Modern Acoustics and Signal Processing

AF183521

More information about this series at http://www.springer.com/series/3754

Series Preface for Modern Acoustics and Signal Processing

In the popular mind, the term "acoustics" refers to the properties of a room or other environment—the acoustics of a room are good or the acoustics are bad. But as understood in the professional acoustical societies of the world, such as the highly influential Acoustical Society of America, the concept of acoustics is much broader. Of course, it is concerned with the acoustical properties of concert halls, classrooms, offices, and factories—a topic generally known as architectural acoustics, but it is also concerned with vibrations and waves too high or too low to be audible. Acousticians employ ultrasound in probing the properties of materials, or in medicine for imaging, diagnosis, therapy, and surgery. Acoustics includes infrasound—the wind-driven motions of skyscrapers, the vibrations of the earth, and the macroscopic dynamics of the sun.

Acoustics studies the interaction of waves with structures, from the detection of submarines in the sea to the buffeting of spacecraft. The scope of acoustics ranges from the electronic recording of rock and roll and the control of noise in our environments to the inhomogeneous distribution of matter in the cosmos.

Acoustics extends to the production and reception of speech and to the songs of humans and animals. It is in music, from the generation of sounds by musical instruments to the emotional response of listeners. Along this path, acoustics encounters the complex processing in the auditory nervous system, its anatomy, genetics, and physiology—perception and behavior of living things.

Acoustics is a practical science, and modern acoustics is so tightly coupled to digital signal processing that the two fields have become inseparable. Signal processing is not only an indispensable tool for synthesis and analysis, it informs many of our most fundamental models about how acoustical communication systems work.

Given the importance of acoustics to modern science, industry, and human welfare Springer presents this series of scientific literature, entitled Modern Acoustics and Signal Processing. This series of monographs and reference books is intended to cover all areas of today's acoustics as an interdisciplinary field. We expect that scientists, engineers, and graduate students will find the books in this series useful in their research, teaching, and studies.

July 2012

William M. Hartmann
Series Editor-in-Chief

Sean F. Wu

The Helmholtz Equation Least Squares Method

For Reconstructing and Predicting Acoustic Radiation

Springer

Sean F. Wu
Department of Mechanical Engineering
Wayne State University
Detroit, MI, USA

ISBN 978-1-4939-3840-7 ISBN 978-1-4939-1640-5 (eBook)
DOI 10.1007/978-1-4939-1640-5
Springer New York Heidelberg Dordrecht London

Printed on acid-free paper

Springer is part of Springer Science+Business Media (www.springer.com)

To my wife Jean

Preface

This is the first and only book on the HELS (Helmholtz equation least-squares) method. While the original contract with Springer to write this book was signed in 2003, it took 10 years for me to actually sit down and complete the writing. This is because during the past decade I have been heavily involved in research projects and teaching, which has constantly distracted me from fulfilling my obligation with the publisher. On the other hand, we have seen tremendous growth and expansion in the HELS theory. Its applications have been extended to many areas that have not been explored such as hybrid near-field acoustical holography (NAH), transient NAH, and NAH-based panel acoustic contributions analyses. Hence, in this sense it was good that I did not write this book 10 years ago. Of course, the HELS method is being further developed and expanded to new frontiers, including reconstruction of the aerodynamically generated sound field generated by an aircraft jet engine and realization of super resolution in discerning acoustic sources by taking input data in space at a rate less than the Nyquist sampling requirement. These new developments will be included in the second edition of this book.

What makes the HELS method unique is its simplicity in mathematical form, efficiency in numerical computation, and flexibility in engineering applications. The idea of using an expansion of certain basis functions to approximate the acoustic field can be traced back to the beginning of the last century. The most famous example was given by Lord Rayleigh to depict the acoustic field scattered from a corrugated surface. The differences and interrelationships between the Rayleigh series and the HELS method are explained in great detail in this book.

The underlying principles of the HELS method are strikingly different from the traditional Fourier acoustics and boundary element method (BEM)-based NAH. The Fourier acoustics-based NAH relies on the Fourier transforms and requires the source surface to contain a level of constant coordinate such as an infinite plate, an infinite cylinder, and a sphere. Moreover, the source must be in free space without the presence of any other source or boundary surface. Although the BEM-based NAH is suitable for arbitrarily shaped surfaces, it also requires the source to be in a source-free region. In addition, both of them require that the hologram surface enclose the entire source surface. If these conditions are not met, then they are invalid theoretically. This makes it difficult for these methods to be adopted in engineering applications because a source-free region is nonexistent and oftentimes a source surface cannot be enclosed by a measurement surface in reality.

In contrast to the traditional NAH implementations, the HELS method does not seek an analytic solution to the acoustic field produced by an arbitrarily shaped structure that cannot be found anyway. Rather, it attempts to obtain the best approximation of an acoustic field through the expansion of certain basis functions. Therefore it significantly simplifies the complexities of the reconstruction process, yet still enables one to acquire a good understanding of the root causes of different noise and vibration problems that involve arbitrarily shaped surfaces in non-free space using much fewer measurement points than both Fourier acoustics and BEM-based NAH do. The examples given in this book illustrate that the HELS method may potentially become a practical and versatile tool for engineers to tackle a variety of complex noise and vibration issues in engineering applications.

Since 2001, I have developed a new course on ME7460 Advanced Acoustic Radiation for graduate students in the Department of Mechanical Engineering at Wayne State University. The main objective of this course is for students to learn the state-of-the-art technology, namely, NAH to diagnose various noise and vibration problems encountered in practice. The major parts of this book are based on my class notes plus new developments in the HELS method accumulated over the past decade. While attending various acoustics conferences sponsored by professional societies such as the Acoustical Society of America, American Society for Mechanical Engineers, and Society for Automobile Engineering, I often have people asking me questions about the HELS method and its implementation. I am happy to report that finally there is a formal textbook on this subject that outlines in great detail this methodology, its implementation steps, and guidelines in practice. In particular, I have provided many examples on how to reconstruct and predict the acoustic fields emitted from different types of sources, and illustrated the intermediate steps in the derivations of various formulations. I sincerely hope that this textbook can serve as a resourceful reference, helpful guidance, and valuable tool for students, engineers, practitioners, and users to understand the HELS-based NAH, how it can be implemented in practice, and why.

This book is structured as follows. Chapter 1 gives a brief history of the major evolution of NAH since its inception in the early 1980s. For a comprehensive review of the development of this technology including its various implementations and extensions over the past three decades, the readers are referred to a review paper by the present author, which was published in the *Journal of the Acoustical Society of America* in 2008.

Chapter 2 reviews the expansion theory using the spherical Hankel functions and spherical harmonics, which form the basis of the HELS method. In particular, many examples are presented that use the expansion theory to reconstruct the acoustic field based on the input data measured on a hologram surface, or to predict the acoustic field based on the boundary condition specified on the source surface.

Chapter 3 discusses the underlying principle of the HELS method and its implementation. In particular, various types of regularization techniques are discussed, including the simplest one that is based on the least-squares minimization and that is very simple to program yet very effective to reconstruct the acoustic pressure field to the most comprehensive hybrid technique that can yield accurate reconstruction for all vibro-acoustic quantities in a nonideal environment. Experimental validations of using the HELS method together with hybrid regularization to reconstruct the vibro-acoustic quantities on the surface of highly non-spherical source geometry are demonstrated. And satisfactory agreements with respect to the benchmark data obtained through direct measurement using a scanning laser probe are obtained.

On the surface the HELS method is very similar to the Rayleigh series or expansion theory. Naturally, people would ask the question whether the HELS method would be subject to the same difficulty as the Rayleigh series did when reconstruction was attempted inside the minimum sphere circumscribing the source for an exterior problem, or beyond the maximum sphere inscribing the source for an interior problem. This question is answered in Chap. 4. In addition, the interrelationships between the HELS method and the Rayleigh series are revealed, and the reasons why the HELS method can be extended beyond the so-called region of validity for the Rayleigh series are presented. Most importantly, rigorous mathematical justifications for the HELS method are provided and its significance is discussed.

Once a solid mathematical foundation for the HELS method has been established, Chap. 5 proceeds to outline the guidelines for implementing this methodology to reconstruct the acoustic fields produced by non-spherical source geometry typically encountered in practice. In particular, detailed steps and formulations to determine the microphone spacing, aperture size, measurement distance, measurement points, etc. are illustrated and explained. In addition, special considerations together with various illustrations and schematic are given to the real-world test configuration and environment for noise and vibration diagnosis.

Chapters 6–10 deal with the extensions of the HELS method to a variety of scenarios that have posed serious challenges to the traditional NAH implementations. Needless to say, there are lots of room for further improvement and new challenges to meet. This is a never-ending process.

This book ends with some of the true stories I have personally experienced in addressing various vehicle-related noise and vibration issues. These stories stress the importance of understanding the physics of sound generation and propagation and how they may help us solve various complex noise and vibration problems in the most cost-effective manner.

I would like to take this opportunity to express my deep gratitudes to my former students, without whom it would not have been possible for me to complete this book. In particular, I would like to thank Dr. Zhaoxi Wang, who was the first student who conducted numerical simulations and validations of the early version of the HELS method; Dr. Nassif Rayess, who was the first one to demonstrate experimentally that the HELS method could indeed be utilized to reconstruct the acoustic field generated by a vehicle buck; Dr. Tatiana Semenova, who revealed the interrelationships between the HELS method and the Rayleigh series and experimented the HELS via various expansions such as localized spherical waves, distributed spherical waves, and distributed point sources; Dr. Manjit Bajwa, who was the first to successfully conduct experiments using transient HELS-based NAH; Dr. Huancai Lu, who was the first to extend the HELS method to reconstruct the normal surface velocity distribution on highly non-spherical source geometry; Dr. Logesh Kumar Natarajan, who was the first to use hybrid regularization and proved experimentally that the HELS method can be used to provide satisfactory reconstruction of the acoustic pressure and the normal surface velocity distribution on highly non-spherical vibrating structures, and to conduct the panel acoustic contributions analyses; Dr. Richard Dziklinski, who was the first to show experimentally that one can violate the Nyquist sampling criterion, namely, take less than two measurement points per wavelength in space, yet still be able to discern two point sources separated by a distance less than one wavelength of the sound emitted by these sources, realizing the so-called super resolution via the HELS method; Dr. Mamohan Singh, who was the first to use the HELS method to visualize the acoustic field generated by an aircraft jet engine and who was instrumental in making HELS codes user-friendly; Mr. Ravi Beniwal, who has made significant contributions toward the graphic user interface of the HELS codes and its real-world applications for diagnosing vehicle noise and vibration problems; and last but not least, my postdoctoral fellow, Dr. Xiang Zhao, who was the first to conduct the numerical simulations using the combined HELS method and hybrid NAH to solve challenging inverse acoustics problems that cannot be done by using other methods.

Finally, I acknowledge that it was Dr. Earl G. Williams who challenged me the validity of using HELS to reconstruct the acoustic quantities on an arbitrarily shaped surface, thus forcing me to look deeper into reasons why the HELS method can be extended beyond the region of validity that the Rayleigh series cannot. However, it was beyond me to prove this validity mathematically. Fortunately, Dr. Victor Isakov came to the rescue and gave a rigorous mathematical justification, for which I am eternally grateful to him for his very important contributions toward the ultimate establishment of the HELS theory.

Detroit, MI, USA Sean F. Wu

Contents

Chapter 1
Introduction

Sound and vibration are intimately related to each other, yet they are two very different physical phenomena. In physics, sound is defined as the disturbance that travels in a compressible medium such as air in terms of a pressure (also called the longitudinal or compressional) wave. In physiology, sound is defined as the sensation of this pressure wave perceived by the brain auditory system of a human being.

Vibration is a mechanical phenomenon that involves oscillations around some equilibrium position. These oscillations can be stationary and periodic such as the membrane of a loudspeaker playing tonal sounds, or non-stationary and random such as the body of an automobile driving on a bumpy road surface. In most engineering applications, sound is regarded as the result of structural vibrations. So vibration is the cause and sound is an effect. The interrelationships between sound and structural vibrations have long been a subject of interest in the acoustics community.

A forward problem is defined as such that the cause is given and effect is to be determined. Conversely, an inverse problem is defined as such that the effect is given via measurement and the cause is to be inferred. In general, a forward problem is much easier to solve than an inverse one because all that is required is to solve some governing equation, given initial values, or boundary conditions, or both. For example, given the geometry and dimensions of a vibrating structure and velocity distribution on its surface, predict the resultant acoustic pressure field. Mathematically, a forward problem is well posed, and its solution is unique and convergent. In the example cited above, there can be one and only one sound field resulting from a particular vibration pattern.

However, the same cannot be said for an inverse problem. Consider the same example as above. Suppose that the acoustic pressures at a finite number of points are measured and geometry and dimensions of the structure are specified. These input data will not be sufficient to uniquely determine the velocity distribution on the surface of the structure. This is because there can be an infinite number of surface velocity distributions that will lead to the same pressure values at these

S.F. Wu, *The Helmholtz Equation Least Squares Method*, Modern Acoustics and Signal Processing, DOI 10.1007/978-1-4939-1640-5_1

measurement points, unless an infinite number of measurements are taken at infinitely close range to the source surface.

Mathematically, all inverse problems are ill posed. In other words, there is an insufficient amount of information on the effect to uniquely determine the cause. Thus an inverse problem is much more difficult to solve than a forward problem.

Unfortunately, most problems encountered in real life are inverse problems, for example, a doctor diagnosing a disease, a detective investigating a crime scene, and an engineer identifying the root cause of a malfunctioning machine. In what follows, we give an overview of the methods commonly used in the industry to diagnose various complex noise and vibration problems, which fall in the category of an inverse problem.

1.1 Conventional Noise and Vibration Diagnoses

Traditionally, noise and vibration diagnoses start from taking measurements of sound and vibration quantities. Typical measurement devices include microphones, accelerometers, intensity probes, and laser vibrometers to gather different physical quantities such as sound pressure level (SPL), sound pressure spectrum, sound intensity, sound power, normal surface velocity, normal surface acceleration, and linear and torsional vibrations. In all these cases, measurements are taken as discrete points in space or on the surface of a vibrating structure. Based on the measured data and one's experiences, noise and vibration abatement strategies are formed. These strategies are trial and error or ad hoc in nature, and the success rate will depend critically on one's experiences in dealing with various noise and vibration problems.

Regardless what measurement devices are used and what physical quantities are measured, there is one thing in common in all these conventional approaches, namely, the measured physical quantities, no matter how many measurement points are taken, are discrete in space and valid only at measurement locations. Moreover, different physical quantities measured at different locations cannot be correlated. For example, the SPL values measured in space cannot be correlated to the accelerations measured on the surface of a vibrating structure. As such, the information gathered is often one dimensional, namely, only the measured physical quantity is seen. There is no overall or big picture of what is going on. The interrelationship between sound and vibration is completely missing in conventional diagnostic approaches. As a result, one might know how to reduce sound and vibration, but not why.

The difficulties encountered in conventional measurement methods for diagnosing noise and vibration problems can be circumvented by a powerful method known as near-field acoustical holography (NAH). In what follows we first define some terminologies in a layman's language.

1.2 Holography

Holography is a technique that makes use of the measured light or sound field to create or reconstruct a three-dimensional (3D) light or sound field after the original light or sound source is removed. The most significant advantage of holography is that it allows one to visualize the entire light or sound field that cannot be obtained by using conventional measurement methodologies.

The earliest holography was done through X-ray in 1920s. The inventor, Dennis Gabor, was awarded the Nobel Prize in Physics in 1971 "for his invention and development of the holographic method" [1]. To date, this technique is still being used in electron microscopy, also known as electron holography. Optical holography had a significant advance after the invention of laser in the 1960s. The first practical laser holography was developed in 1962 by Yuri Denisyuk [2] in the Soviet Union, and by Emmett Leith and Juris Upatnieks [3] at the University of Michigan, USA, which enabled one to see 3D objects with very high spatial resolution.

1.3 Acoustical Holography

The first published acoustical holography through ultrasonic reconstruction was shown by Frederick L. Thurston [4] in 1966. One of the first books on acoustical holography was authored by B.P. Hildebrand and B.B. Brenden [5] in 1971. Acoustical holography employs the same principle as that of laser holography [6], and a 3D sound field is reconstructed based on the measurements of the acoustic pressure in the field. The spatial resolution of acoustical holography is limited to one wavelength of the acoustic wave radiated from the source. In other words, one cannot discern two acoustic sources separated by a distance less than one wavelength. For example, if the frequency is 1,000 Hz and the speed of sound in the air at room temperature (20 °C) is 343 m/s, then the wavelength of this sound wave is $\lambda = c/f = 343/1{,}000 = 0.343$ m. Accordingly, two sources emitting a sound wave of 1,000 Hz is indistinguishable when they are separated by a distance less than 0.34 m.

In general, if the highest frequency of interest is $f_{\max}$ or the shortest wavelength of interest is $\lambda_{\min} = c/f_{\max}$, then the minimal resolvable distance or spatial resolution of acoustical holography is in the order of $R = \lambda_{\min}/2$. The same is true for laser holography, where $R = \lambda_{\min}/2$ represents the separation distance between two (dark and bright) fringes. Since the wavelength of a laser beam is in nanometer range, its spatial resolution is very high. The wavelength of a sound wave, however, is much longer than that of light. Therefore the spatial resolution of acoustical holography is much poorer than that of laser holography.

It is emphasized that acoustical holography can only produce a 3D acoustic pressure image. Moreover, the reconstructed acoustic pressure is accurate in the far field, but not in the near field. It cannot be used to reconstruct the normal surface velocity distribution on a vibrating structure, nor can it be used to reveal the interrelationships between sound and vibration of a vibrating structure. This is because acoustical holography does not consider evanescent waves in its reconstruction. These evanescent waves turn out to be very important in reconstructing vibro-acoustic responses.

1.4 Near-Field Acoustical Holography

The difficulties encountered in acoustical holography can be circumvented by NAH. The major difference between NAH and acoustical holography lies in the fact that the former attempts to capture as much evanescent waves as possible, and the latter does not consider these waves.

By definition, evanescent waves imply the waves whose amplitudes decay exponentially with respect to the measurement distances. For example, at one wavelength of the emitted sound wave, the amplitudes of evanescent waves are reduced down to 0.8 % of their values, or 99.2 % of evanescent waves are lost. At one-half wavelength distance, 95.7 % of evanescent waves will be lost. Therefore, in order to capture evanescent waves the measurement surface must be extremely close to a target source surface. Chapter 5 provides detailed guidelines on setting up measurement distances and microphone spacing for NAH reconstruction in both exterior and interior regions.

Once evanescent waves are captured, the complete vibro-acoustic responses that include the normal component of the particle velocity, acoustic pressure, and acoustic intensity on the source surface and in 3D space can be reconstructed. The interrelationships between sound and vibration can be revealed as well.

Table 1.1 exhibits the major differences between NAH and acoustical holography. Indeed, NAH has been a game changer for analyzing noise and vibration since its inception in early 1980s. The insight one can get from NAH on the underlying mechanisms of sound and vibration cannot be matched by any other methodologies. In theory, NAH can yield an unlimited spatial resolution in reconstruction [7] and produce 3D images for all vibro-acoustic quantities both in space and on the source surface.

NAH can be implemented in many different ways in practice. In what follows, we describe some popular implementations. Each implementation has its advantages and limitations.

Table 1.1 Comparison of the major features and functions of NAH and acoustical holography

Major features and functions	NAH	Acoustical holography
What can it produce?	3D images of all vibro-acoustic responses including the normal component of the particle velocity, acoustic pressure, and acoustic intensity in 3D space and on source surface	3D images of the far-field acoustic pressure
Can it reveal interrelationships between sound and vibration?	It can not only reveal interrelationships between sound and vibration but also uncover which components of structural vibration can produce sound and which ones cannot	No
Can the reconstructed quantities be validated?	Yes, they can be validated against the benchmark values anywhere	Only the acoustic pressure in the far field
What is the spatial resolution?	Unlimited spatial resolution when all evanescent waves are captured	One wavelength of the emitted sound wave
How should the measurement setup be made?	Setups such as measurement distance and microphone spacing are gauged with respect to the critical structural wavelength λ_{cr}	Setup is gauged with respect to the acoustic wavelength λ
What is measurement distance d?	$d < \lambda_{cr}/2$, where λ_{cr} is the critical wavelength of the structural wave	$d < \lambda/2$, where λ is the wavelength of sound wave
What is microphone spacing δ?	$\delta < \lambda_{cr}/2$ where λ_{cr} is the critical wavelength of the structural wave	$\delta < \lambda/2$, where λ is the wavelength of sound wave

1.5 Fourier Transform-Based NAH

The original NAH was implemented through the Fourier transform [8–10], i.e., the acoustic pressures measured on the hologram surface in the space–time domain are transformed to space–frequency domain, followed by another Fourier transform to the wavenumber-frequency domain. These acoustic pressures are then projected toward and away from a source surface in a source-free region in the wavenumber-frequency domain. Once this is done, the inverse Fourier transform is taken to transform the acoustic pressure to the space–frequency domain at new locations, followed by another inverse Fourier transform to take them back to the space–time domain. In this way, the acoustic pressure anywhere in a source-free region is reconstructed. The particle velocity can be reconstructed in a similar manner through a different propagator. Once the acoustic pressures and particle velocities are reconstructed, the acoustic intensity and acoustic energy flow from the surface of a vibrating structure to the surrounding fluid medium are completely determined.

Theoretically, the hologram surface should cover the entire space and measurements must be taken at an equal interval in each dimension to facilitate the Fourier transform. In particular, the hologram surface must have a level of constant coordinate, for example, an infinite planar surface for which the vertical coordinate z is constant; or an infinitely long cylindrical surface for which the radius is constant; or a spherical surface for which the radius is constant.

In practice, however, the hologram surface is always finite, known as a finite measurement aperture. This is inconsistent with the Fourier transform requirement. One way to circumvent this conflict is to apply a spatial window function that keeps the measured acoustic pressure inside the measurement aperture intact, but sets the value to zero outside the aperture. This spatial window in effect extends a finite measurement aperture to an infinite hologram surface.

Note that a finite measurement aperture will induce wraparound errors in convolving the measured acoustic pressures with respect to a propagator. In other words, it introduces artificial wavenumbers that are actually nonexistent in reality. The amount of wraparound errors cannot be determined exactly, but their impacts can be reduced by making the aperture size four times as large as the size of the source surface. Hayek and Luce have demonstrated [11] that it is possible to reduce an aperture to the size of the source surface, yet still allowing reconstruction of the source field with sufficient accuracy.

Recently advancements in the Fourier-based NAH have been made to eliminate the need to have a large measurement aperture. Instead, only the area of the surface on which the source field is interested needs to be scanned. This technique is called patch NAH [12–15], which utilizes an analytic continuation of the patch acoustic pressure and decomposition of the transfer function from measurement and source surfaces.

Another aperture effect is due to a sudden drop in the amplitudes of the acoustic pressures at the edges of a measurement aperture. This discontinuity exists no matter how large an aperture is. As a result, there is a high concentration of very high wavenumbers in the spectrum that poses a serious problem for the inverse Fourier transform. This difficulty can be alleviated by applying a spatial window function that brings down the amplitude of the acoustic pressure along the edge to zero gradually [16]. Examples of such are called the Tukey window or cosine-tapered window [17] that combine the cosine and rectangle windows to reduce the amplitude of the acoustic pressure to zero along the edges of the measurement aperture smoothly without significantly reducing the gain of windowed Fourier transform.

As noted above, the Fourier transform-based NAH is applicable to a surface that contains a level of constant coordinate such as an infinite plane, an infinite circular cylinder, and a sphere. Another limitation is that the Fourier transform-based NAH is suitable for a source-free region. In other words, the region in which reconstruction of an acoustic field is desired must be free from any source and reflecting surface. For an arbitrary source surface, the region of validity is bounded by two parallel infinite planes (see Fig. 1.1), one tangential to the source surface and the other at the infinity. The problem is that such a source-free region is nonexistent in reality. Moreover, the acoustic quantities on arbitrarily shaped surfaces are desired. So other NAH implementations are needed.

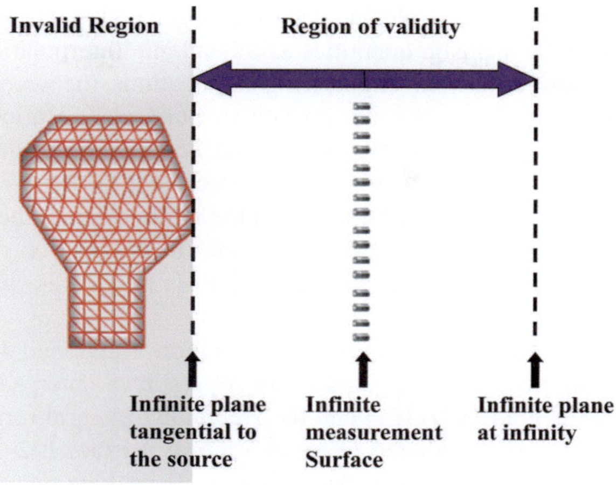

Fig. 1.1 Schematic of region of validity for the Fourier transform-based NAH

1.6 Boundary Element Method-Based NAH

In order to reconstruct the vibro-acoustic quantities on the surface of an arbitrarily shaped structure, different implementations of NAH must be adopted. One such alternative is through the use of the Helmholtz integral theory, which is a reformulation of the Helmholtz equation in terms of an integral formulation (see Sect. 6.1).

The advantage of the Helmholtz integral formulation is a reduction of the dimensionality of an acoustic radiation problem by one. Specifically, the radiated acoustic pressure anywhere in a free field is expressible as an integral of the surface acoustic quantities and free-space Green's function. Hence, all that is required is to determine the acoustic pressure and normal component of the particle velocity on the surface of a vibrating structure. Once this is done, the acoustic field anywhere external to the structure is completely determined.

For an arbitrarily shaped structure, there is no analytic solution to the acoustic quantities on the surface via the Helmholtz integral theory. Therefore, numerical solutions are sought. The most commonly used numerical scheme for the Helmholtz integral theory is the boundary element method (BEM), which discretizes any given surface into a finite number of area segments, each of which contains a fixed number of discrete nodes with specific weighting coefficients. The values of the acoustic quantities anywhere on the source surface can be approximated by a polynomial of the acoustic quantities evaluated at the discrete nodes multiplied by the corresponding weighting coefficients. The number of discrete nodes on an area segment determines the order of polynomial interpolation. The zeroth-order polynomial in BEM approximates the acoustic quantities anywhere on the source surface as a spatial average of this area segment. The first-order polynomial in BEM approximates the acoustic quantities anywhere on the source surface as the linear interpolation of those evaluated at the discrete nodes. The second-order polynomial describes the acoustic quantities as a quadratic interpolation of those evaluated at the discrete nodes. In engineering applications the second-order polynomial is accurate enough to generate satisfactory numerical solutions.

The first attempt to reconstruct the acoustic quantities in an interior region bounded by an arbitrarily shaped surface is depicted by Gardner and Bernhard [18], who use the Helmholtz integral theory to describe the interactions of acoustic sources on the surface and field points in an interior region. Since numerical solutions are obtained by using BEM [19], this approach is known as BEM-based NAH.

Veronesi and Maynard [20] present a more elaborated derivation of BEM-based NAH. Over the next 10 years many papers have emerged that use singular value decomposition (SVD) [21] and the Helmholtz integral formulations to reconstruct the acoustic field radiated from an arbitrary surface [22–32]. SVD is a powerful technique for solving a matrix equation. It enables one to express any complex matrix as a diagonal matrix in the proper bases together with domain and range spaces. If a system of equations is overdetermined, SVD yields a least-squared solution [33]. However, numerical computations involving SVD for a large and complex matrix can be very time consuming, especially at high frequencies.

The advantages of BEM-based NAH are that: (1) it allows for reconstruction of the acoustic quantities on an arbitrarily shaped structure; (2) it has no restriction on locations of measurement points on a hologram surface, so long as they are in a near field; (3) it has no restriction on locations of reconstruction points, whether they are on an arbitrary source surface or in a free field; and (4) it is suitable for both exterior and interior regions.

The disadvantages of BEM-based NAH are that: (1) it fails to yield a unique solution when the frequency is close to one of the characteristic frequencies of a boundary value problem in the interior region [34]; (2) it is valid in an unbounded source-free region only; and (3) numerical computations are very intensive. This is especially the case when additional algorithms are needed to overcome the nonuniqueness difficulties encountered at the characteristic frequencies [34, 35]. The reason for this intensive numerical computation is because the acoustic quantities are reconstructed through spatial discretizations. Typically, six nodes per wavelength are required to ensure a satisfactory reconstruction. Each node contains two unknown variables: the surface acoustic pressure and the normal surface velocity. So the number of measurement points is doubled. For a complex structure such as an engine, the number of discrete nodes needed may be very large, which may lead to an excessive number of measurement points and make reconstruction in the mid-to-high frequency range unrealistically time consuming.

Although BEM-based NAH is based on rigorous integral formulations, its implementation can be a challenge. Another hindrance is that this approach is valid only for an unbounded source-free region, which is nonexistent in practice. Therefore other alternatives are sought.

1.7 Helmholtz Equation Least-Squares Method-Based NAH

An alternative to the aforementioned NAH implementations is HELS (Helmholtz equation least-squares)-based NAH [36, 37]. Unlike the Fourier transform- and BEM-based NAH, HELS-based NAH does not seek an analytic reconstruction of the acoustic field generated by any source, but rather an approximate one by using an expansion of some admissible basis functions with errors minimized by the least-squares method. As such, this approach enables one to tackle a complex source surface with relatively few measurement points, even in a non-source-free region.

The major advantages of HELS-based NAH are its simplicity in formulation, efficiency in computation, and flexibility in application. Since this methodology solves the Helmholtz equation directly, it is immune to the nonuniqueness difficulty inherent in BEM-based Helmholtz integral formulations [34].

The major disadvantage of HELS-based NAH is due to the fact that there is no single set of coordinate system that can yield a good approximation for all surface geometry. For example, the spherical coordinate system may be good for a blunt and convex surface, but not for a highly elongated one; the cylindrical coordinate system is ideal for a slender body, but not for not a flat surface. On the other hand, basis functions are readily available only for the spherical coordinate system. This means that for all practical purposes, the spherical wave functions have to be used in HELS-based NAH to reconstruct the vibro-acoustic responses generated by non-spherical source surface. Consequently, the accuracy in reconstruction is compromised. The farther a source surface differs from a spherical surface, the larger the reconstruction errors on the source surfaces are.

There are other NAH implementations for reconstructing the acoustic fields generated by stationary and moving sources and sources subject to transient excitations, etc. (see [38]). The focus of the present book is on HELS-based NAH and its applications.

Problems

1.1. What is NAH? What does it do?
1.2. What is the difference between NAH and acoustic holography?
1.3. What are the advantages of NAH compared with traditional noise and vibration diagnosis techniques based on measurements of microphones, accelerometers, and intensity probes?
1.4. What are the limitations of NAH compared with traditional noise and vibration diagnosis techniques?
1.5. Why will NAH have these limitations?
1.6. What are the advantages and limitations of the Fourier transform-based NAH?

1.7. What are the advantages and limitations of the BEM-based NAH?
1.8. What can we learn from the reconstruction of the acoustic pressure field?
1.9. What can we learn from the reconstruction of the normal surface velocity distribution?
1.10. What can we learn from the reconstruction of the acoustic intensity field?

Chapter 2
The Spherical Wave Functions

All acoustic radiation problems can be boiled down to solving the wave equation subject to certain initial and boundary conditions. For a constant frequency case, the problem reduces to solving the Helmholtz equation [39], $\nabla^2\hat{p} + k^2\hat{p} = 0$, subject to certain boundary conditions on the source surface. This sounds simple but in reality the analytic solution to the Helmholtz equation exists only for certain types of source geometry that the Helmholtz equation is separable. In most engineering applications the source geometry is arbitrary, so the analytic solution to the Helmholtz equation cannot be found. In these circumstances numerical or approximate solutions are sought.

In this chapter spherical source geometry is considered. Accordingly, the analytic solution to the Helmholtz equation is expressible as the spherical wave functions, which can be obtained by the method of separation of variables. The reasons for choosing the spherical wave functions are (1) the analytic functions are readily available; (2) they are easy to understand; (3) computer codes for the spherical wave functions exist in all software libraries; and (4) they lay the foundation for Chap. 3 of the present book.

2.1 The Helmholtz Equation Under the Spherical Coordinates

Using the spherical coordinates (r, θ, ϕ), we can rewrite the Helmholtz equation as

$$\frac{1}{r^2}\frac{\partial}{\partial r}\left(r^2\frac{\partial\hat{p}}{\partial r}\right) + \frac{1}{r^2\sin\theta}\frac{\partial}{\partial\theta}\left(\sin\theta\frac{\partial\hat{p}}{\partial\theta}\right) + \frac{1}{r^2\sin^2\theta}\frac{\partial\hat{p}^2}{\partial\phi^2} + k^2\hat{p} = 0, \tag{2.1}$$

where $k = \omega/c$ is the acoustic wavenumber; ω and c represent the angular frequency and speed of the sound, respectively; the symbol $\hat{p}$ indicates the complex amplitude

S.F. Wu, *The Helmholtz Equation Least Squares Method*, Modern Acoustics and Signal Processing, DOI 10.1007/978-1-4939-1640-5_2

of the acoustic pressure. Note that a time-harmonic function of the form $e^{-i\omega t}$ is assumed in Eq. (2.1).

The solution to Eq. (2.1) can be obtained by using the separation of variable

$$\hat{p}\,(r,\theta,\phi;\omega) = R(kr)\Theta(\theta)\Phi(\phi), \tag{2.2}$$

where $R(kr)$, $\Theta(\theta)$, and $\Phi(\phi)$ are functions of the spherical coordinates r, θ, and ϕ, respectively.

Substituting Eq. (2.2) into (2.1) then leads to three separate ordinary differential equations:

$$\frac{1}{r^2}\frac{d}{dr}\left(r^2\frac{dR}{dr}\right) + k^2R - \frac{n(n+1)}{r^2}R = 0, \tag{2.3a}$$

$$\frac{1}{\sin\theta}\frac{d}{d\theta}\left(\sin\theta\frac{d\Theta}{d\theta}\right) + \left[n(n+1) - \frac{m^2}{\sin^2\theta}\right]\Theta = 0, \tag{2.3b}$$

$$\frac{d^2\Phi}{d\phi^2} + m^2\Phi = 0, \tag{2.3c}$$

where n and m are integers associated with the radial function $R(kr)$ and the azimuthal function $\Phi(\phi)$, respectively. In particular, $n = 0, 1, 2, \ldots, \infty$, and $m = -n$ to $+n$, which are discussed later. Throughout this book, we use an italic n to indicate an index and a regular n to depict the unit normal.

2.2 Solution to *R*(*kr*)

The solutions to Eq. (2.3a) are expressible as the spherical Bessel functions of the first and second kinds [40], $j_n(kr)$ and $y_n(kr)$, respectively,

$$R(kr) = A_1 j_n(kr) + A_2 y_n(kr), \tag{2.4a}$$

where A_1 and A_2 are arbitrary constants to be determined by boundary conditions. Alternatively, the solutions to Eq. (2.3a) can be written as

$$R(kr) = B_1 h_n^{(1)}(kr) + B_2 h_n^{(2)}(kr), \tag{2.4b}$$

where B_1 and B_2 are arbitrary constants, and $h_n^{(1)}(kr)$ and $h_n^{(2)}(kr)$ are the spherical Hankel functions of the first and second kinds [40], respectively. Since $h_n^{(1)}(kr)$ and $h_n^{(2)}(kr)$ contain the exponential functions of e^{+ikr} and e^{-ikr}, respectively, and since the time-harmonic function in Eq. (2.2) is given by $e^{-i\omega t}$, these spherical Hankel functions depict the outgoing and incoming waves, respectively.

The spherical Hankel functions in Eq. (2.4b) are related to the spherical Bessel functions in Eq. (2.4a). In fact, we can write $h_n^{(1)}(kr)$ and $h_n^{(2)}(kr)$ as

$$h_n^{(1)}(kr) = j_n(kr) + iy_n(kr), \tag{2.5a}$$

$$h_n^{(2)}(kr) = j_n(kr) - iy_n(kr), \tag{2.5b}$$

where $j_n(kr)$ and $y_n(kr)$ are given by

$$j_n(kr) = \sqrt{\frac{\pi}{2kr}} J_{n+1/2}(kr), \tag{2.6a}$$

$$y_n(kr) = \sqrt{\frac{\pi}{2kr}} Y_{n+1/2}(kr), \tag{2.6b}$$

where $J_{n+1/2}(kr)$ and $Y_{n+1/2}(kr)$ are the first and second kinds of Bessel functions, respectively.

Example 2.1 The first four terms of the first and second kinds of the spherical Bessel functions $j_n(kr)$ and $y_n(kr)$ and spherical Hankel functions $h_n^{(1)}(kr)$ and $h_n^{(2)}(kr)$ are given, respectively, by

$$j_0(kr) = \frac{\sin(kr)}{(kr)} \quad \text{and} \quad y_0(kr) = -\frac{\cos(kr)}{(kr)};$$

$$j_1(kr) = -\frac{\cos(kr)}{(kr)} + \frac{\sin(kr)}{(kr)^2} \quad \text{and} \quad y_1(kr) = -\frac{\sin(kr)}{(kr)} - \frac{\cos(kr)}{(kr)^2};$$

$$j_2(kr) = -\frac{\sin(kr)}{(kr)} - \frac{3\cos(kr)}{(kr)^2} + \frac{3\sin(kr)}{(kr)^3} \quad \text{and}$$

$$y_2(kr) = \frac{\cos(kr)}{(kr)} - \frac{3\sin(kr)}{(kr)^2} - \frac{3\cos(kr)}{(kr)^3};$$

$$j_3(kr) = \frac{\cos(kr)}{(kr)} - \frac{6\sin(kr)}{(kr)^2} - \frac{15\cos(kr)}{(kr)^3} + \frac{15\sin(kr)}{(kr)^4} \quad \text{and}$$

$$y_3(kr) = \frac{\sin(kr)}{(kr)} + \frac{6\cos(kr)}{(kr)^2} - \frac{15\sin(kr)}{(kr)^3} - \frac{15\cos(kr)}{(kr)^4};$$

$$h_0^{(1)}(kr) = -i\frac{e^{+ikr}}{kr} \quad \text{and} \quad h_0^{(2)}(kr) = +i\frac{e^{-ikr}}{kr};$$

$$h_1^{(1)}(kr) = -\frac{(kr+i)e^{+ikr}}{(kr)^2} \quad \text{and} \quad h_2^{(2)}(kr) = -\frac{(kr-i)e^{-ikr}}{(kr)^2};$$

$$h_2^{(1)}(kr) = +i\frac{\left[(kr)^2 - 3 + i3(kr)\right]e^{+ikr}}{(kr)^3} \quad \text{and} \quad h_2^{(2)}(kr)$$
$$= -i\frac{\left[(kr)^2 - 3 - i3(kr)\right]e^{-ikr}}{(kr)^3};$$
$$h_3^{(1)}(kr) = +\frac{\left[(kr)^3 - 15(kr) + i6(kr)^2 - i15\right]e^{+ikr}}{(kr)^4} \quad \text{and}$$
$$h_3^{(1)}(kr) = +\frac{\left[(kr)^3 - 15(kr) - i6(kr)^2 + i15\right]e^{-ikr}}{(kr)^4}.$$

In general, we can write

$$j_n(kr) = (-1)^n\frac{d^n}{d(kr)^n}\left[\frac{\sin(kr)}{(kr)}\right] \quad \text{and} \quad y_n(kr) = (-1)^{n+1}\frac{d^n}{d(kr)^n}\left[\frac{\cos(kr)}{(kr)}\right];$$
$$h_n^{(1)}(kr) = (-1)^n\frac{d^n h_0^{(1)}(kr)}{d(kr)^n} \quad \text{and} \quad h_n^{(2)}(kr) = (-1)^n\frac{d^n h_0^{(2)}(kr)}{d(kr)^n}.$$

Example 2.2 The asymptotic forms of the first and second kinds of the spherical Bessel functions, $j_n(kr)$ and $y_n(kr)$, and the spherical Hankel functions, $h_n^{(1)}(kr)$ and $h_n^{(2)}(kr)$, and their derivatives as $kr \to 0$ are given, respectively, by

$$j_n(kr) \approx \frac{(kr)^n}{(2n+1)!!}\left[1 - \frac{(kr)^2}{2(2n+3)} + \cdots\right] \quad \text{and}$$
$$y_n(kr) \approx -\frac{(2n-1)!!}{(kr)^{n+1}}\left[1 - \frac{(kr)^2}{2(1-2n)} + \cdots\right];$$
$$\frac{dj_n(kr)}{d(kr)} \approx \frac{(kr)^{n-1}}{(2n+1)!!}\left[n - \frac{(n+2)(kr)^2}{2(2n+3)} + \cdots\right] \quad \text{and}$$
$$\frac{dy_n(kr)}{d(kr)} \approx \frac{(2n-1)!!}{(kr)^{n+2}}\left[(n+1) - \frac{(n-1)(kr)^2}{2(1-2n)} + \cdots\right];$$
$$h_n^{(1)}(kr) \approx -i\frac{(2n-1)!!}{(kr)^{n+1}} \quad \text{and} \quad h_n^{(2)}(kr) \approx +i\frac{(2n-1)!!}{(kr)^{n+1}};$$
$$\frac{dh_n^{(1)}(kr)}{d(kr)} \approx +i\frac{(n+1)(2n-1)!!}{(kr)^{n+2}} \quad \text{and} \quad \frac{dh_n^{(2)}(kr)}{d(kr)} \approx -i\frac{(n+1)(2n-1)!!}{(kr)^{n+2}};$$

where

$$(2n+1)!! = \frac{(2n+1)!}{2^n n!} \quad \text{and} \quad (2n-1)!! = \frac{(2n-1)!}{2^n n!}.$$

These results show that the first kind of the spherical Bessel functions $j_n(kr)$ and their derivatives are bounded, while the second kind of the spherical Bessel functions $y_n(kr)$ and their derivatives grow without a bound as $kr \to 0$. Because the spherical Hankel functions contain $y_n(kr)$, they are unbounded as well as $kr \to 0$.

Example 2.3 The asymptotic expressions of the spherical Hankel functions of the first and second kinds, $h_n^{(1)}(kr)$ and $h_n^{(2)}(kr)$, and their derivatives as $kr \to \infty$ are given, respectively, by

$$h_n^{(1)}(kr) \approx (-i)^{n+1} \frac{\mathrm{e}^{ikr}}{(kr)} \quad \text{and} \quad h_n^{(2)}(kr) \approx (+i)^{n+1} \frac{\mathrm{e}^{-ikr}}{(kr)};$$

$$\frac{dh_n^{(1)}(kr)}{d(kr)} \approx (-i)^n \frac{\mathrm{e}^{ikr}}{(kr)} \quad \text{and} \quad \frac{dh_n^{(2)}(kr)}{d(kr)} \approx (+i)^n \frac{\mathrm{e}^{-ikr}}{(kr)}.$$

Note that for large real arguments, $kr \to \infty$, one cannot write the true asymptotic forms for Bessel functions of the first and second kinds because they are oscillatory and have zeros all the way to infinity, making it impossible to be matched exactly by any asymptotic expansion.

Example 2.4 The recursion relationships for the first and second kinds of the spherical Bessel functions, $j_n(kr)$ and $y_n(kr)$, and their derivatives are given, respectively, by

$$j_n(kr) = \frac{(2n-1)}{(kr)} j_{n-1}(kr) - j_{n-2}(kr) \quad \text{and}$$

$$y_n(kr) = \frac{(2n-1)}{(kr)} y_{n-1}(kr) - y_{n-2}(kr);$$

where $n \geq 2$. For $n = 0$ and 1, $j_n(kr)$ and $y_n(kr)$ are given in Example 2.1.

$$\frac{dj_n(kr)}{d(kr)} = \left(\frac{n}{kr}\right) j_n(kr) - j_{n+1}(kr) \quad \text{and} \quad \frac{dy_n(kr)}{d(kr)} = \left(\frac{n}{kr}\right) y_n(kr) - y_{n+1}(kr).$$

Example 2.5 The recursion relationships for the first and second kinds of the spherical Hankel functions, $h_n^{(1)}(kr)$ and $h_n^{(2)}(kr)$, and their derivatives are given, respectively, by

$$h_n^{(1)}(kr) = \left(\frac{kr}{2n+1}\right)\left[h_{n-1}^{(1)}(kr) + h_{n+1}^{(1)}(kr)\right] \quad \text{and}$$
$$h_n^{(2)}(kr) = \left(\frac{kr}{2n+1}\right)\left[h_{n-1}^{(2)}(kr) + h_{n+1}^{(2)}(kr)\right];$$
$$\frac{dh_n^{(1)}(kr)}{d(kr)} = h_{n-1}^{(1)}(kr) - \left(\frac{n+1}{kr}\right)h_n^{(1)}(kr) \quad \text{and}$$
$$\frac{dh_n^{(2)}(kr)}{d(kr)} = h_{n-1}^{(2)}(kr) - \left(\frac{n+1}{kr}\right)h_n^{(2)}(kr);$$

where $n \geq 1$. For $n = 0$, $h_0^{(1)}(kr)$ and $h_0^{(2)}(kr)$ are given in Example 2.1, and their derivatives are

$$\frac{dh_0^{(1)}(kr)}{d(kr)} = \left(\frac{kr+i}{kr}\right)\frac{e^{ikr}}{kr} \quad \text{and} \quad \frac{dh_0^{(2)}(kr)}{d(kr)} = \left(\frac{kr-i}{kr}\right)\frac{e^{-ikr}}{kr}.$$

2.3 Solution to $\Theta(\theta)$

The solutions to Eq. (2.3b) can be written as the Legendre functions of the first and second kinds, respectively,

$$\Theta(\theta) = C_1 P_n^m(\cos\theta) + C_2 Q_n^m(\cos\theta), \tag{2.7}$$

where C_1 and C_2 are arbitrary constants. Note that the second kind of the Legendre functions is unbounded at the poles $\cos\theta = \pm 1$, and must be discarded by setting $C_2 = 0$. Also note that n is an integer and the first kind of the Legendre function $P_n^m(\cos\theta) = 0$ whenever $m > n$. Since m ranges from $-n$ to $+n$ and since $P_n^{-m}(\cos\theta)$ is related to $P_n^m(\cos\theta)$ through

$$P_n^{-m}(\cos\theta) = (-1)^m \frac{(n-m)!}{(n+m)!} P_n^m(\cos\theta), \tag{2.8}$$

where m is positive, we only need to be concerned with $P_n^m(\cos\theta)$, which is expressible as [41]

$$P_n^m(\cos\theta) = (-1)^m \sin^m\theta \frac{d^m P_n(\cos\theta)}{d(\cos\theta)^m}, \tag{2.9}$$

where $P_n(\cos\theta)$ is given by an infinite series known as the Legendre polynomials,

$$P_n(\cos\theta) = \frac{(2n-1)!!}{n!}\left[\cos^n\theta - \frac{n(n-1)}{2(2n-1)}\cos^{(n-2)}\theta + \frac{n(n-1)(n-2)(n-3)}{2\cdot 4(2n-1)(2n-3)}\cos^{(n-4)}\theta \right.$$
$$\left. - \frac{n(n-1)(n-2)(n-3)(n-4)(n-5)}{2\cdot 4\cdot 6(2n-1)(2n-3)(2n-5)}\cos^{(n-6)}\theta + \cdots\right], \tag{2.10}$$

where $n = 0, 1, 2, \ldots, \infty$.

The Legendre polynomials $P_n(\cos\theta)$ are orthogonal, namely,

$$\int_{-1}^{+1} P_{n'}(\cos\theta)P_n(\cos\theta)d(\cos\theta) = \left(\frac{2}{2n+1}\right)\delta_{n'n}, \tag{2.11}$$

where $\delta_{n'n}$ is the Kroneker delta function,

$$\delta_{n'n} = \begin{cases} 1, & \text{if } n = n' \\ 0, & \text{if } n \neq n' \end{cases}. \tag{2.12}$$

Since $P_n^m(\cos\theta)$ is related to the Legendre polynomials, a similar orthogonal condition exists for $P_n^m(\cos\theta)$ for any positive value of m,

$$\int_{-1}^{+1} P_{n'}^m(\cos\theta)P_n^m(\cos\theta)d(\cos\theta) = \left(\frac{2}{2n+1}\right)\frac{(n+m)!}{(n-m)!}\delta_{n'n}, \tag{2.13}$$

Example 2.6 The first five terms ($n = 0$, 1, 2, 3, and 4) of the first kind of the Legendre functions are given as follows:

For $n = 0$, $P_0^0(\cos\theta) = 1$.

For $n = 1$, $P_1^{-1}(\cos\theta) = \frac{\sin\theta}{2}$, $P_1^0(\cos\theta) = \cos\theta$, and $P_1^1(\cos\theta) = -\sin\theta$.

For $n = 2$, $P_2^{-2}(\cos\theta) = \frac{\sin^2\theta}{8}$, $P_2^{-1}(\cos\theta) = \frac{\cos\theta\sin\theta}{2}$, $P_2^0(\cos\theta) = \frac{1+3\cos 2\theta}{4}$, $P_2^1(\cos\theta) = -3\cos\theta\sin\theta$, and $P_2^2(\cos\theta) = 3\sin^2\theta$.

For $n = 3$, $P_3^{-3}(\cos\theta) = \frac{\sin^3\theta}{48}$, $P_3^{-2}(\cos\theta) = \frac{\cos\theta\sin^2\theta}{8}$, $P_3^{-1}(\cos\theta) = \frac{(3+5\cos 2\theta)\sin\theta}{16}$, $P_3^0(\cos\theta) = \frac{-3\cos\theta+5\cos^3\theta}{2}$, $P_3^1(\cos\theta) = \frac{3(1-5\cos^2\theta)\sin\theta}{2}$, $P_3^2(\cos\theta) = 15\cos\theta\sin^2\theta$, and $P_3^3(\cos\theta) = -15\sin^3\theta$.

For $n = 4$, $P_4^{-4}(\cos\theta) = \frac{7\sin^4\theta}{2688}$, $P_4^{-3}(\cos\theta) = \frac{7\cos\theta\sin^3\theta}{336}$, $P_4^{-2}(\cos\theta) = \frac{(5+7\cos 2\theta)\sin^2\theta}{96}$, $P_4^{-1}(\cos\theta) = -\frac{(3\cos\theta-7\cos^3\theta)\sin\theta}{2}$, $P_4^0(\cos\theta) = \frac{3-30\cos^2\theta+35\cos^4\theta}{8}$, $P_4^1(\cos\theta) = \frac{5(3\cos\theta-7\cos^3\theta)\sin\theta}{2}$, $P_4^2(\cos\theta) = \frac{15(5+7\cos 2\theta)\sin^2\theta}{4}$, $P_4^3(\cos\theta) = -105\cos\theta\sin^4\theta$, and $P_4^4(\cos\theta) = 105\sin^4\theta$.

Example 2.7 The recursion relations for the Legendre functions can be written in different ways, one of them being

$$P_n^m(\cos\theta) = \frac{(n-m+1)}{(2n+1)\cos\theta} P_{n+1}^m(\cos\theta) + \frac{(n+m)}{(2n+1)\cos\theta} P_{n-1}^m(\cos\theta),$$

where $n \geq 1$. For $n = 0$, $P_0^0(\cos\theta) = 1$ as shown in Example 2.5.

Example 2.8 Similarly, there are different recursion relations for the derivatives of the Legendre functions, one of them being

$$\frac{dP_n^m(\cos\theta)}{d(\cos\theta)} = \frac{(n+1)\cos\theta}{\sin\theta} P_n^m(\cos\theta) - \frac{(n-m+1)}{\sin\theta} P_{n+1}^m(\cos\theta).$$

2.4 Solution to $\Phi(\phi)$

The solutions to Eq. (2.3c) are simply harmonic functions of the azimuthal angle

$$\Phi(\phi) = D_1 \mathrm{e}^{im\phi} + D_2 \mathrm{e}^{-im\phi}, \tag{2.14}$$

where D_1 and D_2 are arbitrary constants.

The angular solutions $\Theta(\theta)$ and $\Phi(\phi)$ given by Eqs. (2.7) and (2.14) can be combined into a single function known as the spherical harmonics $Y_n^m(\theta, \phi)$,

$$Y_n^m(\theta, \phi) = \sqrt{\frac{(2n+1)(n-m)!}{4\pi(n+m)!}} P_n^m(\cos\theta)\mathrm{e}^{im\phi}, \tag{2.15}$$

where $-n \leq m \leq n$.

Since $P_n^{-m}(\cos\theta)$ is related to $P_n^m(\cos\theta)$, so is $Y_n^{-m}(\theta, \phi)$ to $Y_n^m(\theta, \phi)$, namely,

$$Y_n^{-m}(\theta, \phi) = (-1)^m Y_n^{m*}(\theta, \phi), \tag{2.16}$$

where a superscript * implies a complex conjugation. The spherical harmonics are orthonormal,

$$\int_0^{2\pi} d\phi \int_0^{\pi} Y_{n'}^{m'}(\theta, \phi) Y_n^{m*}(\theta, \phi) \sin\theta d\theta = \delta_{n'n}\delta_{m'm}, \tag{2.17}$$

The orthonormal characteristics of the spherical harmonics enable us to use them to represent any arbitrary function $f(\theta, \phi)$ on a spherical surface. For example, we can express $f(\theta, \phi)$ as

$$f(\theta,\phi)=\sum_{n=0}^{\infty}\sum_{m=-n}^{n}A_{nm}Y_n^m(\theta,\phi), \tag{2.18}$$

where A_{nm} are the expansion coefficients that can be obtained by using the orthonormal condition (2.17). Multiplying both sides of Eq. (2.17) by the spherical harmonics $Y_n^m(\theta,\varphi)$ and integrating over the solid angle of a sphere, we obtain

$$\int_0^{2\pi}d\phi\int_0^{\pi}f(\theta,\phi)Y_n^m(\theta,\phi)^*\sin\theta d\theta=\int_0^{2\pi}d\phi\int_0^{\pi}\sum_{n=0}^{\infty}\sum_{m=-n}^{n}A_{nm}Y_n^m(\theta,\phi)Y_{n'}^{m'}(\theta,\phi)^*\sin\theta d\theta$$
$$=\sum_{n=0}^{\infty}\sum_{m=-n}^{n}A_{nm}\int_0^{2\pi}d\phi\int_0^{\pi}Y_n^m(\theta,\phi)Y_{n'}^{m'}(\theta,\phi)^*\sin\theta d\theta=\sum_{n=0}^{\infty}\sum_{m=-n}^{n}A_{nm}\delta_{nm}\delta_{n'm'}=A_{nm}. \tag{2.19}$$

Therefore, the expansion coefficients A_{nm} are given by

$$A_{nm}=\int_0^{2\pi}d\phi\int_0^{\pi}f(\theta,\phi)Y_n^m(\theta,\phi)^*\sin\theta d\theta. \tag{2.20}$$

It is important to point out that the results given by Eqs. (2.18), (2.19), (2.20) are only valid for a spherical source surface.

Example 2.9 The first few terms ($n\le 4$) of the spherical harmonics are given by:

For $n=0$, $Y_0^0(\theta,\varphi)=\frac{1}{2\sqrt{\pi}}$.

For $n=1$, $Y_1^{-1}(\theta,\phi)=\left(\frac{1}{2}\sqrt{\frac{3}{2\pi}}\sin\theta\right)e^{-i\phi}$, $Y_1^0(\theta,\phi)=\frac{1}{2}\sqrt{\frac{3}{\pi}}\cos\theta$, and $Y_1^1(\theta,\phi)$ $=-\left(\frac{1}{2}\sqrt{\frac{3}{2\pi}}\sin\theta\right)e^{i\phi}$.

For $n=2$, $Y_2^{-2}(\theta,\phi)=\left(\frac{3}{4}\sqrt{\frac{5}{6\pi}}\sin^2\theta\right)e^{-i2\phi}$, $Y_2^{-1}(\theta,\phi)=\left(\frac{3}{4}\sqrt{\frac{5}{6\pi}}\sin 2\theta\right)e^{-i\phi}$, $Y_1^0(\theta,\phi)=\frac{1}{4}\sqrt{\frac{5}{\pi}}(3\cos^2\theta-1)$, $Y_2^1(\theta,\phi)=-\left(\frac{3}{4}\sqrt{\frac{5}{6\pi}}\sin 2\theta\right)e^{i\phi}$, and $Y_2^2(\theta,\phi)=\left(\frac{3}{4}\sqrt{\frac{5}{6\pi}}\sin^2\theta\right)e^{i2\phi}$.

For $n=3$, $Y_3^{-3}(\theta,\phi)=\left(\frac{1}{8}\sqrt{\frac{35}{\pi}}\sin^3\theta\right)e^{-i3\phi}$, $Y_3^{-2}(\theta,\phi)=\left(\frac{15}{4}\sqrt{\frac{7}{30\pi}}\cos\theta\sin^2\theta\right)e^{-i2\phi}$, $Y_3^{-1}(\theta,\phi)=\left[\frac{1}{16}\sqrt{\frac{21}{\pi}}(5\cos 2\theta+3)\sin\theta\right]e^{-i\phi}$, $Y_3^0(\theta,\phi)=\frac{1}{4}\sqrt{\frac{7}{\pi}}(5\cos^3\theta-3\cos\theta)$, $Y_3^1(\theta,\phi)=\left[\frac{1}{8}\sqrt{\frac{21}{\pi}}(-5\cos^2\theta+1)\sin\theta\right]e^{i\phi}$, $Y_3^2(\theta,\phi)=\left(\frac{15}{4}\sqrt{\frac{7}{30\pi}}\cos\theta\sin^2\theta\right)e^{i2\phi}$, and $Y_3^3(\theta,\phi)=-\left(\frac{5}{8}\sqrt{\frac{7}{5\pi}}\sin^3\theta\right)e^{i3\phi}$.

For $n = 4$, $Y_4^{-4}(\theta,\phi) = \left(\frac{105}{16}\sqrt{\frac{1}{70\pi}}\sin^4\theta\right)e^{-i4\phi}$,

$$Y_4^{-3}(\theta,\phi) = \left(\frac{105}{8}\sqrt{\frac{1}{35\pi}}\cos\theta\sin^3\theta\right)e^{-i3\phi},$$

$$Y_4^{-2}(\theta,\phi) = \left[\frac{15}{16}\sqrt{\frac{1}{10\pi}}(7\cos 2\theta+5)\sin^2\theta\right]e^{-i2\phi},$$

$$Y_4^{-1}(\theta,\phi) = \left[\frac{15}{8}\sqrt{\frac{1}{5\pi}}(7\cos^3\theta-3\cos\theta)\sin\theta\right]e^{-i\phi},$$

$$Y_4^{0}(\theta,\phi) = \frac{3}{16}\sqrt{\frac{1}{\pi}}(35\cos^4\theta-30\cos^2\theta+3),$$

$$Y_4^{1}(\theta,\phi) = -\left[\frac{15}{8}\sqrt{\frac{1}{5\pi}}(7\cos^3\theta-3\cos\theta)\sin\theta\right]e^{i\phi},$$

$$Y_4^{2}(\theta,\phi) = \left[\frac{15}{16}\sqrt{\frac{1}{10\pi}}(7\cos 2\theta+5)\sin^2\theta\right]e^{i2\phi},$$

$$Y_4^{3}(\theta,\phi) = -\left(\frac{105}{8}\sqrt{\frac{1}{35\pi}}\cos\theta\sin^3\theta\right)e^{i3\phi}, \text{ and}$$

$$Y_4^{4}(\theta,\phi) = \left(\frac{105}{16}\sqrt{\frac{1}{70\pi}}\sin^4\theta\right)e^{i4\phi}.$$

2.5 Solution to $\hat{p}\,(r, \theta, \phi; \omega)$

Combining the radial functions $R(kr)$ and spherical harmonics $Y_n^m(\theta,\phi)$, we can express the solutions to the Helmholtz equation that describe standing waves in an interior region as

$$\hat{p}\,(r,\theta,\phi;\omega) = \sum_{n=0}^{\infty}\sum_{m=-n}^{n}[A_{mn}j_n(kr)+B_{mn}y_n(kr)]Y_n^m(\cos\theta), \tag{2.21a}$$

or the solutions to the Helmholtz equation that describe traveling waves in an exterior region as

$$\hat{p}\,(r,\theta,\phi;\omega) = \sum_{n=0}^{\infty}\sum_{m=-n}^{n}\left[A_{mn}h_n^{(1)}(kr)+B_{mn}h_n^{(2)}(kr)\right]Y_n^m(\cos\theta), \tag{2.21b}$$

where the first term on the right side of Eq. (2.21b) depicts the outgoing spherical waves and second term describes the incoming spherical waves.

Now let us consider the examples of using Eq. (2.21) to predict the acoustic pressure fields generated by a vibrating sphere.

Example 2.10 Consider the case of a vibrating sphere of radius $r=a$ in a free field. Let the acoustic pressure on a spherical surface of radius $r=r^{\text{meas}}$ be specified, $\hat{p}(r,\theta,\phi;\omega)|_{r=r^{\text{meas}}} = \hat{p}(r^{\text{meas}},\theta,\phi;\omega)$. The acoustic pressure field anywhere including the source surface is desired. This problem can be solved by using Eq. (2.21b). Since this is an exterior problem and the field is unbounded, there are only outgoing waves from the vibrating sphere to infinity. Accordingly, Eq. (2.21b) is rewritten as

$$\hat{p}(r,\theta,\phi;\omega) = \sum_{n=0}^{\infty}\sum_{m=-n}^{n} A_{mn} h_n^{(1)}(kr) Y_n^m(\cos\theta).$$

The expansion coefficients A_{mn} can be specified by using the pressure boundary condition on the spherical surface of radius $r=r^{\text{meas}}$.

$$\hat{p}(r^{\text{meas}},\theta,\phi;\omega) = \sum_{n=0}^{\infty}\sum_{m=-n}^{n} A_{mn} h_n^{(1)}(kr^{\text{meas}}) Y_n^m(\cos\theta).$$

Multiplying both sides by the complex conjugate of the spherical harmonics, integrating over the solid angle of a sphere, and using the orthonormal property of the spherical harmonics, we obtain

$$\int_0^{2\pi} d\phi \int_0^{\pi} \hat{p}(r^{\text{meas}},\theta,\phi;\omega) Y_n^{m*}(\theta,\phi)\sin\theta d\theta = A_{nm} h_n^{(1)}(kr^{\text{meas}}).$$

Therefore, the expansion coefficients A_{mn} are given by

$$A_{nm} = \frac{1}{h_n^{(1)}(kr^{\text{meas}})} \int_0^{2\pi} d\phi \int_0^{\pi} \hat{p}(r^{\text{meas}},\theta,\phi;\omega) Y_n^{m*}(\theta,\phi)\sin\theta d\theta.$$

Once A_{mn} are specified, the acoustic pressure anywhere is expressible as

$$\hat{p}(r,\theta,\phi;\omega) = \sum_{n=0}^{\infty} \frac{h_n^{(1)}(kr)}{h_n^{(1)}(kr^{\text{meas}})} \sum_{m=-n}^{n} Y_n^m(\cos\theta) \int_0^{2\pi} d\phi' \int_0^{\pi} \hat{p}\left(r^{\text{meas}},\theta',\phi';\omega\right) Y_n^{m*}\left(\theta',\phi'\right)\sin\theta' d\theta'.$$

Example 2.11 Consider the case of a vibrating sphere of radius $r=a$ in a free field. Assume that the normal surface of this vibrating sphere is given as $\hat{v}_{\text{n}}(a,\theta,\phi;\omega)$. With this boundary condition, we want to predict the radiated acoustic pressure anywhere, including the source surface. From the Euler's equation [42], we can write the boundary condition as

$$\hat{v}_{\mathrm{n}}(a,\theta,\phi;\omega) = \frac{1}{i\omega\rho_0}\frac{\partial\hat{p}\,(r,\theta,\phi;\omega)}{\partial \mathrm{n}}\bigg|_{r=a},$$

where ρ_0 is the ambient density of the medium surrounding the dilating sphere, and a subscript n depicts the unit normal direction.

Since there are only outgoing waves, we can rewrite Eq. (2.21b) as

$$\hat{p}\,(r,\theta,\phi;\omega) = \sum_{n=0}^{\infty}\sum_{m=-n}^{n} A_{mn} h_n^{(1)}(kr) Y_n^m(\cos\theta).$$

Take the normal derivative on both sides of the above expression,

$$\frac{\partial\hat{p}\,(r,\theta,\phi;\omega)}{\partial \mathrm{n}} = \sum_{n=0}^{\infty}\sum_{m=-n}^{n} A_{mn}\frac{dh_n^{(1)}(kr)}{dr} Y_n^m(\cos\theta),$$

where the symbol $\partial/\partial\mathrm{n}$ represents a normal derivative defined as

$$\frac{\partial}{\partial \mathrm{n}} = \vec{\mathrm{n}}\cdot\nabla = \frac{\partial}{\partial r}.$$

Substitute the normal derivative of the acoustic pressure evaluated at $r = a$ yields

$$\hat{v}_{\mathrm{n}}(a,\theta,\phi;\omega) = \frac{1}{i\omega\rho_0}\sum_{n=0}^{\infty}\sum_{m=-n}^{n} C_{mn}\frac{dh_n^{(1)}(kr)}{dr}\bigg|_{r=a} Y_n^m(\cos\theta).$$

Next, we multiply both sides by the complex conjugate of the spherical harmonics, integrate it over the solid angle of the sphere, and use the orthonormal property of the spherical harmonics,

$$\int_0^{2\pi} d\phi \int_0^{\pi} \hat{v}_{\mathrm{n}}(a,\theta,\phi;\omega) Y_n^{m*}(\cos\theta)\sin\theta d\theta = \frac{A_{mn}}{i\omega\rho_0}\frac{dh_n^{(1)}(kr)}{dr}\bigg|_{r=a}.$$

Therefore, the expansion coefficients A_{mn} are found to be

$$A_{mn} = \frac{i\omega\rho_0}{\left.\frac{dh_n^{(1)}(kr)}{dr}\right|_{r=a}}\int_0^{2\pi} d\phi \int_0^{\pi} \hat{v}_{\mathrm{n}}(a,\theta,\phi;\omega) Y_n^{m*}(\theta,\phi)\sin\theta d\theta.$$

Accordingly, the radiated acoustic pressure is given by

$$\hat{p}(r,\theta,\phi;\omega) = i\omega\rho_0 \sum_{n=0}^{\infty} \frac{h_n^{(1)}(kr)}{\left.\frac{dh_n^{(1)}(kr)}{dr}\right|_{r=a}} \sum_{m=-n}^{n} Y_n^m(\cos\theta)$$

$$\int_0^{2\pi} d\phi' \int_0^{\pi} \hat{v}_{\mathrm{n}}\left(a,\theta',\phi';\omega\right) Y_n^{m*}\left(\theta',\phi'\right) \sin\theta' d\theta'.$$

Example 2.12 Let us consider a specific case of a dilating sphere for which $\hat{v}_{\mathrm{n}}(a,\theta,\phi;\omega) \equiv v_0$ is a constant. Accordingly, we set $n = m = 0$ in the above expression. Substituting the derivative of the spherical Hankel functions of the first kind with $n = 0$ (see Example 2.1) yields the expansion coefficient A_{00} as

$$A_{00} = i\rho_0 c v_0 \frac{(ka)^2}{(ka+i)} \mathrm{e}^{-ika}.$$

The corresponding acoustic pressure anywhere is given by

$$\hat{p}(r,\theta,\phi;\omega) = \rho_0 c v_0 \left(\frac{ka}{ka+i}\right)\left(\frac{a}{r}\right)\mathrm{e}^{ik(r-a)},$$

which agrees perfectly with the analytic solution [42].

Example 2.13 Next let us consider the acoustic pressure inside a dilating sphere of radius $r = a$. Once again let that the normal surface of this dilating sphere be constant, $v_{\mathrm{n}} = v_0$. In this case we can use Eq. (2.21a) but have to discard the second term involving the second kind of the spherical Bessel function $y_n(kr)$ because it is unbounded at the center $r = 0$. Accordingly, we have

$$\hat{p}(r,\theta,\phi;\omega) = \sum_{n=0}^{\infty} \sum_{m=-n}^{n} A_{mn} j_n(kr) Y_n^m(\cos\theta) \mathrm{e}^{im\phi},$$

where the expansion coefficients A_{nm} can be determined by the boundary condition together with the orthonormal condition. Take the normal derivative of the acoustic pressure,

$$\frac{\partial \hat{p}(r,\theta,\phi;\omega)}{\partial \mathrm{n}} = \sum_{n=0}^{\infty} \sum_{m=-n}^{n} A_{mn} \frac{dj_n(kr)}{dr} Y_n^m(\cos\theta)\mathrm{e}^{im\phi}.$$

Substitute the normal derivative of the acoustic pressure into the boundary condition,

$$v_0 = \frac{1}{\mathrm{i}\omega\rho_0} \sum_{n=0}^{\infty} \sum_{m=-n}^{n} A_{mn} \left.\frac{dj_n(kr)}{dr}\right|_{r=a} Y_n^m(\cos\theta)\mathrm{e}^{\mathrm{i}m\phi}.$$

Following the same procedures as those outlined in Example 2.11, we obtain

$$\int_0^{2\pi} d\phi \int_0^{\pi} v_0 Y_n^{m*}(\theta,\phi)\sin\theta d\theta = \frac{1}{i\omega\rho_0}\int_0^{2\pi} d\phi \int_0^{\pi} \sum_{n=0}^{\infty}\sum_{m=-n}^{n} A_{nm} \left.\frac{dj_n(kr)}{dr}\right|_{r=a} Y_{n'}^{m'}(\theta,\phi) Y_n^{m*}(\theta,\phi)\mathrm{e}^{im\phi}\sin\theta d\theta.$$

Using the orthogonality property of the spherical harmonics and carrying out the integrations over the solid angle, we obtain

$$v_0 = \frac{1}{i\omega\rho_0}\sum_{n=0}^{\infty}\sum_{m=-n}^{n} A_{nm} \left.\frac{dj_n(kr)}{dr}\right|_{r=a} \mathrm{e}^{im\phi}.$$

The left side in the above equation is constant and independent of the angular variables, and the right side can match this if $n=m=0$. Substituting the derivative of the spherical Bessel function of the first kind given in Example 2.1 with $n=0$, we obtain

$$v_0 = \frac{A_{00}}{i\rho_0 c}\left.\frac{dj_n(kr)}{d(kr)}\right|_{r=a} = \frac{A_{00}[(ka)\cos(ka) - \sin(ka)]}{i\rho_0 c(ka)^2},$$

from which we found the expansion coefficient A_{00} to be

$$A_{00} = \frac{i\rho_0 c v_0 (ka)^2}{(ka)\cos(ka) - \sin(ka)}.$$

Substituting A_{00} into Eq. (2.21a) yields the interior acoustic pressure field,

$$\hat{p}(r,\theta,\phi;\omega) = \frac{i\rho_0 c v_0 (ka)\sin(kr)}{(ka)\cos(ka) - \sin(ka)}\left(\frac{a}{r}\right).$$

In this case resonance will occur inside the spherical surface when the frequency f is equal to one of the eigenfrequencies obtained by solving the following characteristic equation:

$$\tan\left(\frac{2\pi f_n a}{c}\right) = \frac{2\pi f_n a}{c}, \quad n = 1, 2, 3, \ldots,$$

The above equation is a transcendental equation that can only be solved numerically. The first four eigenfrequencies or the roots of this transcendental equation are:

$$f_1 = 0.715148\left(\frac{c}{a}\right) \text{ (Hz)},$$

$$f_2 = 1.229515\left(\frac{c}{a}\right) \text{ (Hz)},$$

$$f_3 = 1.735446\left(\frac{c}{a}\right) \text{ (Hz)},$$

$$f_4 = 2.238705\left(\frac{c}{a}\right) \text{ (Hz)},$$

Examples 2.10 and 2.11 demonstrate that we can use the expansions of the spherical wave functions to describe exactly the acoustic pressure field generated by a vibrating sphere. When the vibration pattern is arbitrary, the number of expansion terms may be infinite. When the vibrating surface is not spherical however, the expansion given by solution Eq. (21) is invalid. In practice, most vibrating surfaces are non-spherical. Therefore, a different methodology is needed to describe the radiated acoustic field. This is the topic of Chap. 3.

Problems

2.1. Use the recursion relations given in Example 2.4 to write the following spherical Bessel function and spherical Hankel function of the first kind: $j_n^{(1)}(kr)$ and $h_n^{(1)}(kr)$, where $n = 4$, 5, and 6.

2.2. Use the recursion relations given in Example 2.4 to write the derivatives for the following spherical Bessel function and spherical Hankel function of the first kind: $dj_n^{(1)}(kr)/d(kr)$ and $dh_n^{(1)}(kr)/d(kr)$, where $n = 4$, 5, and 6.

2.3. Use the recursion relations given in Example 2.6 and definitions of the Legendre functions to write $P_n^m(\cos\theta)$, where $n = 5$ and $m = -n$ to $+n$.

2.4. Continue Problem 2.2 and write down the spherical harmonics $Y_n^m(\theta, \phi)$, where $n = 5$ and $m = -n$ to $+n$.

2.5. Consider a vibrating sphere of radius $r = a$ in free space. Assume that the acoustic pressure on the spherical surface is given as $\hat{p} = \rho_0 c \hat{v}_z(ka)/(ka + i)$. Determine the radiated acoustic pressure anywhere in free space by using the expansion of the spherical Hankel functions and spherical harmonics.

2.6. Consider a vibrating sphere of radius $r = a$ in free space. Assume that the normal surface velocity of this sphere is a constant $\hat{v}_n = V_0$. Find the radiated acoustic pressure anywhere in free space via the expansion of the spherical Hankel functions and spherical harmonics.

2.7. Consider a vibrating sphere of radius $r = a$ in free space. Assume that the acoustic pressure on the spherical surface is given as $\hat{p} = \rho_0 c \hat{v}_z(ka)(ka + i)\cos\theta / \left(k^2a^2 - 2 + i2ka\right)$. Find the radiated acoustic pressure anywhere in free space by using the expansion of the spherical Hankel functions and spherical harmonics.

2.8. Consider a vibrating sphere of radius $r = a$ in free space. Assume that the normal surface velocity of this sphere is given as $\hat{v}_n = V_0 \sin\theta$, where θ is the polar angle. Determine the radiated acoustic pressure in free space by using the expansion of the spherical Hankel functions and spherical harmonics. Will the acoustic pressure contain any resonance frequency? Why?

2.9. Consider the acoustic pressure field inside a sphere of radius $r = a$. Assume that the acoustic pressure on the interior surface is constant $\hat{p} = C$. Determine the acoustic pressure field in the interior region by using the expansion of the spherical Hankel functions and spherical harmonics. Will the acoustic pressure contain any resonance frequency? If so, what are these resonance frequencies?

2.10. Consider the acoustic pressure field inside a sphere of radius $r = a$. Assume that the sphere is oscillating back and forth along the x-axis direction, and the normal surface velocity is given as $\hat{v}_z = V_0 \sin\theta$, where θ is the polar angle. Solve the radiated acoustic pressure field in the interior region using the expansion of the spherical Hankel functions and spherical harmonics. Will the acoustic pressure contain any resonance frequency? If so, what are these resonance frequencies?

Chapter 3
The Helmholtz Equation Least-Squares Method

In engineering applications, most vibrating surfaces are of arbitrary shapes. Moreover, the environments are often nonideal such that the radiated acoustic pressure field cannot be solved by any analytic methods, including expansion theories. Therefore, approximate solutions are sought. The Helmholtz equation least-squares (HELS) method [36, 37] offers such approximate solutions to a wide variety of acoustic radiation problems encountered in practice. Note that HELS can not only be used to reconstruct but also to predict the radiated acoustic field emitted by an arbitrarily shaped vibrating body.

If Fourier transform-based NAH is the first generation and BEM-based NAH the second in the evolution of NAH technology because of their respective significance, the third generation would be HELS-based NAH. Unlike the first two generations, HELS-based NAH does not seek analytic solutions to the acoustic fields generated by arbitrarily shaped structures that cannot be found anyway. Rather, it attempts to find the best approximation of an acoustic field through an expansion of certain basis functions. This approach greatly simplifies reconstruction, yet enables one to tackle a complex problem involving an arbitrarily shaped surface in a non-free field with fewer measurement points than both the Fourier transform and BEM-based NAH do. This makes HELS potentially a practical and versatile tool for diagnosing noise and vibration problems.

The development of HELS method started in the mid-1990s. In HELS the acoustic pressure is written as an expansion of the particular solution to the Helmholtz equation. Using the spherical coordinates, the particular solution is expressible as the spherical wave functions. The coefficients associated with the expansion functions can be determined by solving an overdetermined linear system of equations obtained by matching the assumed-form solution to the measured acoustic pressures, and the errors incurred in this process are minimized by the least-squares method.

S.F. Wu, *The Helmholtz Equation Least Squares Method*, Modern Acoustics and Signal Processing, DOI 10.1007/978-1-4939-1640-5_3

3.1 The HELS Formulations

To put it simply, the HELS method uses an expansion of some basis functions to describe the acoustic pressure generated by an arbitrary source anywhere. The requirements on the basis functions are that they must satisfy the Helmholtz equation and are bounded. Therefore different basis functions may be used, depending on whether the region in which the HELS solution is intended for is external or internal to a source surface. There is no restriction whatsoever on the choice of coordinate system, but the rule of thumb is that for a blunt object whose aspect ratio is close to (1:1:1), then the spherical coordinate system is a natural choice. Similarly, for an elongated object whose aspect ratio is close to (1:1:10), the prolate coordinate would be ideal, and for a discoidal object whose aspect ratio is close to (1:10:10), the oblate would be best. In the matrix form, the HELS formulation can be expressed as follows:

$$\left\{\hat{p}\left(\vec{x};\omega\right)\right\} = \left[\Psi^{(1)}\left(\vec{x};\omega\right)\right]\{C(\omega)\}, \tag{3.1}$$

where $\hat{p}\left(\vec{x};\omega\right)$ implies the complex amplitude of the acoustic pressure at any field point $\vec{x}$, and $\Psi_{ij}^{(1)}\left(\vec{x};\omega\right)$ are the particular solutions to the Helmholtz equation. Using the spherical coordinates, one can write $\Psi_{ij}^{(1)}\left(\vec{x};\omega\right)$ as

$$\Psi_{ij}^{(1)}\left(\vec{x};\omega\right) \equiv \Psi_{nl}^{(1)}(r,\theta,\phi;\omega) = h_n^{(1)}(kr)Y_n^l(\theta,\phi), \tag{3.2}$$

where the first index i in $\Psi_{ij}^{(1)}\left(\vec{x};\omega\right)$ indicates the ith reconstruction point and the second index j implies the jth term of expansion functions. When the spherical coordinates are used, these indices can be replaced by n and l, where n is associated with the radial functions and l is with the angular functions. The indices j, n, and l in Eq. (3.2) are related via $j = n^2 + n + l + 1$ with n starting from 0 to N and l from $-n$ to $+n$. $\Psi_{ij}^{(1)}\left(\vec{x};\omega\right)$ are orthogonal with respect to the angular coordinates

$$\int_0^{2\pi} d\phi \int_0^{\pi} \Psi_{n'l'}^{(1)}(r,\theta,\phi;\omega)\Psi_{nl}^{(1)*}(r,\theta,\phi;\omega)\sin\theta d\theta$$
$$= h_{n'}^{(1)}(kr)h_n^{(1)*}(kr)\int_0^{2\pi} d\phi \int_0^{\pi} Y_{n'}^{l'}(\theta,\phi)Y_n^{l*}(\theta,\phi)\sin\theta d\theta = h_{n'}^{(1)}(kr)h_n^{(1)*}(kr)\delta_{n'n}\delta_{m'm}. \tag{3.3}$$

From Chap. 2 we have learned that the expansion functions $\Psi_{ij}^{(1)}\left(\vec{x};\omega\right)$ form a complete basis from which any function defined on a spherical surface can be uniquely represented.

Using the spherical Hankel functions and spherical harmonics given in Examples 2.1 and 2.9, we can write the first few terms ($n = 0$, 1, 2) of the expansion functions $\Psi_{ij}^{(1)}\left(\vec{x};\omega\right)$ as follows:

For $n = 0$,

$$\Psi_{i1}^{(1)}(r,\theta,\phi;\omega) = -i\frac{1}{2\sqrt{\pi}}\frac{e^{ikr}}{kr}. \tag{3.4}$$

For $n = 1$,

$$\Psi_{i2}^{(1)}(r,\theta,\phi;\omega) = -\frac{1}{2}\sqrt{\frac{3}{\pi}}\frac{(kr+i)\cos\theta}{(kr)^2}e^{ikr}, \tag{3.5}$$

$$\Psi_{i3}^{(1)}(r,\theta,\phi;\omega) = -\frac{1}{2}\sqrt{\frac{3}{2\pi}}\frac{(kr+i)\sin\theta}{(kr)^2}e^{i(kr-\phi)}, \tag{3.6}$$

$$\Psi_{i4}^{(1)}(r,\theta,\phi;\omega) = \frac{1}{2}\sqrt{\frac{3}{2\pi}}\frac{(kr+i)\sin\theta}{(kr)^2}e^{i(kr+\phi)}. \tag{3.7}$$

For $n = 2$,

$$\Psi_{i5}^{(1)}(r,\theta,\phi;\omega) = i\frac{1}{4}\sqrt{\frac{5}{\pi}}\frac{\left[(kr)^2 - 3 + i3(kr)\right](3\cos^2\theta - 1)}{(kr)^3}e^{i(kr+\phi)}, \tag{3.8}$$

$$\Psi_{i6}^{(1)}(r,\theta,\phi;\omega) = i\frac{3}{2}\sqrt{\frac{5}{6\pi}}\frac{\left[(kr)^2 - 3 + i3(kr)\right]\sin\theta\cos\theta}{(kr)^3}e^{i(kr-\phi)}, \tag{3.9}$$

$$\Psi_{i7}^{(1)}(r,\theta,\phi;\omega) = -i\frac{3}{2}\sqrt{\frac{5}{6\pi}}\frac{\left[(kr)^2 - 3 + i3(kr)\right]\sin\theta\cos\theta}{(kr)^3}e^{i(kr+\phi)}, \tag{3.10}$$

$$\Psi_{i8}^{(1)}(r,\theta,\phi;\omega) = i\frac{3}{4}\sqrt{\frac{5}{6\pi}}\frac{\left[(kr)^2 - 3 + i3(kr)\right]\sin^2\theta}{(kr)^3}e^{i(kr-2\phi)}, \tag{3.11}$$

$$\Psi_{i9}^{(1)}(r,\theta,\phi;\omega) = i\frac{3}{4}\sqrt{\frac{5}{6\pi}}\frac{\left[(kr)^2 - 3 + i3(kr)\right]\sin^2\theta}{(kr)^3}e^{i(kr+2\phi)}. \tag{3.12}$$

Physically, the $n = 0$ term (3.4) represents a monopole source, the $n = 1$ terms (3.5)–(3.7) depict the effects of three dipole sources in three mutually orthogonal axes directions, and $n = 2$ terms (3.8)–(3.11) describe the effects of five quadrupole sources, for which the terms given by Eqs. (3.8)–(3.10) portray the effects of longitudinal quadrupoles, and those given by Eqs. (3.11) and (3.12) describe the effects of lateral quadrupoles. In general the number of terms in the nth index is $(2n + 1)$ and that of total expansion terms is $J = (n + 1)^2$.

The coefficients $\{C(\omega)\}$ associated with the expansion functions are obtained by matching Eq. (3.1) to the measured acoustic pressures $\hat{p}\left(\vec{x}_m^{\,meas};\omega\right)$ at $\vec{x}_m^{\,meas}$ on the hologram surface, where $m = 1$ to M, J indicates the number of basis functions, and M the number of measurement points. In practice we take more measurement points than the number of expansion terms, $M > J$. Hence Eq. (3.1) becomes an overdetermined set of equations for the coefficients $\{C(\omega)\}$,

$$\left\{\hat{p}\left(\vec{x}_m^{\,meas};\omega\right)\right\}_{M\times 1} = \left[\Psi\left(\vec{x}_m^{\,meas};\omega\right)\right]_{M\times J}\{C(\omega)\}_{J\times 1}. \tag{3.13}$$

Solving Eq. (3.13) by taking a pseudo inversion, we obtain

$$\begin{aligned}\{C(\omega)\}_{J\times 1} = {} & \left(\left[\Psi\left(\vec{x}_m^{\,meas};\omega\right)\right]^H_{J\times M}\left[\Psi\left(\vec{x}_m^{\,meas};\omega\right)\right]_{M\times J}\right)^{-1} \\ & \left[\Psi\left(\vec{x}_m^{\,meas};\omega\right)\right]^H_{J\times M}\left\{\hat{p}\left(\vec{x}_m^{\,meas};\omega\right)\right\}_{M\times 1},\end{aligned} \tag{3.14}$$

where the superscript H represents Hermitian transpose.

Once the expansion coefficients are specified, the acoustic pressure at $\vec{x}_S^{\,rec}$ on the source surface can be reconstructed by substituting Eq. (3.14) into (3.1),

$$\left\{\hat{p}\left(\vec{x}_s^{\,rec};\omega\right)\right\}_{S\times 1} = \left[G_{pp}\left(\vec{x}_s^{\,rec}\middle|\vec{x}_m^{\,meas};\omega\right)\right]_{S\times M}\left\{\hat{p}\left(\vec{x}_m^{\,meas};\omega\right)\right\}_{M\times 1}, \tag{3.15}$$

where $\left[G_{pp}\left(\vec{x}_s^{\,rec}\middle|\vec{x}_m^{\,meas};\omega\right)\right]_{S\times M}$ represents the transfer matrix that correlates the measured acoustic pressure at $\vec{x}_m^{\,meas}$ to the reconstructed acoustic pressure at $\vec{x}_S^{\,rec}$,

$$\begin{aligned}\left[G_{pp}\left(\vec{x}_s^{\,rec}\middle|\vec{x}_m^{\,meas};\omega\right)\right]_{S\times M} = {} & \left[\Psi\left(\vec{x}_s^{\,rec};\omega\right)\right]_{S\times J}\left(\left[\Psi\left(\vec{x}_m^{\,meas};\omega\right)\right]^H_{J\times M}\left[\Psi\left(\vec{x}_m^{\,meas};\omega\right)\right]_{M\times J}\right)^{-1} \\ & \left[\Psi\left(\vec{x}_m^{\,meas};\omega\right)\right]^H_{J\times M}.\end{aligned} \tag{3.16}$$

The normal component of the surface velocity $\hat{v}_n\left(\vec{x}_s^{\,rec};\omega\right)$ can be obtained from Eq. (3.15) by using the Euler's equation,

$$\left\{\hat{v}_{\mathrm{n}}\left(\vec{x}_s^{\,\mathrm{rec}};\omega\right)\right\}_{S\times 1} = \left[G_{pv}\left(\vec{x}_s^{\,\mathrm{rec}}\middle|\vec{x}_m^{\,\mathrm{meas}};\omega\right)\right]_{S\times M}\left\{\hat{p}\left(\vec{x}_m^{\,\mathrm{meas}};\omega\right)\right\}_{M\times 1}, \tag{3.17}$$

where $\left[G_{pv}\left(\vec{x}_s^{\,\mathrm{rec}}\middle|\vec{x}_m^{\,\mathrm{meas}};\omega\right)\right]_{S\times M}$ represents the transfer matrix that correlates the measured acoustic pressure at $\vec{x}_m^{\,\mathrm{meas}}$ to the reconstructed normal surface velocity at $\vec{x}_S^{\,\mathrm{rec}}$,

$$\begin{aligned}\left[G_{pv}\left(\vec{x}_s^{\,\mathrm{rec}}\middle|\vec{x}_m^{\,\mathrm{meas}};\omega\right)\right]_{S\times M} &= \frac{1}{i\omega\rho_0}\left[\frac{\partial\Psi\left(\vec{x}_s^{\,\mathrm{rec}};\omega\right)}{\partial\mathrm{n}}\right]_{S\times J} \\ &\left(\left[\Psi\left(\vec{x}_m^{\,\mathrm{meas}};\omega\right)\right]_{J\times M}^{H}\left[\Psi\left(\vec{x}_m^{\,\mathrm{meas}};\omega\right)\right]_{M\times J}\right)^{-1}\left[\Psi\left(\vec{x}_m^{\,\mathrm{meas}};\omega\right)\right]_{J\times M}^{H}.\end{aligned} \tag{3.18}$$

Once $\hat{p}\left(\vec{x}_m^{\,meas};\omega\right)$ and $\hat{v}_{\mathrm{n}}\left(\vec{x}_s^{\,\mathrm{rec}};\omega\right)$ are reconstructed, the normal component of the time-averaged acoustic intensity on the source surface can be calculated as

$$\left\{\hat{I}_{av,\mathrm{n}}\left(\vec{x}_s^{\,\mathrm{rec}};\omega\right)\right\}_{S\times 1} = \frac{1}{2}\mathrm{Re}\left\{\hat{p}\left(\vec{x}_s^{\,\mathrm{rec}};\omega\right)\hat{v}_{\mathrm{n}}^{*}\left(\vec{x}_s^{\,\mathrm{rec}};\omega\right)\right\}_{S\times 1}. \tag{3.19}$$

The radiated acoustic power can be obtained by integrating the normal component of the time-averaged acoustic intensity over the source surface,

$$\mathbf{P}_{av}(\omega) = \iint_S \hat{I}_{av,\mathrm{n}}\left(\vec{x}_s^{\,\mathrm{rec}};\omega\right)dS. \tag{3.20}$$

Therefore all acoustic quantities generated by this source are completely determined.

3.2 Reconstructing the Radiated Acoustic Field

Now let us use the HELS formulation (3.1) to reconstruct the acoustic pressures based on the measurements taken on a hologram surface surround a vibrating object.

Example 3.1 For simplicity yet without loss of generality, let us consider a vibrating sphere of the radius $r = a$ at a constant frequency f. Suppose that the acoustic pressures on a spherical hologram surface Γ of radius $r = r_m^{\mathrm{meas}}$ are taken and the acoustic pressures anywhere in the field including the vibrating surface are

Fig. 3.1 Schematic of reconstructing the acoustic field generated by a dilating sphere around radius $r = a$ using HELS-based NAH. The measured acoustic pressures are taken on a concentric spherical surface of radius r_m

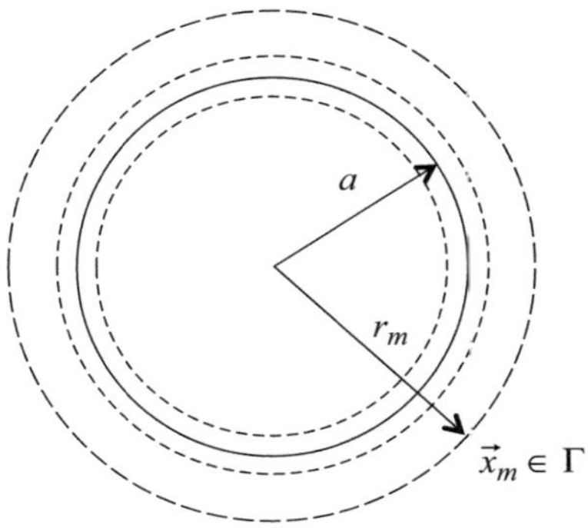

sought (see Fig. 3.1). To illustrate the use of Eq. (3.1), we assume that the number of measurements and that of reconstruction points be equal and set $S = M = 2$, and the reconstructed acoustic pressures are set to equal to the measured ones.

$$\begin{Bmatrix} \hat{p}\left(r_1^{\text{meas}}, \theta_1^{\text{meas}}, \phi_1^{\text{meas}}; \omega\right) \\ \hat{p}\left(r_2^{\text{meas}}, \theta_2^{\text{meas}}, \phi_2^{\text{meas}}; \omega\right) \end{Bmatrix}_{2\times 1}$$

$$= \begin{bmatrix} \Psi_{11}^{(1)}\left(r_1^{\text{meas}}, \theta_1^{\text{meas}}, \phi_1^{\text{meas}}; \omega\right) & \Psi_{12}^{(1)}\left(r_1^{\text{meas}}, \theta_1^{\text{meas}}, \phi_1^{\text{meas}}; \omega\right) \\ \Psi_{21}^{(1)}\left(r_2^{\text{meas}}, \theta_2^{\text{meas}}, \phi_2^{\text{meas}}; \omega\right) & \Psi_{22}^{(1)}\left(r_2^{\text{meas}}, \theta_2^{\text{meas}}, \phi_2^{\text{meas}}; \omega\right) \end{bmatrix}_{2\times 2} \begin{Bmatrix} C_1(\omega) \\ C_2(\omega) \end{Bmatrix}_{2\times 1},$$

where $\Psi_{ij}^{(1)}(r_i^{\text{meas}}, \theta_i^{\text{meas}}, \phi_i^{\text{meas}}; \omega)$ are given by

$$\Psi_{11}^{(1)}\left(r_1^{\text{meas}}, \theta_1^{\text{meas}}, \phi_1^{\text{meas}}; \omega\right) = -i\frac{1}{2\sqrt{\pi}}\frac{e^{ikr_1^{\text{meas}}}}{kr_1^{\text{meas}}},$$

$$\Psi_{12}^{(1)}\left(r_1^{\text{meas}}, \theta_1^{\text{meas}}, \phi_1^{\text{meas}}; \omega\right) = -\frac{1}{2}\sqrt{\frac{3}{\pi}}\frac{\left(kr_1^{\text{meas}} + i\right)\cos\theta_1^{\text{meas}}}{\left(kr_1^{\text{meas}}\right)^2}e^{ikr_1^{\text{meas}}},$$

$$\Psi_{21}^{(1)}\left(r_2^{\text{meas}}, \theta_2^{\text{meas}}, \phi_2^{\text{meas}}; \omega\right) = -i\frac{1}{2\sqrt{\pi}}\frac{e^{ikr_2^{\text{meas}}}}{kr_2^{\text{meas}}},$$

$$\Psi_{22}^{(1)}\left(r_2^{\text{meas}}, \theta_2^{\text{meas}}, \phi_2^{\text{meas}}; \omega\right) = -\frac{1}{2}\sqrt{\frac{3}{\pi}}\frac{\left(kr_2^{\text{meas}} + i\right)\cos\theta_2^{\text{meas}}}{\left(kr_2^{\text{meas}}\right)^2}e^{ikr_2^{\text{meas}}}.$$

The expansion coefficients $\{C(\omega)\}$ can be determined by inverting the square matrix directly:

$$\begin{Bmatrix} C_1(\omega) \\ C_2(\omega) \end{Bmatrix}_{2\times 1} = \begin{bmatrix} \Psi_{11}^{(1)}\left(r_1^{\text{meas}},\theta_1^{\text{meas}},\phi_1^{\text{meas}};\omega\right) & \Psi_{12}^{(1)}\left(r_1^{\text{meas}},\theta_1^{\text{meas}},\phi_1^{\text{meas}};\omega\right) \\ \Psi_{21}^{(1)}\left(r_2^{\text{meas}},\theta_2^{\text{meas}},\phi_2^{\text{meas}};\omega\right) & \Psi_{22}^{(1)}\left(r_2^{\text{meas}},\theta_2^{\text{meas}},\phi_2^{\text{meas}};\omega\right) \end{bmatrix}_{2\times 2}^{-1} \begin{Bmatrix} \hat{p}\left(r_1^{\text{meas}},\theta_1^{\text{meas}},\phi_1^{\text{meas}};\omega\right) \\ \hat{p}\left(r_2^{\text{meas}},\theta_2^{\text{meas}},\phi_2^{\text{meas}};\omega\right) \end{Bmatrix}_{2\times 1}$$

$$= \frac{1}{\det\left[\Psi^{(1)}(r,\theta,\phi;\omega)\right]} \begin{bmatrix} \Psi_{22}^{(1)}\left(r_2^{\text{meas}},\theta_2^{\text{meas}},\phi_2^{\text{meas}};\omega\right) & -\Psi_{12}^{(1)}\left(r_1^{\text{meas}},\theta_1^{\text{meas}},\phi_1^{\text{meas}};\omega\right) \\ -\Psi_{21}^{(1)}\left(r_2^{\text{meas}},\theta_2^{\text{meas}},\phi_2^{\text{meas}};\omega\right) & \Psi_{11}^{(1)}\left(r_1^{\text{meas}},\theta_1^{\text{meas}},\phi_1^{\text{meas}};\omega\right) \end{bmatrix}_{2\times 2} \begin{Bmatrix} \hat{p}\left(r_1^{\text{meas}},\theta_1^{\text{meas}},\phi_1^{\text{meas}};\omega\right) \\ \hat{p}\left(r_2^{\text{meas}},\theta_2^{\text{meas}},\phi_2^{\text{meas}};\omega\right) \end{Bmatrix}_{2\times 1},$$

where the determinant of the transfer matrix is given by

$$\begin{aligned} &\det\left[\Psi^{(1)}(r,\theta,\phi;\omega)\right] \\ &= \Psi_{11}^{(1)}\left(r_1^{\text{meas}},\theta_1^{\text{meas}},\phi_1^{\text{meas}};\omega\right)\Psi_{22}^{(1)}\left(r_2^{\text{meas}},\theta_2^{\text{meas}},\phi_2^{\text{meas}};\omega\right) \\ &-\Psi_{12}^{(1)}\left(r_1^{\text{meas}},\theta_1^{\text{meas}},\phi_1^{\text{meas}};\omega\right)\Psi_{21}^{(1)}\left(r_2^{\text{meas}},\theta_2^{\text{meas}},\phi_2^{\text{meas}};\omega\right). \\ &= \frac{\sqrt{3}e^{ik\left(r_1^{\text{meas}}+r_2^{\text{meas}}\right)}}{4\pi}\left[\frac{r_2^{\text{meas}}\left(1-ikr_1^{\text{meas}}\right)\cos\theta_1^{\text{meas}} - r_1^{\text{meas}}\left(1-ikr_2^{\text{meas}}\right)\cos\theta_2^{\text{meas}}}{k^3\left(r_1^{\text{meas}}r_2^{\text{meas}}\right)^2}\right] \end{aligned}$$

Substituting the determinant and $\Psi_{ij}^{(1)}(r_i,\theta_i,\phi_i;\omega)$ into the elements of $\{C(\omega)\}$, we obtain

$$\begin{aligned} &\frac{\Psi_{22}^{(1)}\left(r_2^{\text{meas}},\theta_2^{\text{meas}},\phi_2^{\text{meas}};\omega\right)}{\det\left[\Psi^{(1)}(r,\theta,\phi;\omega)\right]} \\ &\qquad = -\frac{2\sqrt{\pi}k\left(r_1^{\text{meas}}\right)^2\left(kr_2^{\text{meas}}+i\right)\cos\theta_2^{\text{meas}}e^{-ikr_1^{\text{meas}}}}{r_2^{\text{meas}}\left(1-ikr_1^{\text{meas}}\right)\cos\theta_1^{\text{meas}} - r_1^{\text{meas}}\left(1-ikr_2^{\text{meas}}\right)\cos\theta_2^{\text{meas}}}, \end{aligned}$$

$$-\frac{\Psi_{12}^{(1)}\left(r_1^{\text{meas}},\theta_1^{\text{meas}},\phi_1^{\text{meas}};\omega\right)}{\det\left[\Psi^{(1)}(r,\theta,\phi;\omega)\right]}$$
$$=\frac{2\sqrt{\pi}k\left(r_2^{\text{meas}}\right)^2\left(kr_1^{\text{meas}}+i\right)\cos\theta_1^{\text{meas}}\mathrm{e}^{-ikr_2^{\text{meas}}}}{r_2^{\text{meas}}\left(1-ikr_1^{\text{meas}}\right)\cos\theta_1^{\text{meas}}-r_1^{\text{meas}}\left(1-ikr_2^{\text{meas}}\right)\cos\theta_2^{\text{meas}}},$$
$$-\frac{\Psi_{21}^{(1)}\left(r_2^{\text{meas}},\theta_2^{\text{meas}},\phi_2^{\text{meas}};\omega\right)}{\det\left[\Psi^{(1)}(r,\theta,\phi;\omega)\right]}$$
$$=i2\sqrt{\frac{\pi}{3}}\frac{k^2\left(r_1^{\text{meas}}\right)^2r_2^{\text{meas}}\mathrm{e}^{-ikr_1^{\text{meas}}}}{r_2^{\text{meas}}\left(1-ikr_1^{\text{meas}}\right)\cos\theta_1^{\text{meas}}-r_1^{\text{meas}}\left(1-ikr_2^{\text{meas}}\right)\cos\theta_2^{\text{meas}}},$$
$$\frac{\Psi_{11}^{(1)}\left(r_1^{\text{meas}},\theta_1^{\text{meas}},\phi_1^{\text{meas}};\omega\right)}{\det\left[\Psi^{(1)}(r,\theta,\phi;\omega)\right]}$$
$$=-i2\sqrt{\frac{\pi}{3}}\frac{k^2r_1^{\text{meas}}\left(r_2^{\text{meas}}\right)^2\mathrm{e}^{-ikr_2^{\text{meas}}}}{r_2^{\text{meas}}\left(1-ikr_1^{\text{meas}}\right)\cos\theta_1^{\text{meas}}-r_1^{\text{meas}}\left(1-ikr_2^{\text{meas}}\right)\cos\theta_2^{\text{meas}}}.$$

Accordingly, the reconstructed acoustic pressures are given by

$$\begin{Bmatrix}\hat{p}\left(r_1^{\text{rec}},\theta_1^{\text{rec}},\phi_1^{\text{rec}};\omega\right)\\ \hat{p}\left(r_2^{\text{rec}},\theta_2^{\text{rec}},\phi_2^{\text{rec}};\omega\right)\end{Bmatrix}=-k\begin{bmatrix}i\dfrac{\mathrm{e}^{ikr_1^{\text{rec}}}}{kr_1^{\text{rec}}} & \dfrac{\sqrt{3}\left(kr_1^{\text{rec}}+i\right)\cos\theta_1^{\text{rec}}}{\left(kr_1^{\text{meas}}\right)^2}\mathrm{e}^{ikr_1^{\text{rec}}}\\ i\dfrac{\mathrm{e}^{ikr_2^{\text{rec}}}}{kr_2^{\text{rec}}} & \dfrac{\sqrt{3}\left(kr_2^{\text{rec}}+i\right)\cos\theta_2^{\text{rec}}}{\left(kr_2^{\text{rec}}\right)^2}\mathrm{e}^{ikr_2^{\text{rec}}}\end{bmatrix}$$
$$\times\begin{bmatrix}-\dfrac{\left(r_1^{\text{meas}}\right)^2\left(kr_2^{\text{meas}}+i\right)\cos\theta_2^{\text{meas}}\mathrm{e}^{-ikr_1^{\text{meas}}}}{\Delta} & \dfrac{\left(r_2^{\text{meas}}\right)^2\left(kr_2^{\text{meas}}+i\right)\cos\theta_2^{\text{meas}}\mathrm{e}^{-ikr_2^{\text{meas}}}}{\Delta}\\ i\dfrac{\left(kr_2^{\text{meas}}\right)\left(r_1^{\text{meas}}\right)^2\mathrm{e}^{-ikr_1^{\text{meas}}}}{\sqrt{3}\Delta} & -i\dfrac{\left(kr_1^{\text{meas}}\right)\left(r_2^{\text{meas}}\right)^2\mathrm{e}^{-ikr_2^{\text{meas}}}}{\sqrt{3}\Delta}\end{bmatrix}$$
$$\begin{Bmatrix}\hat{p}\left(r_1^{\text{meas}},\theta_1^{\text{meas}},\phi_1^{\text{meas}};\omega\right)\\ \hat{p}\left(r_2^{\text{meas}},\theta_2^{\text{meas}},\phi_2^{\text{meas}};\omega\right)\end{Bmatrix},$$

where $\Delta=r_2^{\text{meas}}(1-ikr_1^{\text{meas}})\cos\theta_1^{\text{meas}}-r_1^{\text{meas}}(1-ikr_2^{\text{meas}})\cos\theta_2^{\text{meas}}$.

These results are of generality because we have neither specified the measurement points $(r_m^{\text{meas}},\theta_m^{\text{meas}},\phi_m^{\text{meas}};\omega)$, $m=1$ and 2, and the reconstruction points, $(r_s^{\text{rec}},\theta_s^{\text{rec}},\phi_s^{\text{rec}};\omega)$, $m=1$ and 2, nor stipulated the measured acoustic pressures $\hat{p}\left(r_m^{\text{meas}},\theta_m^{\text{meas}},\phi_m^{\text{meas}};\omega\right)$. In practice, however, it will be a good idea to take measurements on a conformal surface very close to the target source surfaces so as to capture as much near-field information as possible.

Once the acoustic pressure is reconstructed, the particle velocity anywhere in the field can be obtained by using the Euler's equation,

$$\vec{\hat{v}}\left(r^{\mathrm{rec}},\theta^{\mathrm{rec}},\phi^{\mathrm{rec}};\omega\right)=\frac{1}{i\omega\rho_0}\nabla\hat{p}\left(r^{\mathrm{rec}},\theta^{\mathrm{rec}},\phi^{\mathrm{rec}};\omega\right).$$

The time-averaged acoustic intensity anywhere in the field is given by

$$\vec{I}_{av}\left(r^{\mathrm{rec}},\theta^{\mathrm{rec}},\phi^{\mathrm{rec}};\omega\right)=\frac{1}{2}\mathrm{Re}\left[\hat{p}\left(r^{\mathrm{rec}},\theta^{\mathrm{rec}},\phi^{\mathrm{rec}};\omega\right)\vec{\hat{v}}^{*}\left(r^{\mathrm{rec}},\theta^{\mathrm{rec}},\phi^{\mathrm{rec}};\omega\right)\right].$$

Therefore, all acoustic quantities in the entire field including the source surface are completely determined.

Example 3.2 Now let us specify the input data $\left\{\hat{p}\left(r_m^{\mathrm{meas}},\theta_m^{\mathrm{meas}},\phi_m^{\mathrm{meas}};\omega\right)\right\}_{M\times 1}$. First, we consider the dilating sphere of radius $r=a$ at a constant frequency f. Suppose that the normal surface velocity on this dilating sphere is $\hat{v}_{\mathrm{n}}$. Then acoustic pressure at any arbitrary field point is given by

$$\hat{p}\left(r_m^{\mathrm{meas}},\theta_m^{\mathrm{meas}},\phi_m^{\mathrm{meas}};\omega\right)=\frac{\rho_0 c\hat{v}_{\mathrm{n}}ka^2\mathrm{e}^{ik\left(r_m^{\mathrm{meas}}-a\right)}}{(ka+i)r_m^{\mathrm{meas}}},\qquad m=\ 1\text{ and }2.$$

Substituting these data into the expansion coefficients given in Example 3.1 yields

$$\begin{Bmatrix}C_1(\omega)\\ C_2(\omega)\end{Bmatrix}=2\sqrt{\pi}k\begin{bmatrix}-\dfrac{\left(r_1^{\mathrm{meas}}\right)^2\left(kr_2^{\mathrm{meas}}+i\right)\cos\theta_2\mathrm{e}^{-ikr_1^{\mathrm{meas}}}}{\Delta} & \dfrac{\left(r_2^{\mathrm{meas}}\right)^2\left(kr_1^{\mathrm{meas}}+i\right)\cos\theta_1\mathrm{e}^{-ikr_2^{\mathrm{meas}}}}{\Delta}\\ i\dfrac{\left(kr_2^{\mathrm{meas}}\right)\left(r_1^{\mathrm{meas}}\right)^2\mathrm{e}^{-ikr_1^{\mathrm{meas}}}}{\sqrt{3}\Delta} & -i\dfrac{\left(kr_1^{\mathrm{meas}}\right)\left(r_2^{\mathrm{meas}}\right)^2\mathrm{e}^{-ikr_2^{\mathrm{meas}}}}{\sqrt{3}\Delta}\end{bmatrix}$$

$$\times\begin{Bmatrix}\rho_0c\hat{v}_{\mathrm{n}}\dfrac{(ka)}{(ka+i)}\left(\dfrac{a}{r_1^{\mathrm{meas}}}\right)\mathrm{e}^{ik\left(r_1^{\mathrm{meas}}-a\right)}\\ \rho_0c\hat{v}_{\mathrm{n}}\dfrac{(ka)}{(ka+i)}\left(\dfrac{a}{r_2^{\mathrm{meas}}}\right)\mathrm{e}^{ik\left(r_2^{\mathrm{meas}}-a\right)}\end{Bmatrix}=\begin{Bmatrix}i2\sqrt{\pi}\rho_0c\hat{v}_{\mathrm{n}}\dfrac{(ka)^2}{(ka+i)}\mathrm{e}^{-ika}\\ 0\end{Bmatrix},$$

where Δ is given in Example 3.1.

Once the expansion coefficients are determined, the acoustic pressure anywhere including the source surface can be reconstructed by using the HELS formulation (3.1).

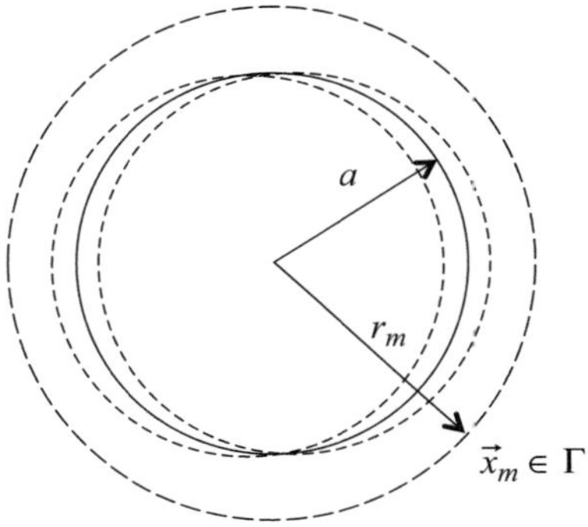

Fig. 3.2 Schematic of reconstructing the acoustic field generated by an oscillating sphere of radius $r = a$ using HELS-based NAH. The measured acoustic pressures are taken on a concentric spherical surface of radius r_m

$$\hat{p}\left(r^{\text{rec}}, \theta^{\text{rec}}, \phi^{\text{rec}}; \omega\right) = \left\{ -i\frac{1}{2\sqrt{\pi}}\frac{e^{ikr^{\text{rec}}}}{kr^{\text{rec}}} \quad -\frac{1}{2}\sqrt{\frac{3}{\pi}}\frac{\left(kr^{\text{rec}}+i\right)\cos\theta^{\text{rec}}}{\left(kr^{\text{rec}}\right)^2}e^{ikr^{\text{rec}}} \right\}$$
$$\left\{ \begin{array}{c} i2\sqrt{\pi}\rho_0 c\hat{v}_{\text{n}}\dfrac{(ka)^2}{(ka+i)}e^{-ika} \\ 0 \end{array} \right\}$$
$$= \rho_0 c\hat{v}_{\text{n}}\left(\frac{ka}{ka+i}\right)\left(\frac{a}{r^{rec}}\right)e^{ik\left(r^{\text{rec}}-a\right)},$$

which matches perfectly the analytic solution [42].

Example 3.3 Next, we reconstruct the acoustic pressure generated by a sphere of radius $r = a$ that oscillates in the z-axis direction at a constant frequency f (see Fig. 3.2). Suppose that the normal surface velocity of this oscillating sphere is $\hat{v}_c \cos\theta$. The acoustic pressures at any field point can be written as

$$\hat{p}\left(r_m^{\text{meas}}, \theta_m^{\text{meas}}, \phi_m^{\text{meas}}; \omega\right) = \frac{\rho_0 c\hat{v}_z a^2\left(kr_m^{\text{meas}}+i\right)\cos\theta_m^{\text{meas}}e^{ik\left(r_m^{\text{meas}}-a\right)}}{\left(k^2a^2-2+i2ka\right)\left(r_m^{\text{meas}}\right)^2}$$
$$= \frac{C_0 a^2\left(kr_m^{\text{meas}}+i\right)\cos\theta_m^{\text{meas}}e^{ikr_m^{\text{meas}}}}{\left(r_m^{\text{meas}}\right)^2},$$

where $m = 1$ and 2, and the constant C_0 is given by

$$C_0 = \rho_0 c\hat{v}_z\left(\frac{ka}{k^2a^2-2+i2ka}\right)e^{-ika}.$$

Accordingly, the expansion coefficients are given by

$$\begin{Bmatrix} C_1(\omega) \\ C_2(\omega) \end{Bmatrix} = 2\sqrt{\pi}k \begin{bmatrix} -\dfrac{\left(r_1^{\text{meas}}\right)^2\left(kr_2^{\text{meas}}+i\right)\cos\theta_2^{\text{meas}}e^{-ikr_1^{\text{meas}}}}{\Delta} & \dfrac{\left(r_2^{\text{meas}}\right)^2\left(kr_1^{\text{meas}}+i\right)\cos\theta_1^{\text{meas}}e^{-ikr_2^{\text{meas}}}}{\Delta} \\ i\dfrac{\left(kr_2^{\text{meas}}\right)\left(r_1^{\text{meas}}\right)^2\text{e}^{-ikr_1^{\text{meas}}}}{\sqrt{3}\Delta} & -i\dfrac{\left(kr_1^{\text{meas}}\right)\left(r_2^{\text{meas}}\right)^2\text{e}^{-ikr_2^{\text{meas}}}}{\sqrt{3}\Delta} \end{bmatrix},$$

$$\times \begin{Bmatrix} \dfrac{C_0a^2\left(kr_1^{\text{meas}}+i\right)\cos\theta_1^{\text{meas}}\text{e}^{ikr_1^{meas}}}{\left(r_1^{\text{meas}}\right)^2} \\ \dfrac{C_0a^2\left(kr_2^{\text{meas}}+i\right)\cos\theta_2^{\text{meas}}\text{e}^{ikr_2^{\text{meas}}}}{\left(r_2^{\text{meas}}\right)^2} \end{Bmatrix} = \begin{Bmatrix} 0 \\ -2\sqrt{\dfrac{\pi}{3}}C_0(ka)^2 \end{Bmatrix}$$

where Δ is given in Example 3.1.

Using Eq. (3.1), the acoustic pressure anywhere including the source surface is given by

$$\begin{aligned} &\hat{p}\left(r^{\text{rec}},\theta^{\text{rec}},\phi^{\text{rec}};\omega\right) \\ &= -\frac{1}{2\sqrt{\pi}}\left\{\frac{i\text{e}^{ikr^{\text{rec}}}}{kr^{\text{rec}}} \quad \frac{\sqrt{3}(kr^{\text{rec}}+i)\cos\theta^{\text{rec}}}{(kr^{\text{rec}})^2}\text{e}^{ikr^{\text{rec}}}\right\}\left\{\begin{matrix} 0 \\ -2\sqrt{\dfrac{\pi}{3}}C_0(ka)^2 \end{matrix}\right\}, \\ &= \rho_0 c\hat{v}_z\frac{(ka)(kr^{\text{rec}}+i)\cos\theta}{(k^2a^2-2+i2ka)}\left(\frac{a}{r^{\text{rec}}}\right)^2\text{e}^{ik(r^{\text{rec}}-a)} \end{aligned}$$

which once again agrees perfectly with the analytic solution [42].

3.3 Predicting the Radiated Acoustic Field

In Examples 3.1–3.3 we have focused on using the HELS formulations to reconstruct the acoustic pressure radiated from a vibrating sphere. Now we demonstrate that the same formulations can be used to predict acoustic radiation from a vibrating object.

To this end, we take the normal derivative of Eq. (3.1), and apply the Euler's equation to express the expansion coefficients $\{C\}_{J\times 1}$ in terms of the normal surface velocity:

$$i\omega\rho_0\left\{\hat{v}_{\text{n}}\left(\vec{x}_s;\omega\right)\right\}_{S\times 1} = \left[\frac{\partial\Psi^{(1)}\left(\vec{x}_s;\omega\right)}{\partial\text{n}}\right]_{S\times J}\{C(\omega)\}_{J\times 1}, \tag{3.21}$$

where $\hat{v}_{\mathrm{n}}\left(\vec{x}_s;\omega\right)$ is the normal component of the particle velocity measured by the source surface at $\vec{x}_s$, $s=1, 2, \ldots, S$, and $\{C\}_{J\times 1}$ are obtained by taking a pseudo inversion of Eq. (3.21).

$$\{C(\omega)\}_{J\times 1} = i\omega\rho_0 \left[\frac{\partial \Psi^{(1)}\left(\vec{x}_s;\omega\right)}{\partial \mathrm{n}}\right]^{\dagger}_{J\times S} \left\{\hat{v}_{\mathrm{n}}\left(\vec{x}_s;\omega\right)\right\}_{S\times 1}, \tag{3.22}$$

where the pseudo inversion in Eq. (3.22) is defined as

$$\begin{aligned}\left[\frac{\partial \Psi^{(1)}\left(\vec{x}_s;\omega\right)}{\partial \mathrm{n}}\right]^{\dagger}_{J\times S} = &\left(\left[\frac{\partial \Psi^{(1)}\left(\vec{x}_s;\omega\right)}{\partial \mathrm{n}}\right]^{H}_{J\times S}\left[\frac{\partial \Psi^{(1)}\left(\vec{x}_s;\omega\right)}{\partial \mathrm{n}}\right]_{S\times J}\right)^{-1} \\ &\left[\frac{\partial \Psi^{(1)}\left(\vec{x}_s;\omega\right)}{\partial \mathrm{n}}\right]^{H}_{J\times S},\end{aligned} \tag{3.23}$$

where a superscript H implies a Hermitian transpose of a matrix.

Substituting Eq. (3.23) into (3.1) leads to the matrix formulation for predicting acoustic radiation from any vibrating structure into three-dimensional space,

$$\left\{\hat{p}\left(\vec{x};\omega\right)\right\}_{N\times 1} = \left[G_{vp}\left(\vec{x}\middle|\vec{x}_s;\omega\right)\right]_{N\times S}\left\{\hat{v}_{\mathrm{n}}\left(\vec{x}_s;\omega\right)\right\}_{S\times 1}, \tag{3.24}$$

where $\left[G_{vp}\left(\vec{x}\middle|\vec{x}_s;\omega\right)\right]_{N\times S}$ indicates the transfer function that correlates the normal surface velocity specified on the source surface to the field acoustic pressure, which is given by

$$\left[G_{vp}\left(\vec{x}\middle|\vec{x}_s;\omega\right)\right]_{N\times S} = i\omega\rho_0\left[\Psi^{(1)}\left(\vec{x};\omega\right)\right]_{N\times J}\left[\frac{\partial \Psi^{(1)}\left(\vec{x}_s;\omega\right)}{\partial \mathrm{n}}\right]^{\dagger}_{J\times S}. \tag{3.25}$$

Equation (3.24) can be utilized to predict the acoustic pressure, given the normal surface velocity on a vibrating object.

Let us consider the case that involves two expansion functions such that Eq. (3.24) may be solved exactly. Moreover, the normal surface velocity is specified at two points on the surface of a vibrating sphere of radius $r=a$ at a constant frequency f, and the acoustic pressures at two arbitrary field points are sought. This scenario leads to a square matrix equation given by

$$\begin{Bmatrix} \hat{p}(r_1,\theta_1,\phi_1;\omega) \\ \hat{p}(r_2,\theta_2,\phi_2;\omega) \end{Bmatrix} = i\omega\rho_0 \begin{bmatrix} \Psi_{11}^{(1)}(r_1,\theta_1,\phi_1;\omega) & \Psi_{12}^{(1)}(r_1,\theta_1,\phi_1;\omega) \\ \Psi_{21}^{(1)}(r_2,\theta_2,\phi_2;\omega) & \Psi_{22}^{(1)}(r_2,\theta_2,\phi_2;\omega) \end{bmatrix}$$

$$\times \begin{bmatrix} +\left.\frac{\partial\Psi_{22}^{(1)}\left(r,\theta_2^{\text{meas}},\phi_2^{\text{meas}};\omega\right)}{\partial r}\right|_{r=a} & -\left.\frac{\partial\Psi_{12}^{(1)}\left(r,\theta_1^{\text{meas}},\phi_1^{\text{meas}};\omega\right)}{\partial r}\right|_{r=a} \\ -\left.\frac{\partial\Psi_{21}^{(1)}\left(r,\theta_2^{\text{meas}},\phi_2^{\text{meas}};\omega\right)}{\partial r}\right|_{r=a} & +\left.\frac{\partial\Psi_{11}^{(1)}\left(r,\theta_1^{\text{meas}},\phi_1^{\text{meas}};\omega\right)}{\partial r}\right|_{r=a} \end{bmatrix} \frac{\begin{Bmatrix} \hat{v}_{\text{n}}\left(a,\theta_1^{\text{meas}},\phi_1^{\text{meas}};\omega\right) \\ \hat{v}_{\text{n}}\left(a,\theta_2^{\text{meas}},\phi_2^{\text{meas}};\omega\right) \end{Bmatrix}}{\det\left[\left.\frac{\partial\Psi^{(1)}(r,\theta,\phi;\omega)}{\partial r}\right|_{r=a}\right]}, \tag{3.26}$$

where the determinant is given by

$$\begin{aligned} \det\left[\left.\frac{\partial\Psi^{(1)}(r,\theta,\phi;\omega)}{\partial r}\right|_{r=a}\right] = & \left.\frac{\partial\Psi_{11}^{(1)}\left(r,\theta_1^{\text{meas}},\phi_1^{\text{meas}};\omega\right)}{\partial r}\right|_{r=a} \left.\frac{\partial\Psi_{22}^{(1)}\left(r,\theta_2^{\text{meas}},\phi_2^{\text{meas}};\omega\right)}{\partial r}\right|_{r=a} \\ & - \left.\frac{\partial\Psi_{21}^{(1)}\left(r,\theta_1^{\text{meas}},\phi_1^{\text{meas}};\omega\right)}{\partial r}\right|_{r=a} \left.\frac{\partial\Psi_{12}^{(1)}\left(r,\theta_2^{\text{meas}},\phi_2^{\text{meas}};\omega\right)}{\partial r}\right|_{r=a}, \end{aligned} \tag{3.27}$$

where the first subscript in the expansion functions implies the order of expansion functions, and the second subscript stands for the sequence of measurement points. The expansion functions and their derivatives are given by

$$\Psi_{11}^{(1)}(r_1,\theta_1,\phi_1;\omega) = -i\frac{1}{2\sqrt{\pi}}\frac{\mathrm{e}^{ikr_1}}{kr_1}, \tag{3.28}$$

$$\Psi_{12}^{(1)}(r_1,\theta_1,\phi_1;\omega) = -\frac{1}{2}\sqrt{\frac{3}{\pi}}\frac{(kr_1+i)\mathrm{e}^{ikr_1}}{k^2r_1^2}\cos\theta_1, \tag{3.29}$$

$$\Psi_{21}^{(1)}(r_2,\theta_2,\phi_2;\omega) = -i\frac{1}{2\sqrt{\pi}}\frac{\mathrm{e}^{ikr_2}}{kr_2}, \tag{3.30}$$

$$\Psi_{22}^{(1)}(r_2,\theta_2,\phi_2;\omega) = -\frac{1}{2}\sqrt{\frac{3}{\pi}}\frac{(kr_2+i)\mathrm{e}^{ikr_2}}{k^2r_2^2}\cos\theta_2 \tag{3.31}$$

$$\left.\frac{\partial\Psi_{11}^{(1)}\left(r,\theta_1^{\text{meas}},\phi_1^{\text{meas}};\omega\right)}{\partial r}\right|_{r=a} = \frac{1}{2\sqrt{\pi}}\frac{(ka+i)\mathrm{e}^{ika}}{ka^2}, \tag{3.32}$$

$$\left.\frac{\partial\Psi_{12}^{(1)}\left(r,\theta_1^{\text{meas}},\phi_1^{\text{meas}};\omega\right)}{\partial r}\right|_{r=a} = \frac{1}{2}\sqrt{\frac{3}{\pi}}\frac{\left[2ka+i\left(2-k^2a^2\right)\right]\mathrm{e}^{ika}}{k^2a^3}\cos\theta_1^{\text{meas}}, \tag{3.33}$$

$$\left.\frac{\partial\Psi_{22}^{(1)}\left(r,\theta_2^{\text{meas}},\phi_2^{\text{meas}};\omega\right)}{\partial r}\right|_{r=a}=\frac{1}{2\sqrt{\pi}}\frac{(ka+i)\mathrm{e}^{ika}}{ka^2},\tag{3.34}$$

$$\left.\frac{\partial\Psi_{22}^{(1)}\left(r,\theta_2^{\text{meas}},\phi_2^{\text{meas}};\omega\right)}{\partial r}\right|_{r=a}=\frac{1}{2}\sqrt{\frac{3}{\pi}}\frac{\left[2ka+i\left(2-k^2a^2\right)\right]\mathrm{e}^{ika}}{k^2a^3}\cos\theta_2^{\text{meas}},\tag{3.35}$$

where θ_m^{meas}, $m=1$ and 2, indicates the polar angles at the measurement points on the sphere.

Substituting the expansion functions and their derivatives into the determinant yields

$$\det\left[\left.\frac{\partial\Psi^{(1)}(r,\theta,\phi;\omega)}{\partial r}\right|_{r=a}\right]=\frac{\sqrt{3}}{4\pi}\frac{(ka+i)\left[2ka+i\left(2-k^2a^2\right)\right]\mathrm{e}^{i2ka}}{k^3a^5}\left(\cos\theta_2^{\text{meas}}-\cos\theta_1^{\text{meas}}\right).\tag{3.36}$$

Substituting Eqs. (3.28)–(3.35) and (3.36) into (3.26), we obtain

$$\begin{aligned}\left\{\begin{array}{l}\hat{p}\left(r_1,\theta_1,\phi_1;\omega\right)\\ \hat{p}\left(r_2,\theta_2,\phi_2;\omega\right)\end{array}\right\}&=i\omega\rho_0\begin{bmatrix}-i\dfrac{\mathrm{e}^{ikr_1}}{kr_1} & -\sqrt{3}\dfrac{(kr_1+i)\mathrm{e}^{ikr_1}}{k^2r_1^2}\cos\theta_1\\ -i\dfrac{\mathrm{e}^{ikr_2}}{kr_2} & -\sqrt{3}\dfrac{(kr_2+i)\mathrm{e}^{ikr_2}}{k^2r_2^2}\cos\theta_2\end{bmatrix}\\ &\times\begin{bmatrix}\dfrac{ka^2\mathrm{e}^{-ika}}{(ka+i)}\left(\dfrac{\cos\theta_2^{\text{meas}}}{\cos\theta_2^{\text{meas}}-\cos\theta_1^{\text{meas}}}\right) & -\dfrac{ka^2\mathrm{e}^{-ika}}{(ka+i)}\left(\dfrac{\cos\theta_1^{\text{meas}}}{\cos\theta_2^{\text{meas}}-\cos\theta_1^{\text{meas}}}\right)\\ -\dfrac{k^2a^3\mathrm{e}^{-ika}}{\left[2ka+i\left(2-k^2a^2\right)\right]\left(\cos\theta_2^{\text{meas}}-\cos\theta_1^{\text{meas}}\right)} & \dfrac{k^2a^3\mathrm{e}^{-ika}}{\left[2ka+i\left(2-k^2a^2\right)\right]\left(\cos\theta_2^{\text{meas}}-\cos\theta_1^{\text{meas}}\right)}\end{bmatrix}\\ &\times\left\{\begin{array}{l}\hat{v}_{\mathrm{n}}\left(a,\theta_1^{\text{meas}},\phi_1^{\text{meas}};\omega\right)\\ \hat{v}_{\mathrm{n}}\left(a,\theta_2^{\text{meas}},\phi_2^{\text{meas}};\omega\right)\end{array}\right\},\end{aligned}\tag{3.37}$$

Equation (3.37) is of generality for a spherical source surface because neither the normal surface velocity nor the measurement and field points are specified.

Example 3.4 Consider the case of a dilating sphere for which the normal surface velocity $\hat{v}_{\mathrm{n}}$ is a constant (see Fig. 3.3). Substituting $\hat{v}_{\mathrm{n}}$ into Eq. (3.37), we obtain

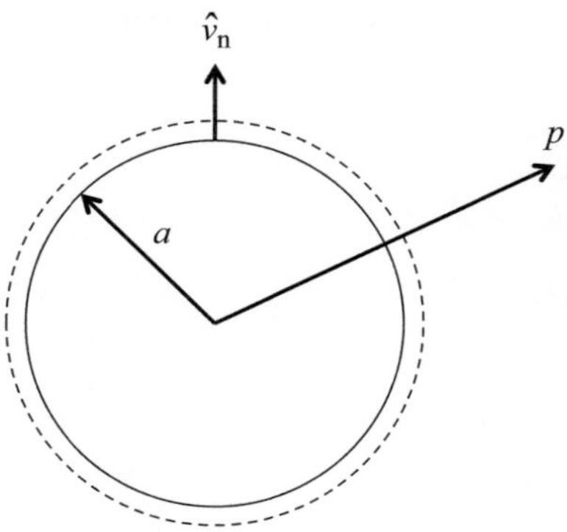

Fig. 3.3 Schematic of predicting the acoustic field generated by a dilating sphere of radius $r = a$ by using HELS-based NAH. The normal surface velocity is given at two arbitrary points on the surface and the field acoustic pressure is predicted

Fig. 3.4 Schematic of predicting the acoustic field generated by a sphere of radius a oscillating along the z-axis direction by using HELS-based NAH. The normal surface velocity is specified at two arbitrary points on the surface and the field acoustic pressure is predicted

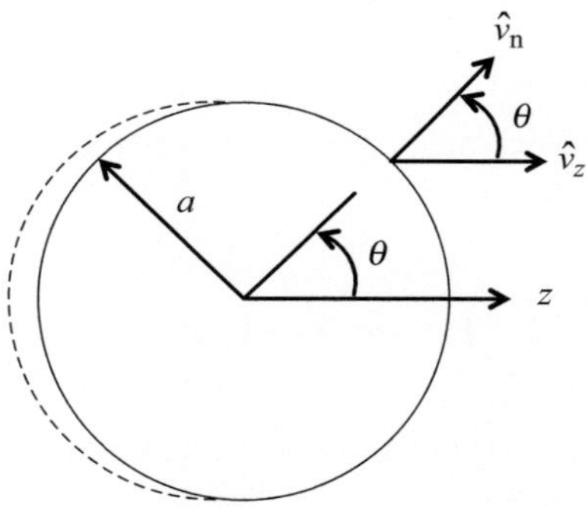

$$\begin{Bmatrix} \hat{p}(r_1, \theta_1, \phi_1; \omega) \\ \hat{p}(r_2, \theta_2, \phi_2; \omega) \end{Bmatrix} = \begin{bmatrix} -i\dfrac{e^{ikr_1}}{kr_1} & -\sqrt{3}\dfrac{(kr_1 + i)e^{ikr_1}}{k^2 r_1^2} \cos\theta_1 \\ -i\dfrac{e^{ikr_2}}{kr_2} & -\sqrt{3}\dfrac{(kr_2 + i)e^{ikr_2}}{k^2 r_2^2} \cos\theta_2 \end{bmatrix}$$

$$\begin{Bmatrix} i\rho_0 c\hat{v}_n \dfrac{(ka)^2}{(ka + i)} e^{-ika} 0 \end{Bmatrix} = \begin{Bmatrix} \rho_0 c\hat{v}_n \left(\dfrac{ka}{ka + i}\right)\left(\dfrac{a}{r_1}\right) e^{ik(r_1 - a)} \\ \rho_0 c\hat{v}_n \left(\dfrac{ka}{ka + i}\right)\left(\dfrac{a}{r_2}\right) e^{ik(r_2 - a)} \end{Bmatrix},$$

which is exactly the same as the analytic solution [42].

Example 3.5 Next, we consider a sphere of radius $r = a$ that oscillates back and forth in the z-axis direction at the velocity $\hat{v}_z$ (see Fig. 3.4). The normal surface velocity measured at two arbitrary points on the surface can be written as $\hat{v}_{\mathrm{n}}(a,\theta,\phi;\omega)=\hat{v}_z\cos\theta_m^{\mathrm{meas}}$, $m=1$ and 2,

$$\begin{Bmatrix} \hat{p}\left(r_1,\theta_1,\phi_1;\omega\right) \\ \hat{p}\left(r_2,\theta_2,\phi_2;\omega\right) \end{Bmatrix} = \begin{bmatrix} \dfrac{\mathrm{e}^{ikr_1}}{\sqrt{3}kr_1} & -\dfrac{i(kr_1+i)\mathrm{e}^{ikr_1}}{k^2r_1^2}\cos\theta_1 \\ \dfrac{\mathrm{e}^{ikr_2}}{\sqrt{3}kr_2} & -\dfrac{i(kr_2+i)E^{ikr_2}}{k^2r_2^2}\cos\theta_2 \end{bmatrix} \begin{Bmatrix} 0 \\ \rho_0 c\hat{v}_z\dfrac{(ka)^3}{2ka+i\left(2-k^2a^2\right)}\mathrm{e}^{-ika} \end{Bmatrix}$$

$$= \begin{Bmatrix} \rho_0 c\hat{v}_z\dfrac{(ka)(kr_1+i)E^{ik(r_1-a)}}{\left(k^2a^2-2\right)+i2ka}\left(\dfrac{a}{r_1}\right)^2\cos\theta_1 \\ \rho_0 c\hat{v}_z\dfrac{(ka)(kr_2+i)\mathrm{e}^{ik(r_2-a)}}{\left(k^2a^2-2\right)+i2ka}\left(\dfrac{a}{r_2}\right)^2\cos\theta_2 \end{Bmatrix},$$

which again matches the analytic solution perfectly [42].

3.4 Error Analyses

To acquire an in-depth understanding of the HELS solutions (3.16) and (3.18), we carry out an error analysis to see the impact of the measurement errors on reconstruction results.

There are different types of errors in the measured data that can be caused by a number of reasons due to aliasing, aperture size, bias, random fluctuations, etc. Most of these errors may be corrected by using various techniques that have been developed in the past. For example, temporal and spatial aliasing can be overcome by using sufficiently high sampling rates in either temporal or spatial special domains; errors due to a finite measurement aperture can be reduced by applying a spatial window to the measured data; biased errors can be adjusted by calibrating the system, and random fluctuations can be suppressed by taking time averages of the measurements. In this error analysis, however, we do not distinguish the causes of errors. Moreover, we assume that they are uncorrelated to the true values of the measured data. Accordingly, Eq. (3.16) can be written as

$$\left\{\hat{p}\left(\vec{x}_S^{\,\mathrm{rec}};\omega\right)\right\}_{S\times1} = \left[G_{pp}\left(\vec{x}_S^{\,\mathrm{rec}}\middle|\vec{x}_m^{\,\mathrm{meas}};\omega\right)\right]_{S\times M}^{-1}\left\{\hat{p}\left(\vec{x}_m^{\,\mathrm{meas}};\omega\right)+\varepsilon_m\right\}_{M\times1}, \tag{3.38}$$

where ε_m represents the measurement errors that are Gaussian white and spatially uncorrelated.

As an example, we consider the two-term expansion ($n=1$) and assume that the acoustic pressure is specified on a spherical surface of radius $r=r_m$, $m=1$ and 2. Substituting the expansion functions $\Psi_{ij}^{(1)}(r_i,\theta_i,\phi_i;\omega)$ into Eq. (3.38) yields the following explicit solutions:

$$\begin{Bmatrix} \hat{p}\left(r_1^{\text{rec}},\theta_1^{\text{rec}},\phi_1^{\text{rec}};\omega\right) \\ \hat{p}\left(r_2^{\text{rec}},\theta_2^{\text{rec}},\phi_2^{\text{rec}};\omega\right) \end{Bmatrix} = \begin{Bmatrix} G_{11}\hat{p}\left(r_1^{\text{meas}},\theta_1^{\text{meas}},\phi_1^{\text{meas}};\omega\right) - G_{12}\hat{p}\left(r_2^{\text{meas}},\theta_2^{\text{meas}},\phi_2^{\text{meas}};\omega\right) + O_1(\varepsilon_1,\varepsilon_2) \\ G_{21}\hat{p}\left(r_1^{\text{meas}},\theta_1^{\text{meas}},\phi_1^{\text{meas}};\omega\right) - G_{22}\hat{p}\left(r_2^{\text{meas}},\theta_2^{\text{meas}},\phi_2^{\text{meas}};\omega\right) + O_2(\varepsilon_1,\varepsilon_2) \end{Bmatrix}, \tag{3.39}$$

where G_{ij} are defined as

$$G_{11} = \frac{\begin{matrix}\Psi_{11}^{(1)}\left(r_1^{\text{rec}},\theta_1^{\text{rec}},\phi_1^{\text{rec}};\omega\right)\Psi_{22}^{(1)}\left(r_2^{\text{meas}},\theta_2^{\text{meas}},\phi_2^{\text{meas}};\omega\right) \\ -\Psi_{12}^{(1)}\left(r_1^{\text{rec}},\theta_1^{\text{rec}},\phi_1^{\text{rec}};\omega\right)\Psi_{21}^{(1)}\left(r_2^{\text{meas}},\theta_2^{\text{meas}},\phi_2^{\text{meas}};\omega\right)\end{matrix}}{\det\left[\Psi^{(1)}\left(r^{\text{meas}},\theta^{\text{meas}},\phi^{\text{meas}};\omega\right)\right]}, \tag{3.40}$$

$$G_{12} = \frac{\begin{matrix}\Psi_{11}^{(1)}\left(r_1^{\text{rec}},\theta_1^{\text{rec}},\phi_1^{\text{rec}};\omega\right)\Psi_{12}^{(1)}\left(r_1^{\text{meas}},\theta_1^{\text{meas}},\phi_1^{\text{meas}};\omega\right) \\ -\Psi_{12}^{(1)}\left(r_1^{\text{rec}},\theta_1^{\text{rec}},\phi_1^{\text{rec}};\omega\right)\Psi_{11}^{(1)}\left(r_1^{\text{meas}},\theta_1^{\text{meas}},\phi_1^{\text{meas}};\omega\right)\end{matrix}}{\det\left[\Psi^{(1)}\left(r^{\text{meas}},\theta^{\text{meas}},\phi^{\text{meas}};\omega\right)\right]}, \tag{3.41}$$

$$G_{21} = \frac{\begin{matrix}\Psi_{21}^{(1)}\left(r_2^{\text{rec}},\theta_2^{\text{rec}},\phi_2^{\text{rec}};\omega\right)\Psi_{22}^{(1)}\left(r_2^{\text{meas}},\theta_2^{\text{meas}},\phi_2^{\text{meas}};\omega\right) \\ -\Psi_{22}^{(1)}\left(r_2^{\text{rec}},\theta_2^{\text{rec}},\phi_2^{\text{rec}};\omega\right)\Psi_{21}^{(1)}\left(r_2^{\text{meas}},\theta_2^{\text{meas}},\phi_2^{\text{meas}};\omega\right)\end{matrix}}{\det\left[\Psi^{(1)}\left(r^{\text{meas}},\theta^{\text{meas}},\phi^{\text{meas}};\omega\right)\right]}, \tag{3.42}$$

$$G_{22} = \frac{\begin{matrix}\Psi_{21}^{(1)}\left(r_2^{\text{rec}},\theta_2^{\text{rec}},\phi_2^{\text{rec}};\omega\right)\Psi_{12}^{(1)}\left(r_1^{\text{meas}},\theta_1^{\text{meas}},\phi_1^{\text{meas}};\omega\right) \\ -\Psi_{22}^{(1)}\left(r_2^{\text{rec}},\theta_2^{\text{rec}},\phi_2^{\text{rec}};\omega\right)\Psi_{11}^{(1)}\left(r_1^{meas},\theta_1^{\text{meas}},\phi_1^{\text{meas}};\omega\right)\end{matrix}}{\det\left[\Psi^{(1)}\left(r^{\text{meas}},\theta^{\text{meas}},\phi^{\text{meas}};\omega\right)\right]}, \tag{3.43}$$

where the determinant is given by

$$\begin{aligned}
&\det\left[\Psi^{(1)}\left(r^{\text{meas}}, \theta^{\text{meas}}, \phi^{\text{meas}}; \omega\right)\right] \\
&\quad = \Psi_{11}^{(1)}\left(r_1^{\text{meas}}, \theta_1^{\text{meas}}, \phi_1^{\text{meas}}; \omega\right)\Psi_{22}^{(1)}\left(r_2^{\text{meas}}, \theta_2^{\text{meas}}, \phi_2^{\text{meas}}; \omega\right) \\
&\qquad - \Psi_{12}^{(1)}\left(r_1^{\text{meas}}, \theta_1^{\text{meas}}, \phi_1^{\text{meas}}; \omega\right)\Psi_{21}^{(1)}\left(r_2^{\text{meas}}, \theta_2^{\text{meas}}, \phi_2^{\text{meas}}; \omega\right).
\end{aligned} \tag{3.44}$$

The error terms in Eq. (3.39) are defined as

$$\begin{Bmatrix} O_1(\varepsilon_1, \varepsilon_2) \\ O_2(\varepsilon_1, \varepsilon_2) \end{Bmatrix} = \begin{Bmatrix} G_{11}\varepsilon_1 - G_{12}\varepsilon_2 \\ G_{21}\varepsilon_1 - G_{22}\varepsilon_2 \end{Bmatrix}, \tag{3.45}$$

where G_{ij} are given in Eq. (3.40)–(3.43).

Equation (3.39) has generality for a spherical source since neither the measured acoustic pressures $\hat{p}\left(r_m^{\text{meas}}, \theta_m^{\text{meas}}, \phi_m^{\text{meas}}; \omega\right)$ nor the measurement points $(r_m, \theta_m, \phi_m; \omega)$, $m = 1$ and 2, as well as the reconstruction points $(r_s^{\text{rec}}, \theta_s^{\text{rec}}, \phi_s^{\text{rec}}; \omega)$, $s = 1$ and 2, are specified.

Substituting the expressions for individual expansion functions $\Psi_{ij}^{(1)}$ as given in Example 3.1 to Eq. (3.40-3.43) and further to Eq. (3.45), we can write the leading order terms of errors as

$$\begin{Bmatrix} O_1(\varepsilon_1, \varepsilon_2) \\ O_2(\varepsilon_1, \varepsilon_2) \end{Bmatrix} = \begin{Bmatrix} a_1\left(\frac{r_1^{\text{meas}}}{r_1^{\text{rec}}}\right)^2\varepsilon_1 + a_2\left(\frac{r_2^{\text{meas}}}{r_1^{\text{rec}}}\right)^2\varepsilon_2 \\ b_1\left(\frac{r_1^{\text{meas}}}{r_2^{\text{rec}}}\right)^2\varepsilon_1 + b_2\left(\frac{r_2^{\text{meas}}}{r_2^{\text{rec}}}\right)^2\varepsilon_2 \end{Bmatrix}, \tag{3.46}$$

where a_m and b_m, $m = 1$ and 2, are independent of the radial distance r. Equation (3.46) show that when $r_s^{\text{rec}} < r_m^{\text{meas}}$, the measurement errors ε_m can be amplified quadratically.

It is easy to show that if we consider a single-term expansion, namely, $n = 0$ in (3.38), the error term will be proportional to $(r_m^{\text{meas}}/r_s^{\text{rec}})$, where $r_s^{\text{rec}} < r_m^{\text{meas}}$,

$$O(\varepsilon) = a\left(\frac{r^{\text{meas}}}{r^{\text{rec}}}\right)\varepsilon, \tag{3.47}$$

where a is independent of the radial distance.

These results can be extended to any order of expansion in $\Psi_{ij}^{(1)}$. As shown in Example 2.2, the asymptotic behaviors of the spherical Hankel function of the first kind as $kr \to 0$, namely, the reconstruction point approaches the source surface, are given by

$$h_n^{(1)}(kr) \to \frac{1}{(kr)^{n+1}}. \tag{3.48}$$

So for the nth-order expansion of $\Psi_{ij}^{(1)}$ in Eq. (3.38), the reconstruction errors are dominated by

$$O_s(\varepsilon_1, \varepsilon_2, \ldots, \varepsilon_M) = \sum_{m=1}^{M} a_m \left(\frac{r_m^{\text{meas}}}{r_s^{\text{rec}}} \right)^{n+1} \varepsilon_m, \quad r_s^{\text{rec}} < r_m^{\text{meas}} \tag{3.49}$$

where a_m, $m = 1, 2, \ldots, M$, are independent of the radial distance. Since $r_s^{\text{rec}} < r_m^{\text{meas}}$, $(r_m^{\text{meas}}/r_s^{\text{rec}}) > 1$. Therefore, the measurement errors ε_m are amplified by $(r_m^{\text{meas}}/r_s^{\text{rec}})^{n+1}$ times in reconstruction.

Equation (3.49) demonstrates the importance of keeping the measurement surface close to the source surface in order to ensure the accuracy in reconstruction.

However, when the HELS formulations (3.24) are utilized to predict the radiated acoustic pressure, given the normal surface velocity of a vibrating structure, errors in the predicted results will be bounded.

Following the procedures outlined above, it can be shown that the errors in Eq. (3.24) are proportional to

$$O(\varepsilon) \propto \sum_{n=0}^{N} \sum_{m=1}^{M} \frac{h_n^{(1)}(kr^{\text{rec}})}{dh_n^{(1)}(kr_m^{\text{meas}})/d(kr_m^{\text{meas}})} \varepsilon_m. \tag{3.50}$$

Example 2.3 illustrates that the asymptotic behaviors of the spherical Hankel functions and their derivatives in the far-field ($kr \to \infty$) are given by

$$h_n^{(1)}(kr) \to \frac{1}{kr} \text{and} \frac{dh_n^{(1)}(kr)}{d(kr)} \to \frac{1}{kr}. \tag{3.51}$$

Substituting Eq. (3.51) into (3.50) yields the upper bound of the errors $O(\varepsilon)$ in Eq. (3.24),

$$O(\varepsilon) \propto N \sum_{m=1}^{M} b_m \left(\frac{r_m^{\text{meas}}}{r^{\text{rec}}} \right) e^{ikr^{\text{rec}}} \varepsilon_m \leq N \sum_{m=1}^{M} b_m \left(\frac{r_m^{\text{meas}}}{r^{\text{rec}}} \right) |\varepsilon_m|, \tag{3.52}$$

where N represents the total number of the spherical Hankel functions involved in the expansion, and b_m, $m = 1$ and 2, are independent of the radial distance.

Since $r^{\text{meas}} < r^{\text{rec}}$, $(r_m^{\text{meas}}/r_s^{\text{rec}}) < 1$. So the errors in prediction are always bounded. In fact, the further the prediction distance r^{rec} is, the smaller the errors in prediction become.

As the field point r^{rec} approaches the source surface ($kr \to 0$), the asymptotic behaviors of the spherical Hankel functions and their derivatives are given by (see Example 2.2),

$$h_n^{(1)}(kr) \to \frac{1}{(kr)^{n+1}} \text{and} \frac{dh_n^{(1)}(kr)}{d(kr)} \to \frac{1}{(kr)^{n+2}}. \tag{3.53}$$

Substituting Eq. (3.53) into (3.50) leads to the errors $O(\varepsilon)$ in Eq. (3.24),

$$O(\varepsilon) \propto \sum_{n=0}^{N} \sum_{m=1}^{M} b_m \left(kr_m^{\text{meas}}\right) \left(\frac{r_m^{\text{meas}}}{r^{\text{rec}}}\right)^{n+1} \varepsilon_m. \tag{3.54}$$

Since $(r_m^{\text{meas}}/r_s^{\text{rec}}) < 1$, the errors in prediction using Eq. (3.24) are bounded by

$$O(\varepsilon) \leq \sum_{m=1}^{M} b_m \left(kr_m^{\text{meas}}\right) \left(\frac{r_m^{\text{meas}}}{r^{\text{rec}}}\right) |\varepsilon_m|. \tag{3.55}$$

Therefore, errors in measurements ε_m will not affect the acoustic pressure in prediction as much as they do in reconstruction.

3.5 Regularization

The examples discussed in Sects. 3.2–3.4 all dealt with a spherical source surface. In practice, a vibrating structure is seldom spherical. Moreover, the measured acoustic pressures are neither accurate nor complete. So the acoustic quantities reconstructed by using Eqs. (3.15) and (3.17) may be quite unsatisfactory. This is because reconstruction is an inverse process, namely, we want to identify the cause (e.g., structural vibrations) based on the effect (the resultant acoustic pressure field). Such a problem is mathematically ill posed. Accordingly, any slight error in the input data may be amplified significantly as it has been demonstrated in the preceding error analysis section.

The solution strategies to an inverse problem encountered in all fields can be summarized in one word: regularization [43]. The underlying principle in all regularization techniques is to smooth the dependence of the output data on the input data [44–46].

The simplest regularization for Eq. (3.15) is to use an optimal number of expansion terms J_{op} in reconstruction. This is because Eq. (3.15) employs the superposition of the spherical wave functions to describe the acoustic pressure. The more expansion terms are used in Eq. (3.15), the more details in the reconstructed acoustic pressure are included. Theoretically, if the input data are accurate and complete, the expansion solution given by Eq. (3.15) for a spherical source is exact as $J \to \infty$ [47]. However, this is not the case in reality. Sect. 3.4 has shown that errors embedded in measurement ε_m can be amplified by $(r_m^{\text{meas}}/r_s^{\text{rec}})^{n+1}$ times in reconstruction on the source surface. Therefore, it is necessary to find the optimal number of expansion J_{op} in reconstruction.

In what follows, we consider an arbitrarily shaped source surface and the measured data are neither accurate nor complete. For convenience sake, we express the basis functions in terms the spherical Hankel functions and spherical harmonics since they are available in many mathematical libraries. Use of these spherical wave functions may be very effective for a surface with an aspect ratio close to unity, $x : y : z \to 1$. For elongated, flat, or both elongated and flat objects, the prolate, oblate, or elliptic coordinates, respectively, can be used to provide faster convergence in numerical computations. Unfortunately, the analytic forms of the spheroidal functions in these corresponding coordinates are nonexistent and direct numerical computations of these spheroidal functions must be carried out, which can be extremely time consuming.

3.6 Regularization Through Truncation of the Expansion Functions

In this section, we want to develop a simple yet effective methodology to reconstruct the radiated acoustic pressure field. To achieve this goal, we must devise a procedure that can yield the desired accuracy with relatively few measurements. The questions that must be addressed are:

1. Where and how the measurements should be taken?
2. How many measurement points are necessary to achieve the desired resolution?
3. Given the number of measurements, what is the optimal number of expansion functions?

Unfortunately, there are no definite answers to these questions because of the uncertainties involved in an inverse problem. What we can do is to develop guidelines with which satisfactory reconstruction of the radiated acoustic pressure fields can be obtained. The detailed guidelines are presented in Chap. 5.

Suppose that the input data are collected by an array of microphones (see Fig. 3.5). For simplicity yet without loss of generality, we use a rectangular array that consists of $M_1 \times M_2 = M$ microphones and collect M data points.

Fig. 3.5 An array of $M_1 \times M_2$ microphones used to reconstruct the acoustic pressures radiated from a source. *Red dots*, input data points; *green dots*, validation points

These measured data can be substituted into Eq. (3.15) to reconstruct the acoustic pressure anywhere.

To find an optimal number of expansion terms in reconstruction, we only use one-half the data points and the other half for validation. Detailed procedures are as follows:

1. Use every other data points (red dots) as indicated in Fig. 3.5 as input to Eq. (3.15) to reconstruct the acoustic pressures at other points on the hologram surface (green dots) with $N=1$ and $J=(N+1)^2=4$ terms for any selected frequency f. The corresponding expansion functions include $h_0(kr_m)P_{0,0}(\cos\theta_m)$, $h_1(kr_m)P_{1,0}(\cos\theta_m)$, $h_1(kr_m)P_{1,1}(\cos\theta_m)\sin\phi_m$, and $h_1(kr_m)P_{1,1}(\cos\theta_m)\cos\phi_m$.
2. Calculate the L2-norm errors of the reconstructed acoustic pressures at the green dots with respect to the measured data.

$$\|L\|^2 = \sum_{m=1}^{M/2} |\hat{p}^{\,\mathrm{rec}}(r_m, \theta_m, \phi_m; \omega) - \hat{p}^{\,\mathrm{meas}}(r_m, \theta_m, \phi_m; \omega)|^2, \tag{3.56}$$

where $m=1$ to $M/2$.

3. Increase the number of expansion terms by one, $J=J+1$ in Eq. (3.15) and repeat the same processes as outlined in 1 and 2 and calculate the L2-norm errors again.
4. Find the minimum value of L2-norm errors. Notice that because of uncertainties involved in the measurements, L2-norm errors may fluctuate with respect to the value of J. However the general trend of this L2-norm error curve is U-shaped. Accordingly, its slope becomes zero at the global minimum of the L2-norm error curve, and the corresponding value of J is optimal. The resultant expansions will provide the best approximation of the reconstructed acoustic pressure on the source surface. In practice, an exact zero slope may not occur. So we can set a criterion for the slope to be less than certain value, say, $\tan\alpha \le \varepsilon_0$, where ε_0 is a preselected small value. The corresponding number of expansion terms is the optimal value J_{op}, which can be interpreted as being optimized with respect to a particular set of the measured acoustic pressures at a particular frequency f. Obviously, different frequency and measurement conditions will yield a different optimal value of J_{op}.

In general, the higher the frequency f is, the larger the value of J_{op} becomes and the longer the computation takes. This slowdown in computation speed is inherent in all expansion theories, including the standard finite element or boundary element methods. Therefore other methods such as asymptotic approximations should be utilized to reconstruct the radiated acoustic pressure field in the high-frequency regime.

The value of J_{op} thus determined may be used to reconstruct acoustic pressures anywhere, including the source surface.

Example 3.6 Figure 3.6 displays a typical example [37] of L2-norm errors curves for reconstructing the acoustic pressures with respect to benchmark values on a

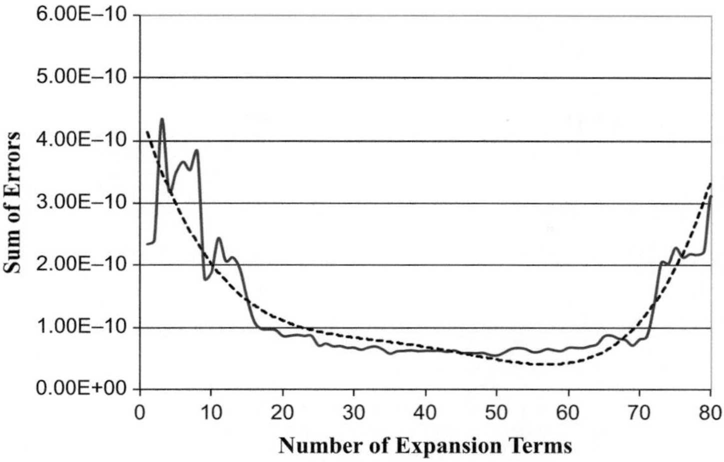

Fig. 3.6 The L2-norm errors curves with respect to the expansion terms for the reconstructed acoustic pressure on a source surface. *Solid line*, errors; *dashed line*, curve-fitted errors

source surface. Results indicate that the accuracy in reconstruction on the source surface increases at first with the number of expansion terms to certain level, and then decreases monotonically thereafter. The presence of this minimum L2-norm error is reasonable. Physically, the higher-order terms represent the small-scale or near-field effects, which may have been lost in the measured data and cannot be recovered anyway. On the other hand, the lower-order terms describe the large-scale or propagating wave effects, which may be captured in the measured data and can be reconstructed. As more expansion functions are used in reconstruction, more details are added to the reconstructed acoustic pressure and therefore, more accurate the result is. However the higher the order of expansions is, the more the amplifications of measurement errors in the reconstructed acoustic pressure become (see Sect. 3.4). Eventually, the errors embedded in the high-order terms will be so large that they may completely distort the reconstructed acoustic pressures.

Consequently, it is necessary to find the optimal number of expansion terms in order to get a satisfactory reconstruction of the radiated acoustic pressure field in both near and far fields in a cost-effective manner in practice.

On the other hand, the L2-norm error curves for the acoustic pressures reconstructed on the measurement surface shows a monotonic decay with respect to the expansion terms. In other words, the accuracy in reconstruction on the measurement surface increases monotonically with the number of expansion terms (Fig. 3.7). This is expected because errors in reconstruction on a measurement surface have been minimized by the least-squares method.

It is emphasized that the above optimization process is effective for producing a satisfactory reconstruction for the acoustic pressure, but not enough for reconstructing the normal component of the particle velocity on a vibrating surface.

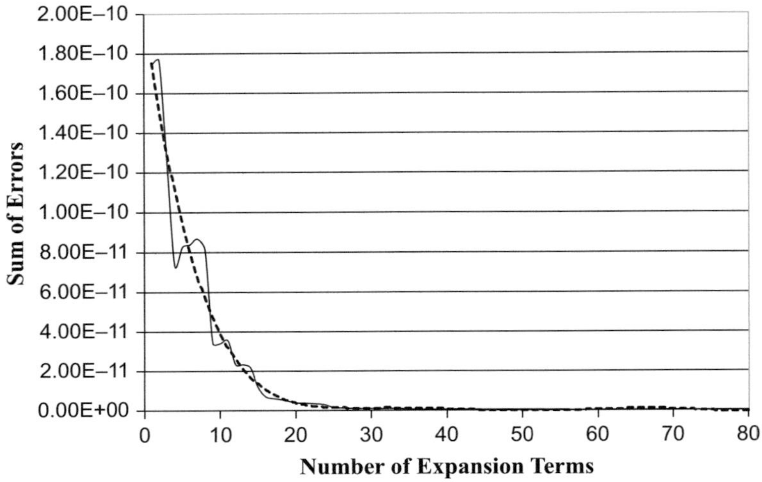

Fig. 3.7 The L2-norm errors curves with respect to the expansion terms for the reconstructed acoustic pressure on a measurement surface. *Solid line*, errors; *dashed line*, curve-fitted errors

This is because the normal surface velocity usually contains much more details (evanescent waves) than the acoustic pressure do. Accordingly, different methods must be used to ensure a satisfactory reconstruction of the normal surface velocity.

3.7 Other Regularization Techniques

Consider Eqs. (3.15) and (3.17), which represent the explicit solutions for the reconstructed surface acoustic pressure and normal surface velocity. Since in practice the measured data will not be error free and the source surface may not be spherical, the transfer matrices on the right sides of Eqs. (3.15) and (3.17), $\left[G_{pp}\left(\vec{x}_s^{\,\text{rec}}\middle|\vec{x}_m^{\,\text{meas}};\omega\right)\right]_{S\times M}^{-1}$ and $\left[G_{pv}\left(\vec{x}_s^{\,\text{rec}}\middle|\vec{x}_m^{\,\text{meas}};\omega\right)\right]_{S\times M}^{-1}$, may be singular and errors embedded in the measured data $\hat{p}\left(\vec{x}_m^{\,\text{meas}};\omega\right)$ may grow without a bound. One way to obtain a bounded solution is to apply a truncated singular value decomposition (TSVD) to Eqs. (3.15) and (3.17) [22]

$$\left\{\hat{p}\left(\vec{x}_s^{\,\text{rec}};\omega\right)\right\}_{S\times 1} = [V_P]_{S\times S}\left[\sum\nolimits_P\right]_{S\times M}^{-1}[U_P]_{M\times M}^{H}\left\{\hat{p}\left(\vec{r}_m^{\,\text{meas}};\omega\right)\right\}_{M\times 1}, \quad (3.57)$$

$$\left\{\hat{v}_{\mathrm{n}}\left(\vec{x}_s^{\,\mathrm{rec}};\omega\right)\right\}_{S\times 1}=[V_v]_{S\times S}\left[\sum\nolimits_v\right]^{-1}_{S\times M}[U_v]^{H}_{M\times M}\left\{\hat{p}\left(\vec{r}_m^{\,\mathrm{meas}};\omega\right)\right\}_{M\times 1},\tag{3.58}$$

where $[V_p]$ and $[U_p]$ in Eq. (3.57) are the right and left unitary orthonormal matrices, respectively, of the matrix $\left[G_{pp}\left(\vec{x}_s^{\,\mathrm{rec}}\middle|\vec{x}_m^{\,\mathrm{meas}};\omega\right)\right]^{-1}_{S\times M}$ in Eq. (3.15), $[\Sigma_p]^{-1}$ is the diagonal matrix that contains inverted singular values of the corresponding matrix; $[V_v]$ and $[U_v]$ in Eq. (3.58) are the right and left unitary orthonormal matrices, respectively, of the matrix $\left[G_{pv}\left(\vec{x}_s^{\,\mathrm{rec}}\middle|\vec{x}_m^{\,\mathrm{meas}};\omega\right)\right]^{-1}_{S\times M}$ in Eq. (3.17), and $[\Sigma_v]^{-1}$ is the diagonal matrix containing inverted singular values of the corresponding matrix.

The simplest regularization for the HELS method is to set the number of expansion terms in Eqs. (3.57) and (3.58) at the optimal value J_{op} for any given set of the input data, as discussed above. Mathematically, this optimization process can be written as

$$\min_J\sum_{i=1}^{M}\left\|\hat{p}\left(\vec{x}_{m,i}^{\,\mathrm{rec}};\omega\right)-\hat{p}\left(\vec{x}_{m,i}^{\,\mathrm{meas}};\omega\right)\right\|_2^2\rightarrow J_{\mathrm{op}},\tag{3.59}$$

where $\hat{p}\left(\vec{x}_{m,i}^{\,\mathrm{rec}};\omega\right)$ and $\hat{p}\left(\vec{x}_{m,i}^{\,\mathrm{meas}};\omega\right)$ represent, respectively, the reconstructed and measured acoustic pressures at the ith measurement point $\vec{x}_{m,i}$. As pointed out above, Eq. (3.59) may be effective for reconstructing the acoustic pressure, but not enough for the normal surface velocity.

The effectiveness of using regularization methods such as Tikhonov regularization (TR) and L-curve. to improve the accuracy in reconstruction by HELS has been examined previously [48]. Results have indicated that the modified Tikhonov regularization [49] (MTR) with its regularization parameter determined by the generalized cross-validation [50] (GCV) can provide the most accurate reconstruction for HELS.

Recently, hybrid regularization [51] has been developed to determine the optimal number for the basis functions. In this hybrid regularization the MTR and GCV method are used to regularize the reconstruction of acoustic pressures on a hologram surface for each value of J first, and the least-squares method to minimize residual by matching the reconstructed acoustic pressure with respect to the measured one. Mathematically, this process is expressible as

$$\min_J\sum_{m=1}^{M}\left\|\hat{p}\left(\vec{x}_m^{\,\mathrm{rec},\alpha};\omega\right)-\hat{p}\left(\vec{x}_m^{\,\mathrm{meas}};\omega\right)\right\|_2^2\rightarrow J_{\mathrm{op,MTR}},\tag{3.60}$$

where $\hat{p}\left(\vec{x}_m^{\,\mathrm{rec},\alpha};\omega\right)$ is the reconstructed acoustic pressure at the ith measurement point $\vec{x}_m$ by using MTR and GCV, and the Tikhonov functional, J_α, for regularizing $\hat{p}\left(\vec{x}_m^{\,\mathrm{rec},\alpha};\omega\right)$ in Eq. (3.60) may be written as

$$\begin{aligned}&\mathbf{J}_\alpha\left(\left\{\hat{p}\left(\overrightarrow{x}_m^{\text{rec},\alpha};\omega\right)\right\}_{M\times 1}\right)\\&=\left\|[U]_{M\times M}[\Sigma]_{M\times M}[V]_{M\times M}^H\left\{\hat{p}\left(\overrightarrow{x}_m^{\text{rec},\alpha};\omega\right)\right\}_{M\times 1}-\left\{\hat{p}\left(\overrightarrow{x}_m^{\text{meas}};\omega\right)\right\}_{M\times 1}\right\|_2^2\\&\quad+\alpha\left\|\left[F_h^\alpha\right]_{M\times M}[V]_{M\times M}^H\left\{\hat{p}\left(\overrightarrow{x}_m^{\text{rec}};\omega\right)\right\}_{M\times 1}\right\|_2^2,\end{aligned}\tag{3.61}$$

where the regularization parameter α is determined by GCV through a minimization process,

$$\min_\alpha\left(\frac{\left\|\left[F_h^\alpha\right]_{M\times M}[U]_{M\times M}^H\left\{\hat{p}\left(\overrightarrow{x}_m^{\text{meas}};\omega\right)\right\}_{M\times 1}\right\|_2^2}{\left(\text{Trace}\left[F_h^\alpha\right]_{M\times M}\right)^2}\right),\tag{3.62}$$

where $[F_h^\alpha]_{M\times M}$ is the high-pass filter defined as

$$\left[F_h^\alpha\right]_{M\times M}=\text{diag}\left([\mathbf{I}]_{M\times M}-[F^\alpha]_{M\times M}\right),\tag{3.63}$$

where $[F^\alpha]_{M\times M}$ is a low-pass filter for regularizing the measured acoustic pressure and is a diagonal matrix containing the singular values σ_i of the matrix $[\Sigma]$ in Eq. (3.61), where $i = 1$ to M,

$$[F^\alpha]_{M\times M}=\text{diag}\left(\ldots,\frac{\sigma_i^2\left(\alpha+\sigma_i^2\right)^2}{\alpha^3+\sigma_i^2\left(\alpha+\sigma_i^2\right)^2},\ldots\right).\tag{3.64}$$

The reconstructed acoustic pressure $\left\{\hat{p}\left(\overrightarrow{x}_m^{\text{rec},\alpha};\omega\right)\right\}_{M\times 1}$ can then be rewritten as

$$\left\{\hat{p}\left(\overrightarrow{x}_m^{\text{rec},\alpha};\omega\right)\right\}_{M\times 1}=[V]_{M\times M}[F^\alpha]_{M\times M}[\Sigma]_{M\times M}^{-1}[U]_{M\times M}^H\left\{p\left(\overrightarrow{x}_m^{\text{meas}};\omega\right)\right\}_{M\times 1}.\tag{3.65}$$

The differences between Eq. (3.15) and Eq. (3.65) are that the former relies on the least-squares method to determine the optimal HELS expansion number J_{op} to reconstruct the acoustic pressure, whereas the latter uses a hybrid regularization that consists of MTR, SVD, and the least-squares minimization to determine the optimal HELS expansion number $J_{\text{op,MTR}}$ and to reconstruct the acoustic pressure.

Using $J_{\text{op,MTR}}$, the acoustic quantities anywhere can be reconstructed. For example, we can express the reconstructed surface acoustic pressure and normal surface velocity as

$$\left\{\hat{p}\left(\overset{\rightarrow \text{rec},\alpha}{x_s};\omega\right)\right\}_{S\times 1}=\left[V_p\right]_{S\times S}\left[F_p^{\beta}\right]_{S\times S}\left[\Sigma_p\right]_{S\times M}^{-1}\left[U_p\right]_{M\times M}^{H}\left\{p\left(\overset{\rightarrow \text{meas}}{x_m};\omega\right)\right\}_{M\times 1}, \tag{3.66}$$

$$\left\{\hat{v}_{\text{n}}\left(\overset{\rightarrow \text{rec},\alpha}{x_s};\omega\right)\right\}_{S\times 1}=\left[V_v\right]_{S\times S}\left[F_v^{\gamma}\right]_{S\times S}\left[\sum\nolimits_v\right]_{S\times M}^{-1}\left[U_v\right]_{M\times M}^{H}\left\{\hat{p}\left(\overset{\rightarrow \text{meas}}{x_m};\omega\right)\right\}_{M\times 1}, \tag{3.67}$$

where $[F_p^{\beta}]$ and $[F_v^{\gamma}]$ are the diagonal matrices containing the singular values of the matrix $[\Sigma_p]$ in Eq. (3.66) and those of $[\Sigma_v]$ in Eq. (3.67), respectively,

$$\left[F_p^{\beta}\right]_{S\times S}=\text{diag}\left(\ldots,\frac{\varepsilon_i^2\left(\beta+\varepsilon_i^2\right)^2}{\beta^3+\varepsilon_i^2\left(\beta+\varepsilon_i^2\right)^2},\ldots\right), \tag{3.68}$$

$$\left[F_v^{\gamma}\right]_{S\times S}=\text{diag}\left(\ldots,\frac{\eta_i^2\left(\gamma+\eta_i^2\right)^2}{\gamma^3+\eta_i^2\left(\gamma+\eta_i^2\right)^2},\ldots\right), \tag{3.69}$$

where β and γ are obtained by GCV through a minimization process given by, respectively,

$$\min_{\beta}\left(\frac{\left\|\left[F_{h,p}^{\beta}\right]_{M\times M}\left[U_p\right]_{M\times M}^{H}\left\{\hat{p}\left(\overset{\rightarrow \text{meas}}{x_m};\omega\right)\right\}_{M\times 1}\right\|_2^2}{\left(\text{Trace}\left[F_{h,p}^{\beta}\right]_{M\times M}\right)^2}\right), \tag{3.70}$$

$$\min_{\gamma}\left(\frac{\left\|\left[F_{h,p}^{\gamma}\right]_{M\times M}\left[U_v\right]_{M\times M}^{H}\left\{\hat{p}\left(\overset{\rightarrow \text{meas}}{x_m};\omega\right)\right\}_{M\times 1}\right\|_2^2}{\left(\text{Trace}\left[F_{h,p}^{\gamma}\right]_{M\times M}\right)^2}\right), \tag{3.71}$$

where $[F_{h,p}^{\beta}]$ and $[F_{h,v}^{\gamma}]$ are the high-pass filters obtained by subtracting the low-pass filters $[F_p^{\beta}]$ and $[F_v^{\gamma}]$ from the identity matrix, respectively, and by filling its null space with unity,

$$\left[F_{p,h}^{\beta}\right]_{M\times M}=\text{diag}\left[\left([\mathbf{I}]_{S\times S}-\left[F_p^{\beta}\right]_{S\times S}\right),[\mathbf{I}]_{(M-S)\times(M-S)}\right], \tag{3.72}$$

$$\left[F_{v,h}^{\gamma}\right]_{M\times M}=\text{diag}\left[\left([\mathbf{I}]_{S\times S}-\left[F_v^{\gamma}\right]_{S\times S}\right),[\mathbf{I}]_{(M-S)\times(M-S)}\right]. \tag{3.73}$$

Experimental results have confirmed that Eqs. (3.66) and (3.67) can yield more robust results than Eqs. (3.15) and (3.17) do.

Example 3.7 Consider a thin square plate mounted on a large baffle. The addition of a baffle allows for rigorous examinations of the reconstructed vibro-acoustic quantities with respect to the analytic solutions. Also, a plate represents a class of structures that are commonly encountered in practice, as many structures consist of flat or slightly curved panels. Hence it will have significant impacts if we can show

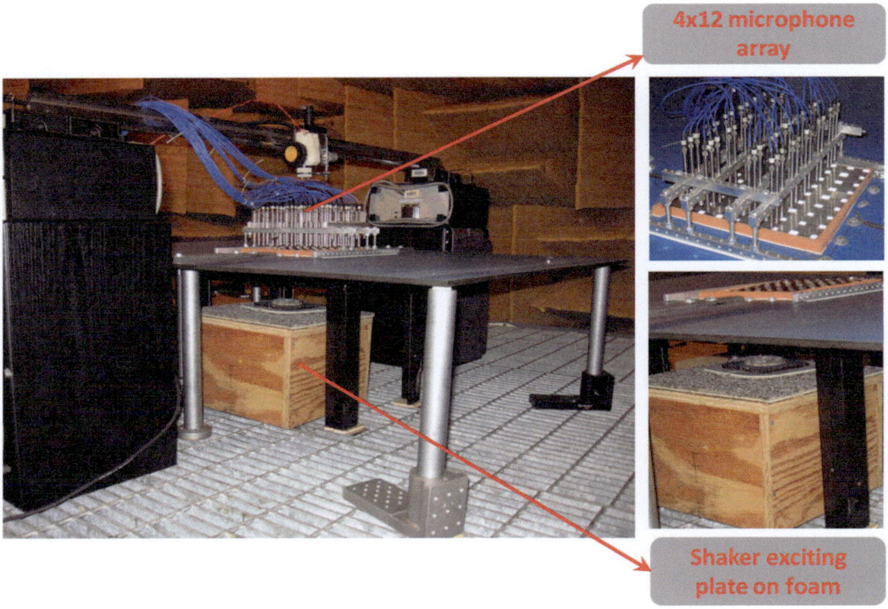

Fig. 3.8 Test setup for collecting the near-field acoustic pressures above a baffled square plate

the capability of using computationally simple HELS codes based on the spherical wave functions to produce satisfactory reconstruction of the vibro-acoustic quantities on a highly non-spherical panel surface.

For convenience yet without loss of generality, we consider a square plate subject to free-free boundary conditions. The reason for choosing a square plate is that its symmetry can pose a challenge for traditional Experimental Modal Analysis (EMA) to distinguish coupled modes. It is interesting to examine if the HELS-based NAH can discern these coupled modes.

Figure 3.8 shows the test setup of a square steel plate of dimensions 220×220 mm^2 with a thickness of 1.25 mm, which was mounted on a large baffle and excited by a mechanical shaker using random signals. The plate edges were supported on soft foam to mimic free-free boundary conditions. The excitation point on the thin plate was selected to be away from the nodal lines of the first 10 natural modes. The radiated acoustic pressures were measured by using a 12×4 array that contained 48 microphones (PCB T130D21, PCB Piezotronics, Inc. Depew, New York). Measurements were taken on three patches, resulting in total 144 data points at the standoff distance of 10 mm.

The tests were conducted inside a fully anechoic chamber at the Acoustics, Vibration, and Noise Control Laboratory at Wayne State University. The excitation signal level was adjusted to ensure that the signal to noise ratio (SNR) was maintained at least 10 dB or higher over the entire frequency of interest. The relative phases of the acoustic pressures were obtained by taking cross correlations of the measured acoustic

Fig. 3.9 Test setup for reconstructing the vibro-acoustic responses of a baffled square plate using the HELS-based NAH

pressures with respect to the excitation force as a reference signal. The measured data were used as input to the HELS codes to reconstruct the normal surface velocity distributions on the 9×9 grid that coincided with the laser scanning measurement points (see Fig. 3.9). A point laser vibrometer (Polytec OFV 551 fiber optic interferometer, Polytec, Irvine, California) was used to collect the benchmark data to validate the reconstructed normal surface velocity. The measured normal surface velocities were taken as input to EMA to identify the natural modes of the plate.

Figure 3.10 shows the comparison of the acoustic pressures reconstructed at four randomly selected locations on the hologram plane using hybrid regularization with the optimal number of expansion terms $J_{\mathrm{op,MTR}}$ determined by Eq. (3.60). For comparison purposes, we also employed Eq. (3.15) with the optimal number of expansion terms J_{op} given by the least-squares minimization process Eq. (3.59) to reconstruct the acoustic pressures at the same locations. Results confirm that even with a straight application of the HELS method using the value of J_{op}, the accuracy in the reconstructed acoustic pressure is still very high. By using the hybrid regularization strategies, we can get more accurate reconstruction of the acoustic pressures over a much wider frequency span than those obtained by using the least-squares minimization alone.

Figure 3.11 shows the comparison of the minimized residuals in reconstructing the acoustic pressure over the entire frequency range of interest on the hologram surface by using J_{op} and those by using $J_{\mathrm{op,MTR}}$. These results clearly show the advantage of using $J_{\mathrm{op,MTR}}$ in reconstruction.

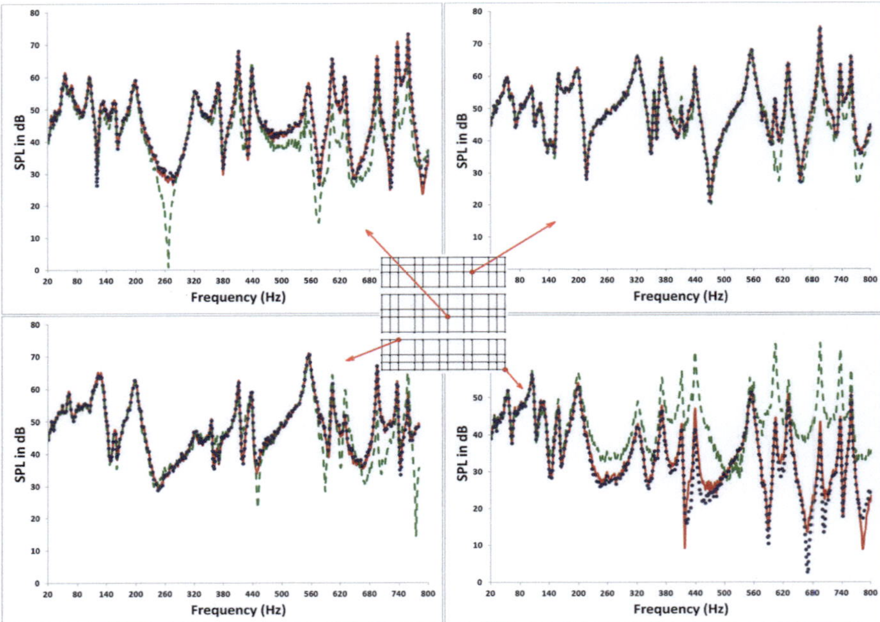

Fig. 3.10 Comparisons of the reconstructed acoustic pressures at four randomly selected points vs. the benchmark results on the hologram surface. *Solid line*, benchmark results; *dotted line*, reconstructed by using the least-squares minimization alone; *crosses*, reconstructed by using the hybrid regularization

Figure 3.12 depicts the comparisons of the normal surface velocity spectra reconstructed at randomly selected points on the plate surface, including its edge and corner, and benchmark data. Once again, we utilized the optimal number of expansion terms J_{op} given by Eq. (3.59) (dashed lines) and that of expansion terms $J_{\mathrm{op},MTR}$ provided by Eq. (3.60) (dotted lines) to reconstruct the normal surface velocities. The results indicate that a straight application of the HELS formulations with J_{op} yields satisfactory reconstruction in the normal surface velocity distributions at the lower-order modes; however, errors in reconstructions increase with the mode order. This is because the number of expansion terms J_{op} obtained by the least-squares minimization alone can be optimal for reconstructing the acoustic pressure, but not enough for reconstructing the normal surface velocity. By using the optimal number of expansion terms $J_{\mathrm{op},MTR}$ obtained through a hybrid regularization scheme, satisfactory reconstruction in the normal surface velocity distribution can be obtained up to the operational deflection shape (ODS) dominated by the natural mode of the target structural wavelength λ_{cr}. The accuracy is improved even along the edges and at corners of the plate.

Figures 3.13 and 3.14 exhibit comparisons of the first 18 natural modes and corresponding natural frequencies of this square plate under free-free boundary conditions from 100 to 420 Hz and from 420 to 800 Hz, respectively.

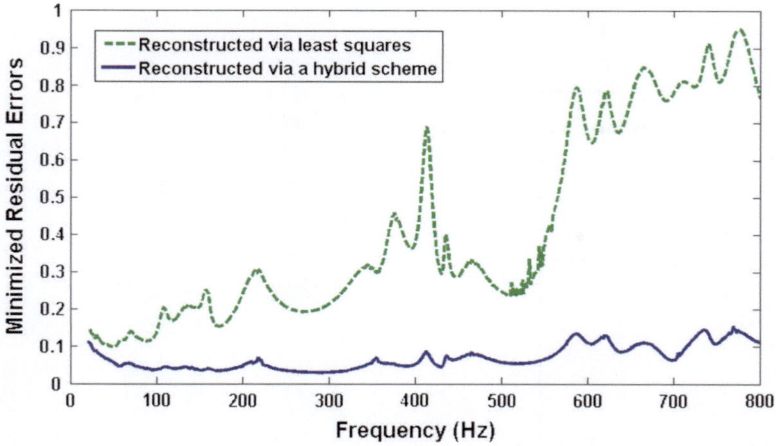

Fig. 3.11 Comparison of the minimized residuals in reconstructing the acoustic pressures by using J_{op} and $J_{op,MTR}$. *Dashed line*, residuals resulting from J_{op}; solid line, residuals resulting from $J_{op,MTR}$

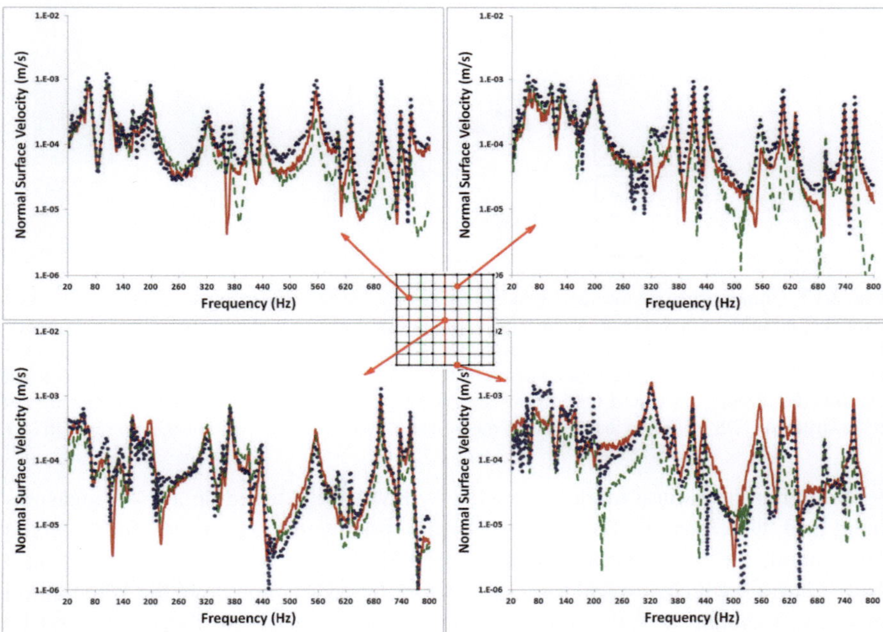

Fig. 3.12 Comparisons of the reconstructed normal surface velocity spectra at some randomly selected points on the baffled square plate surface with the benchmark data. Solid line, benchmark; dotted line, reconstructed using least-squares minimization alone; crosses, reconstructed using a hybrid regularization

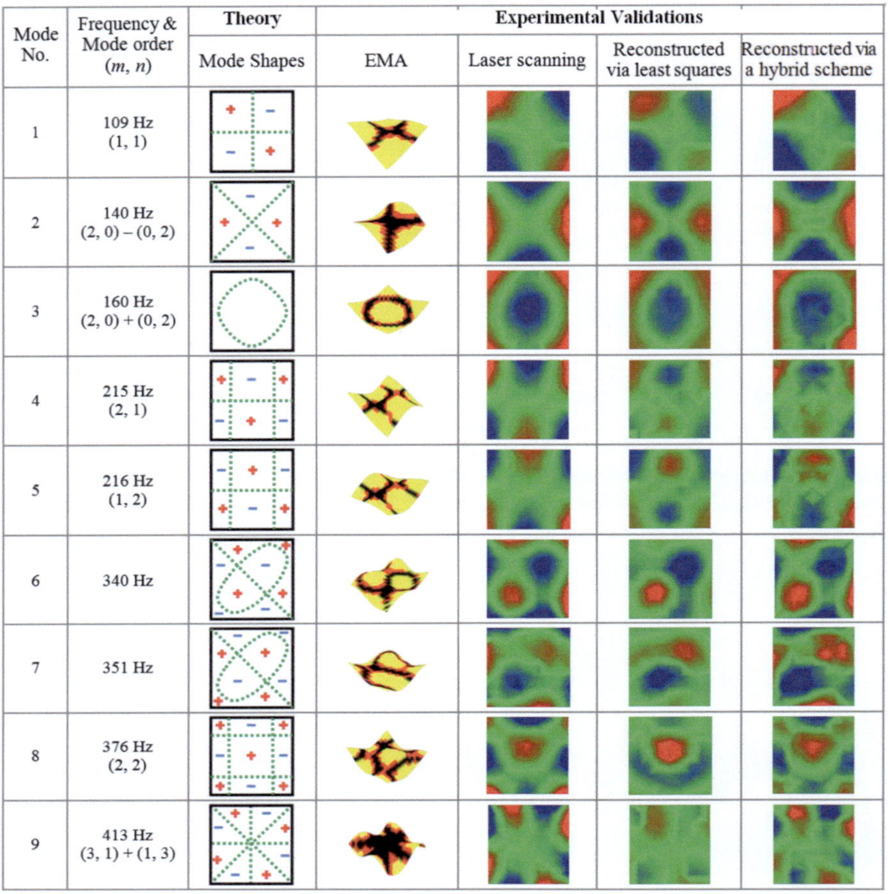

Fig. 3.13 Comparisons of the reconstructed ODS of a baffled square plate subject to free-free boundary conditions and measured ones from 100 to 420 Hz. The first column indicates the mode number, second column implies the mode order and natural frequency, third column shows the theoretical mode shape, fourth column displays the mode shape extracted from EMA, fifth column exhibits the directly measured ODS from laser scanning, and sixth and seventh columns are the reconstructed ODS using the least-squares method and hybrid regularization scheme, respectively

Note that the natural modes extracted by EMA were based on a 9×9 measurement grid, which enabled one to get the 15th mode at 695 Hz successfully. On the other hand, the microphone grid used for HELS reconstruction was based on a 12×12 array, which allowed for a satisfactory reconstruction up to the 18th natural mode. It is emphasized that the theoretical natural modes and those extracted by using EMA were displayed in Figs. 3.13 and 3.14 as a reference, but not for validation. The reconstructed normal surface velocity distributions were validated against the ODS dominated by the 18th natural mode that were obtained by scanning a laser vibrometer on the plate surface. For comparisons beyond the

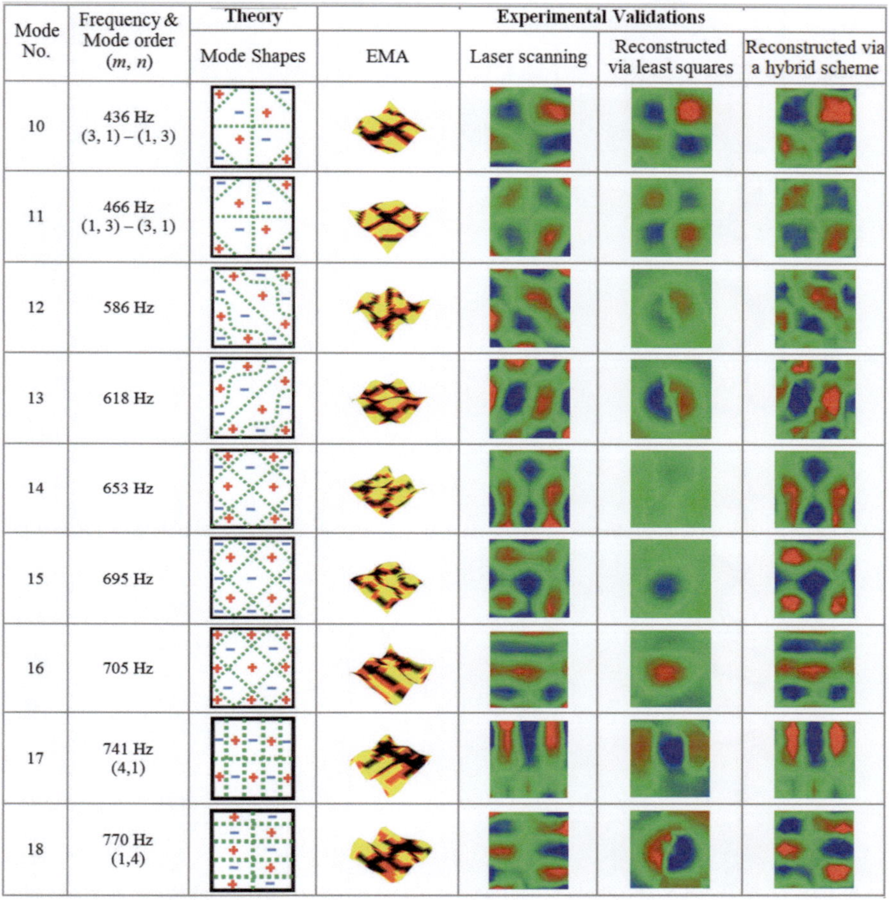

Fig. 3.14 Comparisons of reconstructed ODS of a baffled square plate subject to the free-free boundary condition and measured ones from 420 to 800 Hz. The first column indicates the mode number, second column implies the mode order and natural frequency, third column shows the theoretical mode shape, fourth column displays the mode shape extracted from EMA, fifth column exhibits the directly measured ODS from laser scanning, and sixth and seventh columns are the reconstructed ODS obtained by using the least-squares method alone and by a hybrid regularization scheme, respectively

18th natural mode, a denser measurement grid for the laser vibrometer and a finer measurement grid for the HELS-based NAH must be adopted.

The results shown in Figs. 3.13 and 3.14 confirm that: (1) the HELS based NAH can be utilized to reconstruct the ODSs dominated by the natural modes of a panel structure; (2) a direct application of the HELS formulations using J_{op} alone can yield satisfactory reconstructions of the ODSs dominated by lower-order modes, for example, up to the 11th natural mode in this case; and (3) by using the hybrid regularization scheme, it is possible to obtain satisfactory reconstructions

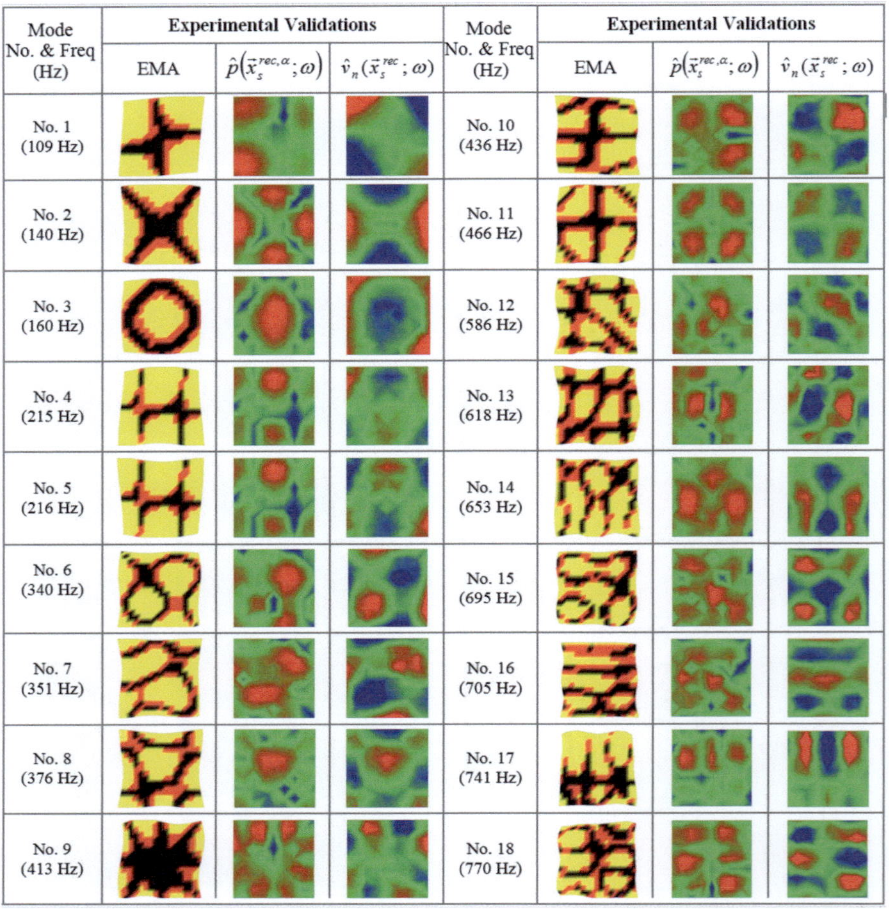

Fig. 3.15 Comparisons of the reconstructed vibro-acoustic quantities on the surface of a baffled square plate up to the 18th natural mode at 770 Hz. The first column is the mode number and frequency; second depicts the mode shape extracted from EMA; third and forth columns demonstrate the surface acoustic pressure and normal surface velocity distributions reconstructed by using the HELS-based NAH with $J_{\text{op,MTR}}$ determined by hybrid regularization scheme, respectively

of all ODSs up to the target one that is dominated by the 18th natural mode of the square plate.

Note that it is not possible to measure the surface acoustic pressure directly. So validations of the reconstructed acoustic pressures are performed on the hologram surface (see Fig. 3.9). It is important, however, to display the capabilities of reconstructing all vibro-acoustic quantities by using the HELS-based NAH at any frequency. Figure 3.15 depicts the natural modes extracted by EMA together with the surface acoustic pressure and normal surface velocity distributions at the natural frequencies of a square plate up to the 18th natural mode. Note that the acoustic pressure distributions showed the same patterns as those of the normal surface velocity distributions, but without the presence of phase changes. This is because the acoustic pressure is a scalar quantity.

The HELS-based NAH has been successfully used to reconstruct acoustic radiation from arbitrarily shaped objects in both exterior [52, 53] and interior regions [54, 55]. Test results in these cases demonstrate that HELS can provide satisfactory reconstruction of the acoustic quantities even on an arbitrarily shaped surface with relatively few measurements in the low-to-mid frequencies and in particular, allow a piecewise or patch measurement and reconstruction.

Note that in conducting patch measurement at least one reference point should be selected. This reference point must not be moved in moving from one patch measurement to another.

It is worth noticing that since the expansion functions in the HELS formulations consist of the spherical wave functions, the solutions provided by Eqs. (3.66) and (3.67) converge very fast when measurement and reconstruction points are outside the minimum spherical surface that circumscribes an arbitrarily shaped source surface.

Inside this minimum spherical surface, however, the situation is unclear in the beginning. While numerous numerical and experimental results have confirmed the correctness of the HELS results inside this minimum spherical surface, the validity of HELS solutions was not established theoretically at first. Naturally, the validity of the HELS formulations inside the spherical surface that circumscribes arbitrary source geometry in an exterior region, or outside a maximum sphere that inscribes arbitrary source geometry in an interior region has been challenged when the HELS method was first introduced to the acoustics community.

Problems

3.1. Use the knowledge learned in Chap. 2 for the spherical Hankel functions and spherical harmonics and Eq. (3.4)–(3.12) to write the HELS expansion functions for $n = 3$ and 4.

3.2. Use Eq. (3.15) to reconstruct the acoustic pressure emitted by a dilating sphere of radius a. Suppose that the acoustic pressures measured at two arbitrary points r_1 and r_2 in space are $\hat{p}_1 = \rho_0 c \hat{v}_0 k a^2 e^{ik(r_1 - a)} / [(ka + i) r_1]$ and $\hat{p}_2 = \rho_0 c \hat{v}_0 k a^2 e^{ik(r_2 - a)} / [(ka + i) r_2]$, respectively. Find the acoustic pressure $\hat{p}\,(r, \theta, \phi; \omega)$ at any field point. Hint: use a one-term expansion in the HELS to reconstruct the radiated acoustic pressure field.

3.3. Continue Problem 3.2. Use Eq. (3.17) to reconstruct the normal component of the particle velocity $\hat{v}_n(r, \theta, \phi; \omega)$ at any field point based on the same measured acoustic pressures.

3.4. Use a two-term HELS expansion in Eq. (3.15) to reconstruct the acoustic pressure emitted by an oscillating sphere of radius a. Suppose that the input acoustic pressure at (r_1, θ_1, ϕ_1) is $\hat{p}_1 = \rho_0 c \hat{v}_0 k a^3 (k r_1 + i) e^{ik(r_1 - a)}$

$\cos\theta_1/\left[\left(k^2a^2-2+i2ka\right)r_1^2\right]$ and that at (r_2, θ_2, ϕ_2) is given by $\hat{p}_2=\rho_0 c\hat{v}_0 ka^3(kr_2+i)e^{ik(r_2-a)}\cos\theta_2/\left[\left(k^2a^2-2+i2ka\right)r_2^2\right]$. Solve the acoustic pressure $\hat{p}(r,\theta,\phi;\omega)$ at any field point.

3.5. Continue Problem 3.4. Use Eq. (3.17) to reconstruct the normal component of the particle velocity $\hat{v}_n(r,\theta,\phi;\omega)$ at any field point based on the same measured acoustic pressures.

3.6. Use a two-term HELS expansion in Eq. (3.37) to predict the acoustic pressure emitted by an oscillating sphere of radius a. Suppose that the normal surface velocities at two arbitrary points on the surface of the sphere are given by $\hat{v}_{\mathrm{n}}(r_m,\theta_m,\phi_m;\omega)=V_z\cos\theta_m$, where $m=1$ and 2 and V_z is known.

3.7. Show that the errors in reconstructing the acoustic quantities on the source surface based on the acoustic pressures measured in the near field by using the HELS formulation (3.15) are of the order as given by Eq. (3.49).

3.8. Show that the errors in reconstructing acoustic quantities in the far field based on the near-field acoustic pressure measurements are of the order as given by Eq. (3.52).

3.9. Show that the errors in predicting the acoustic quantities on the source surface based on the near-field measurements by using the HELS formulation (3.44) are of the order as given by Eq. (3.55).

3.10. Show that the errors in predicting the acoustic quantities in the far field by using the HELS formulation (3.44) are of the order as given by Eq. (3.55).

3.11. Write a simple code for the HELS formulations to reconstruct the acoustic pressure fields based on the acoustic pressure input data. In this program the number of optimal terms in the HELS expansion should be specified automatically by the least-squares minimization process as given by Eq. (3.56).

3.12. Write a simple code for the HELS formulations to reconstruct the acoustic pressure fields based on the normal surface velocity input data. Once again, the number of optimal terms in the HELS expansion is specified automatically by using the least squares minimization process as given by Eq. (3.56).

Chapter 4
Validity of the HELS Method

The validity challenges came at the joint meetings of the 136th Meeting of the Acoustical Society of America (ASA), the 2nd Convention of the European Acoustics Association (EAA), and the 25th German Annual Conference on Acoustics (DAGA) held in Berlin, Germany, 1999 [56]. The major questions were as follows: "How can the acoustic field on the surface of any non-spherical structure be described by the spherical wave functions?" "Is this a Rayleigh hypothesis in NAH that pushes a solution formulation beyond its region of validity?"

4.1 Rayleigh Hypothesis

At the turn of the last century, Rayleigh used a series expansion of plane waves to depict the acoustic pressure field resulting from an incident time-harmonic acoustic plane wave scattered on a one-dimensional periodic, impenetrable corrugated surface S (see Fig. 4.1). The corrugations can be expressed mathematically as [57, 58]

$$\varsigma = b \cos(2\pi x/\lambda_x), \tag{4.1}$$

where b and λ_x are, respectively, the amplitude and wavelength of corrugation of the surface S and θ is the angle of incidence with respect to the unit normal of the surface S.

Above the highest point of the corrugations of surface S ($z > b$), the complex amplitude of the total acoustic pressure (with the time-harmonic function $e^{-i\omega t}$ omitted for brevity) as given by Rayleigh was

S.F. Wu, *The Helmholtz Equation Least Squares Method*, Modern Acoustics and Signal Processing, DOI 10.1007/978-1-4939-1640-5_4

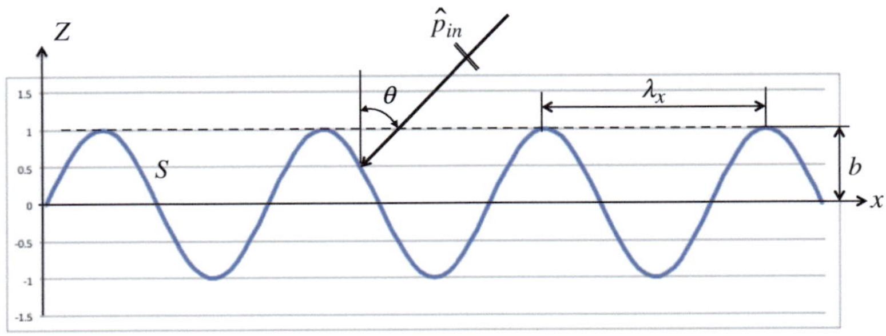

Fig. 4.1 Schematic of an incident time-harmonic acoustic pressure plane wave on a periodic corrugated surface S

$$\hat{p}(x,z;\omega) = \hat{p}_{\text{in}}\left[e^{ikz\cos\theta} + R(\omega)e^{-ikz\cos\theta}\right] + \sum_{n=-\infty}^{\infty} A_n e^{i(nk_x x + kx\sin\theta - kz\cos\theta_n)}, \quad (4.2)$$

where $\hat{p}_{in}$ is the complex amplitude of the incident acoustic pressure, $R(\omega)$ is the acoustic pressure reflection coefficient, A_n represent the expansion coefficients that are determined by the boundary conditions on the corrugated surface S, $k_x = 2\pi/\lambda_x$ is the spatial wavenumber of the corrugations, and $\cos\theta_n = \sqrt{1 - \sin^2\theta_n}$, here $\sin_n = \sin\theta + n(k_x/k)$. The first and second terms on the right side of Eq. (4.2) represent, respectively, the incident and reflected acoustic pressure waves acting on a smooth surface and the infinite series imply the acoustic pressure scattered from the corrugated surface as shown in Fig. 4.1.

In an attempt to use the boundary conditions on S, Rayleigh assumed that the infinite series (4.42) was valid everywhere, including the corrugated surface S. This is known as the Rayleigh hypothesis. This hypothesis was tested on various acoustic scattering problems and had aroused many controversies over the next 60 years. Sometimes the results given by the Rayleigh series (4.2) were correct, but most of times they were completely wrong.

The validity of Rayleigh hypothesis may be examined through analyticity of the solution. If the solution to the acoustic pressure can be analytically continued from the field to the surface, then the expansion coefficients may be determined by the boundary conditions, and the Rayleigh hypothesis is correct [59]. Therefore to answer the question of the validity of Rayleigh hypothesis, it is necessary to find the distribution of singularities using the analytic continuation of the acoustic field across the surface of any scatterer.

These controversies were eventually settled by Millar [59–61] who proved that the Rayleigh hypothesis was neither completely right nor completely wrong. In fact, the validity of a Rayleigh series was governed by the locations of the singularities of the analytic continuation of the exterior scattered field across a scattering surface. For example, in the case of scattering from the gratings and periodic corrugated structures, the Rayleigh series solution would be valid if singularities lay below the lowest point of a corrugated surface. If the singularities lay above the lowest point of a corrugated surface, the series solution is valid only in the region above the highest singularities.

For the case of a periodic corrugated surface as depicted in Fig. 4.1, Millar showed that the Rayleigh hypothesis would be wrong and Eq. (4.2) be invalid when $\lambda_x b > 0.0448$, and neither wrong nor right when $\lambda_x b \leq 0.0448$ [62]. For example, consider a corrugated surface of a wavelength $\lambda_x = 1$ m and corrugation height $b = 0.05$ m. Because $\lambda_x b = 0.05 > 0.0448$, it would be wrong to use Eq. (4.2) to depict the scattered acoustic pressure on the corrugated surface S. On the other hand, if $\lambda_x = 1$ m and $b = 0.045$ m, then $\lambda_x b = 0.045 < 0.0448$ and it might be acceptable to use Eq. (4.2) to describe the scattered acoustic pressure on and above the corrugated surface S.

Millar gave the formal proof of the method for determining the singularities of the acoustic field for a periodically corrugated surface [60]. Hill and Celli offered a heuristic method to estimate the singularities of a periodic corrugated surface [63]. van der Berg and Fokkema studied the acoustic scattering from a nonperiodic corrugated surface [64].

Similarly, in a two-dimensional acoustic scattering scenario, we can use a Rayleigh series in terms of the outgoing cylindrical waves to describe the scattered acoustic pressure field,

$$\hat{p}_{\text{scattered}}(r, \phi; \omega) = \sum_{n=-\infty}^{\infty} A_n H_n^{(1)}(kr) e^{in\phi}, \tag{4.3}$$

where $H_n^{(1)}(kr)$ represents the nth-order cylindrical Hankel functions of the first kind.

Once again, the validity of Eq. (4.3) will be correlated to the distribution of singularities in the analytic continuation of the acoustic pressure field across the surface of a two-dimensional scatterer. Figure 4.2 depicts an arbitrary circle S, which is the cross section of an infinite cylinder. When the singularities all lie inside the maximum circle S_{max} that inscribes the circle S, the series solution (4.3) converges absolutely and uniformly in the compact subsets of the exterior of S_{min} that circumscribes the scatterer. When the singularities lie on or outside the maximum circle S_{max}, the series solution (4.3) will be valid to depict the scattered acoustic pressure above the highest singularities, but invalid below these singularities, because Eq. (4.3) only converges absolutely and uniformly outside the circle defined by the locations of the singularities.

A number of people have looked into the problem of locating possible singularities of the analytic continuation of solutions to the Helmholtz equation for a two-dimensional scatterer with analytic data across analytic boundaries [65–67]. In particular, Maystre and Cadilhac developed a method for determining possible singularities [68], and Keller gave the proof of its validity [69].

Note that in general there is no way of determining the locations of the singularities in the analyticity of solution to the Helmholtz equation because analytic solutions for arbitrary geometry do not exist. In an attempt to determine possible

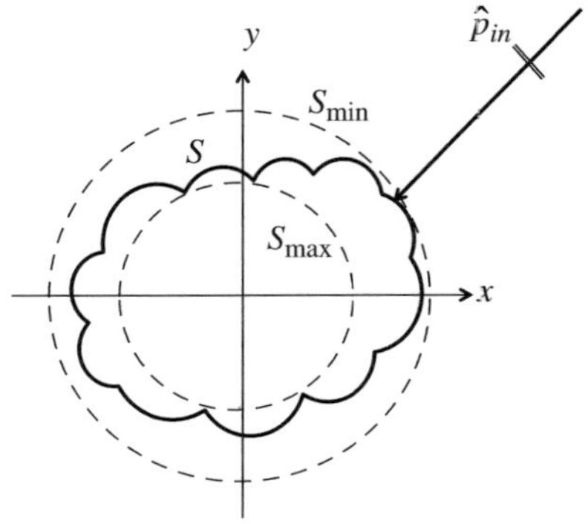

Fig. 4.2 Schematic of an acoustic scattering problem in two-dimensional space. The Rayleigh series solution is valid outside the minimum circle S_{min} circumscribing the scatterer S when singularities all lie inside the maximum circle S_{max} inscribing the scatterer S

singularities of the analytic continuation without the explicit knowledge of the solution, Millar made use of the Schwarz function [70], which utilized of the geometric properties of the boundary. By locating the singularities of the Schwarz function, possible singularities in the analytic continuation of the solution might be determined. However, there is no way of knowing if these possible singularities are indeed the actual singularities. Thus in practice the Rayleigh series solution (4.3) is utilized for domains that are free of singularities. Examples of these include separable geometry such as a sphere and an infinite cylinder.

In three-dimensional acoustic scattering problems, the Rayleigh series can be expanded in terms of the spherical Hankel functions and spherical harmonics, with their expansion coefficients determined by the orthogonality properties of the spherical harmonics.

$$\hat{p}\,(r,\theta,\phi;\omega) = \sum_{n=0}^{\infty}\sum_{m=-n}^{n} A_{mn}h_n^{(1)}(kr)Y_n^m(\cos\theta). \tag{4.4}$$

In Examples 2.10 and 2.11 it has been shown that the infinite series expansion (4.4) can be used to predict acoustic radiation from a vibrating sphere, given the normal surface velocity on a spherical source surface as the boundary condition; or reconstruct the acoustic pressure anywhere including the spherical source surface, given the acoustic pressure on a spherical surface at some distance away from the source surface.

Note that there is a major difference between prediction and reconstruction problems. The former represents a forward problem, whereas the latter stands for an inverse problem. A forward problem is mathematically well defined and errors in input data are bounded in prediction. On the other hand, an inverse problem is mathematically ill posed and errors in input data may increase without a bound in reconstruction. To get a bounded reconstruction, regularization must be used.

Another complication for the infinite series solution may arise in practice when the source is non-spherical. Figure 4.3 demonstrates the schematic of acoustic scattering from an arbitrarily shaped source in three-dimensional space. Analyses have shown the infinite series solution (4.4) is only valid outside the minimum

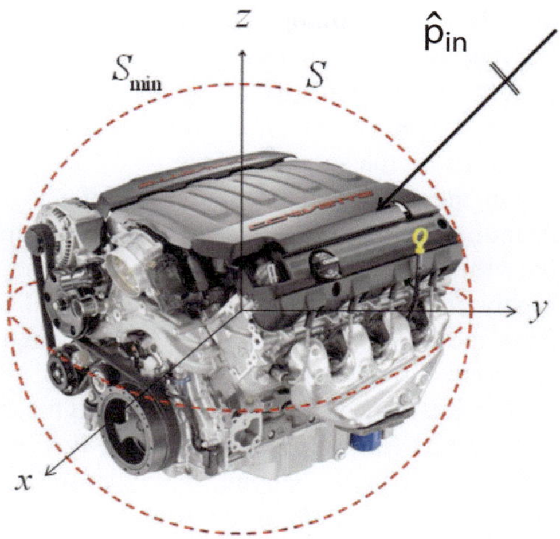

Fig. 4.3 Schematic of an acoustic scattering problem in the three-dimensional space. The Rayleigh series solution (4.4) is valid outside the minimum sphere S_{min} circumscribing the scatterer but invalid inside S_{min}

sphere S_{min} that circumscribes an arbitrary source surface S, but invalid inside the minimum sphere S_{min} in general [71].

On the surface, it looks as though the infinite series solution (4.4) is quite similar to the HELS formulation (3.1), which is expressible as

$$\hat{p}\,(r,\theta,\phi;\omega) = \sum_{n=0}^{N}\sum_{l=-n}^{n} C_{ln}h_n^{(1)}(kr)Y_n^l(\cos\theta). \tag{4.5}$$

Equation (4.5) shows that the acoustic pressure can be described by a superposition of the spherical Hankel functions and spherical harmonics, which is the same as the Rayleigh series (4.4) in three dimensions. Therefore a natural question is as follows: "Will HELS expansions be subject to the same restriction as the Rayleigh series does? Specifically, will Eq. (4.5) be valid only inside the region bounded by spheres?" Moreover, "How will the HELS formulations be related to the Rayleigh series?" These questions are answered in the next section.

4.2 The Rayleigh Series Versus HELS Formulations

Section 4.1 has discussed in detail that the Rayleigh hypothesis is valid and the Rayleigh series converges absolutely and uniformly when the singularities of the analytic continuation of the solution lie inside the maximum sphere that inscribes the source surface of interest. Since in general the analytic solution to the Helmholtz equation for an arbitrarily shaped surface does not exist, there is no way of knowing if the Rayleigh series (4.4) is a valid solution, and where the region of validity is. One thing for sure is that the infinite series will diverge when Eq. (4.4) is evaluated on an arbitrarily shaped source surface. Even if this series is truncated, the solution can still diverge when input data are noisy. So a safe tactic is to use the Rayleigh series to predict the scattered acoustic pressure outside the minimum that circumscribes an arbitrarily shaped surface. The trouble with this approach is that the expansion coefficients cannot be determined because the boundary conditions are given on the source surface.

On the other hand, Chap. 3 has demonstrated examples of using the HELS formulations (4.5) to reconstruct very accurately all the acoustic quantities on a flat vibrating panel, which is way inside the minimum sphere circumscribing this highly non-spherical surface. Thus the HELS formulations (4.5) must be different from the Rayleigh series (4.4), even though they both use the expansion of the spherical wave functions.

The differences between the Rayleigh series (4.4) and the HELS formulations (4.5) are as follows. First of all, the Rayleigh series is infinity, while the HELS expansion is finite. Second, the expansion coefficients in the Rayleigh series are specified by using the orthonormal property of the spherical harmonics and integrating over the solid angle of a sphere, while those in the HELS formulation are specified by matching the expansion (4.5) to the measured data, and the errors involved in this process are minimized by using the least-squares method. Last but not the least, the orthonormal property of the spherical harmonics holds true for a spherical surface, but not an arbitrary surface. So the Rayleigh series is bound to fail when applying it to an arbitrarily shaped surface. In contrast, the HELS formulations always utilize an optimal number of expansion terms to best approximate the reconstructed acoustic quantities. In other words, the HELS formulations always attempt to produce the best approximation for the acoustic quantities radiated from a non-spherical source surface under any given set of input data.

The interrelationships between HELS and Rayleigh series are revealed by Semenova and Wu [72] in reconstructing the acoustic field generated by an arbitrary surface in the exterior region. For simplicity, Semenova and Wu consider infinite cylinders with arbitrary cross sections. They discover that outside the minimum circle that circumscribes the singularities of the cylinder, the Rayleigh series yields an identical result as HELS does when the input data are error free. This is because the high-order terms are negligibly small, so the differences between the Rayleigh series and HELS solutions (a truncated expansion) are minuscule.

When the input data are noisy, the results are different. The normalized errors are the same for all expansion terms in the Rayleigh series because in calculating the coefficients of the series solution by integration, the noise affects all the coefficients equally. In order to obtain a bounded solution, the Rayleigh series must be truncated. Meanwhile, the normalized errors change with the number of expansion terms for the HELS formulations, and are minimal at the optimal number of expansion terms. This is because errors embedded in measurements affect the higher-order terms more than the lower-order ones, as demonstrated in Eq. (3.49) in Sect. 3.4.

Of particular concern is the difference between Rayleigh series and HELS solution inside the minimum circle circumscribing a source. Semenova and Wu illustrates that the Rayleigh series diverges once it is extended inside the minimum circle bounded by the singularities. This confirms Millar's theory on the validity of the Rayleigh hypothesis. In contrast, HELS formulations are not subject to this restriction and may provide satisfactory reconstruction everywhere. Of course, the further the reconstruction point is extended into the minimum circle, the larger the reconstruction errors may become. Note that even if the Rayleigh series is truncated at the same order as that of the HELS expansion, its reconstruction errors inside the minimum circle are still much larger than those of HELS.

These results suggest that the HELS formulations are different from the Rayleigh series as far as back propagating an acoustic field to an arbitrary surface is concerned. However, knowing this difference is not enough to justify the validity of HELS inside a minimum sphere. Moreover, previous results have demonstrated that the accuracy of reconstruction on a non-spherical surface using HELS decreases continuously as the aspect ratio of a source surface and dimensionless frequency ka increase, where a is the characteristic dimension [73].

Therefore, a rigorous mathematical justification of the validity of the HELS formulations to reconstruct the acoustic quantities on an arbitrary surface is needed, which are given rigorously in the next section.

4.3 Justification of the HELS Formulations

Since its first publication in 1997, the HELS-based NAH method has been successfully used to reconstruct the acoustic pressure fields generated by arbitrarily shaped vibrating structures in both exterior and interior regions. Of course, in these cases the structures are not highly elongated, but nonetheless arbitrary. From the acoustics point of view, one can claim that the HELS method may yield satisfactory reconstruction of the acoustic field by using superposition of the spherical wave functions, which explains many phenomena observed in the previous studies. However, this is contradictory to the belief that the expansion solutions using the spherical wave functions and spherical harmonics are valid only inside the regions bounded by spheres and invalid outside these regions.

In this section we present rigorous mathematical justifications for the HELS formulations [74]. Basically, we show that for reconstructing acoustic radiation from an arbitrary source surface, the solutions given by the HELS formulations are approximate; but nonetheless, reconstruction errors are bounded.

Consider reconstruction of acoustic radiation from a finite, arbitrarily shaped object, which includes the acoustic pressure and the normal component of the velocity on the source surface and those in the field. Mathematically, this is equivalent to solving the Helmholtz equation in a three-dimensional domain Ω bounded by the source surface Γ and a surface at infinity Γ_∞.

The acoustic field u with the acoustic wavenumber k satisfies the Helmholtz equation in Ω,

$$\nabla^2 u(r,\theta,\phi;\omega) + k^2 u(r,\theta,\phi;\omega) = 0 \quad \text{in } \Omega\big(\text{or in } \Omega_e = R^3 \backslash \overline{\Omega}\big). \tag{4.6}$$

In practice, the domain Ω can be either the exterior or interior region of a passenger vehicle or an aircraft cabinet. For the exterior problems, solutions to Eq. (4.6) satisfy the Sommerfeld radiation condition,

$$\lim_{r\to\infty} r\left[\frac{\partial u(r,\theta,\phi;\omega)}{\partial r} - iku(r,\theta,\phi;\omega)\right] = 0, \quad as \ r \to \infty. \tag{4.7}$$

Such a solution u is called a radiating solution. In what follows the arguments of u are omitted for brevity.

To reconstruct the acoustic field, we need to measure the acoustic pressures u around the source. Suppose that the acoustical sensors are placed on a surface Γ_0 either inside or outside the source surface. These measured data are utilized to reconstruct u on the source surface Γ and in Ω and, in particular, the normal surface velocity v_n defined by

$$v_n = \frac{1}{i\omega\rho_0}\partial_r u \quad \text{on } \Gamma, \tag{4.8}$$

where the subscript n indicates the unit outward normal on Γ and the symbol ∂_r indicates a partial derivative with respect to r. Note that here we assume that there are no sources other than the one under consideration.

The steps involved in our mathematical justifications are outlined as follows. First, we show that any radiating solution to the Helmholtz equation outside a bounded Lipschitz domain Ω with a connected complement can be approximated by a family of certain known special solutions, for example, the spherical wave functions. Next, by using this approximation together with conditional stability estimates in the Cauchy problem for an elliptic equation, namely, the Helmholtz equation, we demonstrate that these special solutions are bounded on Ω_e and their convergence on Γ_0 implies convergence in Ω_e. Finally, we derive estimates of the convergence of Hölder type at a distance from Ω and that of logarithmic type in Ω_e. These results justify mathematically the validity of the HELS formulations, in which the measured acoustic pressures on Γ_0 are approximated by a linear combination of the special solutions. For an exterior problem this linear combination is well defined everywhere except at the origin, and gives an approximate solution in the exterior region.

Note that we use Ω_e to denote the complement $R^3 \backslash \overline{\Omega}$ and fix a (large) ball B that contains $\overline{\Omega}$. Also we use $H_{(\ell)}(\Omega)$ to imply the Sobolev space of functions in Ω, whose partial derivatives up to the order n are square integrable, and use

$$||u||_{(\ell)}(\Omega) = \sqrt{\sum_{\alpha \le \ell} \int_{\Omega} |\partial^{\alpha} u|^2}, \tag{4.9}$$

to denote the standard norm in this space. Note that in Eq. (4.9), we let $\| \|_2 = \| \|_{(0)}$ be the norm in the space $L^2(\Omega)$ and use the symbol ∂^{α} to indicate the αth-order partial derivative. Accordingly, $\| \|_{\ell+\lambda}$, where $0<\lambda<1$ means the norm in the Hoelder space $\Upsilon^{\ell+\lambda}$ of the functions whose partial derivatives up to order ℓ are Hoelder continuous of an exponent λ, where Υ is a generic constant depending only on Ω, Γ_0, and k.

Now we focus on the approximation of u through the simplest solutions. Our purpose is to interpolate the measured data on Γ_0 for solutions to some integral equations, which can be crucial for higher acoustic wavenumbers k.

Theorem 4.1 *Let Ω be a bounded Lipschitz domain in $\boldsymbol{R}^3$ with connected Ω_e. Let $u \in H_{(1)}(B \backslash \overline{\Omega})$ be a radiating solution to the Helmholtz equation (4.6) in Ω_e. Let $B_0 \subset \overline{B}_0 \subset \Omega$ for a ball B_0. Then for any positive ε there is a radiating solution u_{ε} to the Helmholtz equation outside B_0 such that*

$$||u - u_{\varepsilon}||_{(\ell)}(\Omega_e) < \varepsilon. \tag{4.10}$$

In the proofs we will use the following Green's formula:

$$\int_{\Omega_e} [(\nabla^2 u + k^2 u)u^* - u(\nabla^2 u^* + k^2 u^*)] = \int_{\partial\Omega_e} (u\partial_{\gamma} u^* - u^* \partial_{\gamma} u), \tag{4.11}$$

for u *and* $u^* \in H_{(2)}(B \backslash \Omega)$, *which are radiating solutions to the Helmholtz equation in* B_{1e} *for some ball* $B_1 \subset \overline{B}_1 \subset \Omega$. *Also, we need the following Runge property of radiating solutions.*

Lemma 4.1 *Let Ω_1 and Ω_2 be two bounded domains that contain B_0 with connected Ω_{1e} and Ω_{2e}, $\overline{\Omega}_1 \subset \Omega_2$. Let u_1 be a radiating solution to the Helmholtz equation $(\nabla^2 + k^2)u_1 = 0$ in Ω_{1e}. Then for any $\varepsilon > 0$ there is a radiating solution u outside B_0 such that $//u_1 - u//_{(\ell)}(\Omega_{2e} \cap B) < \varepsilon$.*

Proof Due to interior Schauder-type estimates for elliptic equations, it suffices to consider $\ell = 0$. By shrinking Ω_2 we can achieve that $\partial\Omega_2 \in \Upsilon^{\infty}$.

Let us assume the opposite. Let $\Omega^* = \Omega_{2e} \cap B$. Then by the Hahn-Banach theorem there is a function $f^* \in L^2(\Omega^*)$ such that

$$\int_{\Omega^*} uf^* = 0, \tag{4.12}$$

for all functions u, but

$$\int_{\Omega^*} u_1 f^* \neq 0, \tag{4.13}$$

for some functions u_1. We will extend f^* outside Ω^* as zero.

To obtain a contradiction, we introduce a ball B^* centered at the origin and contained in Ω_1. Since there is a unique radiating solution $u^* \in H_2$ to the equation $(\nabla^2 + k^2)u^* = f^*$ in B_e^* with zero Dirichlet data $u^* = 0$ on ∂B^*, we can find radiating solutions from the Green's formula (4.11),

$$-\int_{\Omega^*} u f^* = \int_{\partial B^*} u \partial_\nu u^*. \tag{4.14}$$

Using Eq. (4.11) and completeness of u in $L^2(\partial B^*)$, we conclude that $\partial_\nu u^* = 0$ on ∂B^*.

Since u^* solves the Helmholtz equation (4.6) in the connected open set $\Omega_2 \backslash \overline{B}^*$, we can conclude from the uniqueness in the Cauchy problem for elliptic equations [75] that $u^* = 0$ on $\Omega_2 \backslash \overline{B}^*$. Now applying again the Green's formula (4.11) to the radiating solutions u_1 and u^*, we obtain

$$\int_{\Omega^*} u_1 f^* = \int_{\partial \Omega_{2e}} (u^* \partial_\nu u_1 - u_1 \partial_\nu u^*) = 0 \tag{4.15}$$

which contradicts Eq. (4.13).

Proof of Theorem 4.1 By extension theorems for Sobolev space in Lipschitz domains, there is an extension $u^* \in H_{(1)}(B)$ of u onto $\mathbf{R}^3$. Let $f^* = \nabla^2 u^* + k^2 u^*$. Then $f^* \in H_{(-1)}(\mathbf{R}^3)$ and supp$f^* \subset \overline{\Omega}$. It is known that $f^* = f_0 + \sum_j \partial_j f_j$ for some $f_0, \ldots, f_3 \in L^2(\mathbf{R}^3)$ that are supported in $\overline{\Omega}$. Let χ_n be a sequence of measurable functions with values 0 or 1 supported in Ω and pointwise convergent to 1 on Ω. Then f_n^* defined as $f_0 \chi_n + \sum_j \partial_j \left(f_j \chi_n \right)$ will converge to f in $H_{(-1)}(\mathbf{R}^3)$ with sup p$f_n^* \subset \Omega$. From the theories of elliptic equations and scattering [76], it follows that radiating solutions to the Helmholtz equation $(\Delta + k^2)u_n^* = f_n^*$ in $\mathbf{R}^3$ converge to u in $H_{(1)}(B \backslash \Omega)$ for any ball B. So one can write u_n such that

$$||u - u_n||_{(\ell)}(B \backslash \Omega) < \frac{\varepsilon}{2}. \tag{4.16}$$

By the Runge property for scattering solutions in $\mathbf{R}^3 \backslash \overline{B}_0$ (Lemma 1), there is a radiating solution u_ε to the Helmholtz equation outside $\overline{B}_0$ such that

$$\left|\left|u_{\rm n} - u_{\varepsilon}\right|\right|_{(\ell)}(B\backslash\Omega) < \frac{\varepsilon}{2}. \tag{4.17}$$

From Eqs. (4.16) and (4.17) we obtain Eq. (4.10).

The proof is complete.

In practice it is very helpful to use a special family of radiating solutions to the Helmholtz equation ε_{mn}, which are expressible as

$$e_{mn}(r,\theta,\phi;\omega) = h_n^{(1)}(kr)Y_n^m(\cos\theta), \tag{4.18}$$

where $h_n^{(1)}$ represent the spherical Hankel functions of the first kind, and $Y_n^m(\cos\theta)$ stand for the spherical harmonics orthonormal in $L^2(S^2)$ on a unit sphere S^2. It is convenient to approximate the solution u to the Helmholtz equation by a linear combination of

$$u_{\rm e}(r,\theta,\phi;\omega) = \sum_{n=0}^{N}\sum_{m=-n}^{n} C_{mn}e_{mn}(r,\theta,\phi;\omega), \tag{4.19}$$

where C_{mn} are the expansion coefficients to be determined.

Corollary 4.1 *Let* $0 \in \Omega$.

For any positive ε *there is* u_ε *such that*

$$\left|\left|u - u_{\rm e}\right|\right|_{(1)}(B\backslash\Omega) < \varepsilon. \tag{4.20}$$

Proof By Theorem 4.1 there is a radiating solution u_ε to the Helmholtz equation in $\mathbf{R}^3\backslash\overline{B}_0$ so that $||u - u_\varepsilon||_{(1)}(B\backslash\Omega) < \varepsilon/2$. Let B_1 be a ball of radius r_1 ($r_1 > r_0$) centered at 0 such that $\overline{B}_1 \subset \Omega$. The spherical harmonics $Y_n^m(\cos\theta)$ form an orthonormal basis in $L^2(S^2)$. Expanding the function u_ε at r_1 with respect to this basis, we can conclude that the partial sums of the corresponding series are convergent in $L^2(\partial B_1)$ and therefore, due to the known results of these series (Theorem 2.14 in Ref. [77]), these partial sums are convergent to u_ε on $B\backslash\Omega$ in $H_1(B\backslash\Omega)$. Consequently, we can find a partial sum u_ε such that $||u_\varepsilon - u_{\rm e}||_{(1)}(B\backslash\Omega) < \varepsilon/2$, and the claim follows from the triangle inequality.

A similar result is valid for interior problems.

Theorem 4.2 *Let* $u \in H_1(\Omega)$ *be a solution to the Helmholtz equation (4.6) in* Ω. *Then for a positive* ε, *there is a solution* u_ε *to the Helmholtz equation in* $\boldsymbol{R}^3$ *such that*

$$\left|\left|u - u_{\rm e}\right|\right|_{(1)}(\Omega) < \varepsilon. \tag{4.21}$$

For interior problems a partial family of useful solutions can be spanned by the functions

$$E_{mn}(r,\theta,\phi;\omega) = j_n(kr)Y_n^m(\cos\theta). \tag{4.22}$$

Now we discuss how to use these results to approximate u *via* u_ε.

Let $\varepsilon = 1$ *in Eq. (4.21). Since for* $\varepsilon < 1$*, there are approximate functions* u_ε *such that*

$$||u_e||_{(1)}(\Omega_0\backslash\Omega) \leq M_1 = ||u||_{(1)}(B\backslash\Omega) + 1. \tag{4.23}$$

Replacing u_ε *by its definition, we have*

$$\int_{\Omega_0\backslash\Omega} \left(\left| \sum_{n=0}^{N} \sum_{m=-n}^{n} C_{mn} e_{mn}(r,\theta,\phi;\omega) \right|^2 + \left| \sum_{n=0}^{N} \sum_{m=-n}^{n} C_{mn} \nabla e_{mn}(r,\theta,\phi;\omega) \right|^2 \right) d(r,\theta,\phi) \leq M_1^2, \tag{4.24}$$

where $d(r, \theta, \phi)$ represents integrations over the source region in Ω_0. Since input data are given on Γ_0, we can approximate u via u_ε by solving a minimization problem,

$$\min_{u_e} ||u - u_e||_{(0)}(\Gamma_0) \tag{4.25}$$

subject to the constraint (4.23). By solving this problem for sufficiently large $N = N(\delta)$, we find $u_\varepsilon(;\delta)$ such that

$$||u - u_e(;\delta)||_{(0)}(\Gamma_0) < \delta, \tag{4.26}$$

so that the constraint (4.23) holds.

Lemma 4.2 *Let* Ω_0 *be a bounded domain,* $\overline{\Omega} \subset \Omega_0$. *Let either* $\Gamma_0 = \partial\Omega_0$ *or* $\partial\Omega_0$ *be analytic and* Γ_0 *be a non-void open part of* $\partial\Omega_0$. *Let* $\Omega_1 \subset \Omega_0$. *Then there is a function* $\omega(\delta) \to 0$ *as* $\delta \to 0$ *such that*

$$||u - u_e(;\delta)||_{(0)}(B\backslash\Omega_1) < \omega(\delta). \tag{4.27}$$

In addition, if $\overline{\Omega} \subset \Omega_1$*, then one can choose* $\omega(\delta) = (C/d^2)M_1^{1-\theta}\delta^\theta$ *[*$\theta \in (0, 1), \theta > d/\Upsilon$*, where d is the distance from* $\partial\Omega_1$ *to* Ω*]; and if* $\Omega_1 = \Omega_0$*, then one can let* $\omega(\delta) = M(-\Upsilon/\log\delta)^{1/4}$.

Proof We use the Carleman-type estimates (Sect. 3.3 in Ref. [75]) for the Helmholtz operator,

$$\sum_{\alpha\leq 1}\int_B \tau\left|\partial^\alpha u_0\right|^2 e^{2\tau\phi} \leq C\int_B \left|\nabla^2 u_0 + k^2 u_0\right|^2 e^{2\tau\phi} \tag{4.28}$$

for any $H_{(2)}(B)$-function u_0 compactly supported in B and $0<\tau$. Here $\varphi\in\Upsilon^2(\mathbf{R}^3)$ is the so-called strongly pseudo-convex function for the Laplace operator in $\mathbf{R}^3$ (see Sects. 3.2 and 3.3 in Ref. [75]). In particular $\nabla\varphi\neq 0$ on $\overline{\Omega}$. We can show that there is such a function satisfying the conditions $\varphi=0$ on $\partial\Omega$, $\varphi>0$ on $B\backslash\Omega$ (see Sect. 3.3 in Ref. [75]).

Let $\delta=\|u-u_e\|_{(0)}(\Gamma_0)$. First, we consider the case in which $\Gamma_0=\partial\Omega_0$. Let Ω_2 be a domain containing $\overline{\Omega}_0$ with $\overline{\Omega}_2\subset B$. By using the Green's function for the exterior Dirichlet problem for the Helmholtz equation (4.6) in Ω_0 we conclude that

$$\left|\left|u-u_e(;\delta)\right|\right|_{(1)}(B\backslash\Omega_2) < C\delta. \tag{4.29}$$

To obtain an interior bound, we introduce the cutoff function $\chi\in\Upsilon^\infty$, which is 1 on $\Omega_2\backslash\Omega(d)$ and which is supported in $B\backslash\overline{\Omega}$. We utilize χ with $0\leq\chi\leq 1$, $|\nabla^j\chi|\leq d^{-j}$, when $j=1$ and 2, where $\Omega(d)=B\cap\{\varphi<d\}$. Observe that due to our choice of φ, $d(r,\theta,\phi)/\Upsilon<\varphi(r,\theta,\phi)<\Upsilon d(r,\theta,\phi)$, thus we can obtain χ with the above bounds. Let $u_0=\chi(u-u_e)$. Then u_0 is compactly supported in B. Using Eq. (4.28) and the equality $(\nabla^2+k^2)u_0=2\nabla(u-u_e)\cdot\nabla\chi+(u-u_e)\nabla^2\chi$, we have

$$\begin{aligned}
&\sum_{\alpha\leq 1}\tau\int_{\Omega_0\backslash\Omega(2d)}\left(|\nabla(u-u_e)|^2+|u-u_e|^2\right)e^{2\tau\phi}\\
&\leq C\int_{\left(\Omega(d)\backslash\Omega\right)\cap\left(B\backslash\Omega_2\right)}\left|2\nabla(u-u_e)\cdot\nabla\chi+(u-u_e)\nabla^2\chi\right|^2 e^{2\tau\phi}\\
&\leq Ce^{2\tau d}\int_{\Omega_0\backslash\Omega}\left(\frac{|\nabla u|^2}{d^2}+\frac{|u|^2}{d^4}\right)+Ce^{2\tau\Phi}\int_{B\backslash\Omega_2}\left(|\nabla(u-u_e)|^2+|u-u_e|^2\right),
\end{aligned} \tag{4.30}$$

where $\Phi=\max\varphi$ over $\overline{B}$ and the inequality $\varphi<d$ on $\Omega(d)\backslash\Omega$ is used. In addition, using $2d<\varphi$ on $\Omega(2d)$ and replacing φ on the left side of Eq. (4.30) by $2d$, we obtain

$$e^{4\tau d}\left|\left|u-u_e\right|\right|^2_{(1)}(\Omega_0\backslash\Omega(2d))\leq C\left(\frac{e^{2\tau d}}{d^4}\left|\left|u-u_e\right|\right|^2_{(1)}(\Omega_0\backslash\Omega)+e^{2\tau\Phi}\left|\left|u-u_e\right|\right|^2_{(1)}(B\backslash\Omega_2)\right). \tag{4.31}$$

Dividing the both parts by $e^{4\tau d}$ and using that due to Eq. (4.23) $\|u-u_e\|_{(1)}(B\backslash\Omega)$, we find

$$||u-u_e||^2_{(1)}(\Omega_0\backslash\Omega(2d)) \leq C\left(\frac{e^{-2\tau d}}{d^4}M_1^2 + e^{2\tau(\Phi-2d)}\varepsilon^2\right). \tag{4.32}$$

Minimizing the right side of Eq. (4.32) with respect to $\tau > 0$ yields the minimum point,

$$2\tau = \frac{1}{\Phi - d}\ln\left[\frac{M_1^2}{d^3(\Phi-2d)\varepsilon^2}\right], \tag{4.33}$$

where the value of the minimized function is less than $Cd^{-4}M_1^{2(1-\theta)}\varepsilon^{2\theta}$, here $\theta = \frac{d}{\Phi - d}$.

This proves the interior bound.

To obtain the logarithmic bound, we will split the norm

$$||u-u_e||^2_{(0)}(\Omega_0\backslash\Omega) = ||u-u_e||^2_{(0)}(\Omega\backslash\Omega(d)) + ||u-u_e||^2_{(0)}(\Omega(d)\backslash\Omega). \tag{4.34}$$

By the Hoelder inequality the second term on the right side of Eq. (4.34) is

$$\int\limits_{\Omega(d)\backslash\Omega} 1|u-u_e|^2 \leq \left(\int\limits_{\Omega(d)\backslash\Omega} 1\right)^{\frac{2}{3}}\left(\int\limits_{\Omega(d)\backslash\Omega} |u-u_e|^6\right)^{\frac{1}{3}}$$
$$\leq Cd^{\frac{2}{3}}||u-u_e||^2_{(1)}(\Omega_0\backslash\Omega), \tag{4.35}$$

by Sobolev embedding theorems (see the Appendix of Ref. [75]). Now using the interior bound, we obtain

$$||u-u_e||^2_{(0)}(\Omega_0\backslash\Omega) \leq C\left[\frac{1}{d^4}M_1^2\left(\frac{\delta}{M_1}\right)^{\frac{d}{C}} + M_1^2 d^{\frac{2}{3}}\right]. \tag{4.36}$$

Letting $d = [-\ln(\delta/M_1)]^{-3/4}$, we conclude that Eq. (4.36) can be rewritten as

$$||u-u_e||^2_{(0)}(\Omega_0\backslash\Omega) \leq CM_1^2\left(L^3 e^{-\frac{L}{C}} + L^{-2}\right), \tag{4.37}$$

where $L = [-\ln(\delta/M_1)]^{1/4}$. Using $L^3 e^{-L/C} \leq CL^{-2}$, we complete the proof of a logarithmic bound.

The case of analytic $\partial\Omega_0$ is similar. We only have to observe that due to known conditional stability estimates of the analytic continuation for the analytic (in some two-dimensional complex neighborhood of the analytic surface $\partial\Omega_0$) function $u - u_e$ from Γ_0 to $\partial\Omega_0$ (see Corollary 1.2.2 of Ref. [78]), we have

$$||u - u_e||_{(0)}(\partial\Omega_0) \leq C||u - u_e||_{(0)}^{\theta_1}(\Gamma_0). \tag{4.38}$$

Here, as in Sect. 6.3 of Ref. [75], the neighborhood of $\partial\Omega_0$ and a bound of the complex-analytic continuation onto this neighborhood depend only on Ω, $\partial\Omega_0$, and M_1. The bound in Ref. [75] is given for a plane domain and for a function of a single complex variable, but by using the local analytic coordinates and continuation in each of two coordinate variables, we obtain the same bound on $\partial\Omega_0$. After that we proceed as above.

The proof is complete.

By some standard but more complicated argument, we can replace the exponent 1/4 in the logarithmic bound by any value smaller than 1. Also we can demonstrate that the interior bound of Lemma 4.2 holds when the bound (4.23) in $H_{(1)}(\Omega\backslash\Omega_0)$ is weakened to the following,

$$||u_e||_{(0)}(\Omega_0\backslash\Omega) \leq M_0, \tag{4.39}$$

and correspondingly the constraint (4.24) is weakened to the bound,

$$\int_{\Omega_0\backslash\Omega} \left|\sum_n \sum_m u_{n,m} e_{n,m}(x)\right|^2 dx \leq M_0^2, \quad m = 0, \ldots, 2n+1 \text{ and } n = 0, \ldots, N. \tag{4.40}$$

The approximation and stability results in this section suggest the following strategies for finding the approximate solution u_e. First, guess N, which is the number of the expansion term for radial functions. Next, find the convex constraint minimization, namely, Eqs. (4.25) and (4.24). Note that sometimes it might be easier to solve Eqs. (4.25) and (4.39) instead.

4.4 Significance of the Justification

The rigorous mathematical justification of the HELS formulations provided by Isakov and Wu is significant in that:

1. It demonstrates that any radiating solution to the Helmholtz equation outside a bounded Lipschitz domain with connected complement can be approximated by using a family of special solutions.
2. Using these approximations and conditional stability estimates in the Cauchy problem for the Helmholtz equation, these special solutions are proven to be bounded outside a vibrating surface and converge to the exact solution, provided that they converge to the exact solution on the measurement surface.

3. Moreover, the estimates of convergence of Hölder and logarithmic types in different regions are derived.

Isakov and Wu's work has provided definitive answers to the question of the validity of the HELS formulations: one can indeed use the spherical wave functions to approximate an acoustic field on a non-spherical surface. This conclusion also holds for the interior region.

The most significant impact of Isakov and Wu's work on HELS method is the suggestion of an effective regularization to overcome ill-posedness difficulty inherent in all inverse acoustic problems. Specifically, Isakov and Wu propose a regularization technique using quasi-solutions [79] by imposing a limit on the growth of reconstructed acoustic quantities in the entire exterior region, including the source surface S. Using the same symbols as those defined in Chap. 3, we obtain

$$||\hat{p}_j(\vec{x}_S;\omega)||_2^2 \equiv \iint_S |\hat{p}_J(\vec{x}_S;\omega)|^2 \leq K^2, (r_s, \theta_s, \phi_s;\omega) \in S, \tag{4.41}$$

where $j = 1$ to J_{op}. The constant on the right side of Eq. (4.41) has been shown to be correlated to the time-averaged acoustic power [71], which is a constant for any given source and is independent of measurement locations, or be correlated to the propagating component of the acoustic pressure,

$$K = \max_{(r_m,\theta_m,\phi_m)\in\Gamma;\ m=1,2,\ldots,M} |\hat{p}(r_m, \theta_m, \phi_m;\omega)|\left(\frac{r_m}{a}\right), \tag{4.42}$$

where a is the characteristic radius of the source surface.

Using this constraint on the source surface together with an iteration scheme to obtain J_{op}, Semenova and Wu [72] illustrate unambiguously that reconstruction errors remain finite everywhere including the source surface, whereas errors in reconstruction using HELS with the least-squares minimization alone can grow to an unsatisfactory level as the reconstruction point approaches the source surface. This explains why sometimes the accuracy of reconstruction on the surface of an arbitrarily shaped structure may be unsatisfactory.

By the way, the validity of using the spherical wave functions and spherical harmonics to reconstruct the acoustic quantities on a non-spherical surface was investigated by Prager [80] as well. In particular, Prager proposed a method to approximate the sound field not fulfilling the Rayleigh hypothesis by transforming a non-converging spherical wave function expansion to a converging one.

Isakov and Wu's theory has laid a solid foundation for the HELS method, answered any questions surrounding its validity in reconstructing acoustic radiation from an arbitrary object and provided the stability estimates for regularizing an ill-posed inverse acoustic problem. The work described in [71, 72] further reveals the interrelationship between a Rayleigh series and HELS solution and most significantly demonstrates that HELS solutions are convergent with bounded errors whenever a surface constraint condition is imposed.

Problems

4.1. What is the Rayleigh hypothesis? What does it attempt to do?

4.2. Consider the solution to the Helmholtz equation that describes the standing waves inside a spherical surface as given by Eq. (2.21a)

$$\hat{p}(r,\theta,\phi;\omega) = \sum_{n=0}^{\infty}\sum_{m=-n}^{n}\left[A_{mn}j_n(kr) + B_{mn}y_n(kr)\right]Y_n^m(\cos\theta)$$

Does this solution subject to the Rayleigh hypothesis? In other words, will this formulation work if the interior surface is corrugated, namely, not exactly spherical?

4.3. Consider the solution to the Helmholtz equation that describes the traveling waves outside a spherical surface as given by Eq. (2.21b). Suppose that this infinite series is truncated to a finite one as follows,

$$\hat{p}(r,\theta,\phi;\omega) = \sum_{n=0}^{N}\sum_{m=-n}^{n}\left[A_{mn}h_n^{(1)}(kr) + B_{mn}h_n^{(2)}(kr)\right]Y_n^m(\cos\theta),$$

and regularization is applied to the expansion. Will this modified solution subject to the Rayleigh hypothesis? Will it be applicable to a corrugated, namely, not exactly spherical surface?

4.4. What are the differences between the Rayleigh hypothesis and HELS formulations?

4.5. Will the HELS formulations be subject to the same restrictions as the Rayleigh hypothesis does?

4.6. How are the HELS formulations related to the Rayleigh hypothesis?

4.7. What does the mathematical justification prove for the HELS formulations?

4.8. What is the significance of this mathematical justification?

Chapter 5
Implementation of the HELS-Based NAH

In this chapter we present the general guidelines for setting up measurement to get the desired accuracy and spatial resolution for the HELS-based NAH. There are several parameters that may influence the reconstruction results such as the number of measurement points, standoff distance, measurement aperture size versus source surface area, microphone spacing, and SNR. These parameters are generic for all NAH applications. The strategies for setting up the optimal measurement scheme are basically the same.

It is the hope of the present author that the proposed reconstruction guidelines and hybrid regularization strategy would help potential users to get an accurate reconstruction of the normal surface velocity for a non-spherical structure. This is because a straight application of the original HELS formulations can result in not-so-satisfactory reconstruction of the normal surface velocity. We believe that the HELS method possesses certain advantages for engineering applications that the Fourier acoustics- and BEM-based NAH do not. This is because the Fourier acoustics- and BEM-based NAH are based on the exact theories. They can provide an exact reconstruction of acoustic fields when the required conditions are satisfied, for example, separable source geometry for the Fourier acoustics and a source-free field for both Fourier acoustics and BEM-based NAH, but are invalid when these conditions are not met. Unfortunately, in engineering practice these conditions are seldom met. On the other hand, the restrictions on the HELS method are significantly relaxed because it only seeks approximate reconstruction and is suitable for patch reconstruction. Needless to say, the results obtained by using HELS are approximate and their accuracy depends on that of input data. So it is important to follow the proposed guidelines and strategies to obtain satisfactory reconstruction of all the acoustic quantities including the acoustic pressures, particle velocities, and acoustic intensities everywhere in three-dimensional space.

S.F. Wu, *The Helmholtz Equation Least Squares Method*, Modern Acoustics and Signal Processing, DOI 10.1007/978-1-4939-1640-5_5

5.1 Guidelines for Implementing the HELS Method

It is emphasized that the HELS method imposes no restrictions whatsoever on the use of the coordinate systems and the corresponding wave functions. The spherical coordinates and the spherical wave functions can yield good approximate solutions for a blunt radiator. The prolate, oblate, and elliptic coordinates and the corresponding spheroids can generate good approximate solutions for a slender, a flat, and an arbitrarily shaped radiator, respectively. Regardless of the coordinate systems selected, the expansion coefficients in HELS are specified by matching the assumed-form solution to the measured acoustic pressures, and the errors are minimized by the least-squares method and regularization. In practice, it is not easy to utilize the prolate, oblate, or elliptic spheroids because the corresponding analytic solutions do not exist. On the other hand, the spherical wave functions are readily available in any software library, making programming very straightforward and numerical computations very fast.

Although our ultimate goal is to extend HELS to reconstructing vibro-acoustic responses of an arbitrarily shaped structure, we begin investigation from a simple yet highly non-spherical surface, such as a baffled thin plate, and examine the reconstruction accuracy using the spherical wave functions that are readily available in most software tools (MATLAB, LabVIEW, etc.) [81, 82]. The choice of a baffled plate also allows for rigorous examinations on the HELS results because the corresponding analytic solutions are readily available.

Since the HELS-based NAH utilizes the expansion of certain basis functions, it is ideal if the geometry of a target source surface fits naturally with the basis functions. For example, if the spherical coordinates are used in the basis functions, then the HELS-based NAH will be naturally fit for reconstructing the acoustic quantities generated by a spherical source or blunt object, whose aspect ratio is close to 1:1:1. The accuracy in reconstruction will be quite high. If source geometry is different from a sphere or its aspect ratio is not 1:1:1, the HELS-based NAH is still applicable, but the accuracy in reconstruction may be compromised. The farther the source geometry is from the coordinates of the basis functions, the larger the errors will be in reconstruction. In practice, a vibrating object is usually of an arbitrary shape. In order to obtain satisfactory reconstruction using HELS, it will be a good idea to follow some tested guidelines.

In what follows, we consider a class of structures that are commonly used in practice, i.e., a plate, which is highly non-spherical and represents a serious challenge to the suitability of using HELS to reconstruct the resultant acoustic field in three-dimensional space. This is because there are additional factors that may affect reconstruction results, for example, location of the origin of the coordinate system. Since the thickness of a plate is usually negligible, there is no way of placing the origin at its geometric center. Thus it must be placed outside the planar surface, but where? How far should the origin of the coordinate system be? How far should the measurement aperture be? How large should the measurement aperture be?

It is emphasized that there are no analytic solutions to these questions because HELS is an approximation not an exact theory. Reference [74] gives a mathematical proof of the HELS method. For the exterior problems solutions are bounded in the three-dimensional domain Ω enclosed by the source surface Γ and a surface at infinity Γ_∞ (excluding the origin of the coordinate system), where the Helmholtz equation is satisfied. Moreover, HELS solutions converge logarithmically in Ω. There is no restriction on where the origin of the coordinate system should be placed. For a blunt object, it is natural to place the origin at the geometric center. For a thin plate, however, this is not possible. Our studies have illustrated that there is an optimum position for the origin of the coordinate system on the opposite side of the plate, which can produce satisfactory reconstruction results [83]. Experiments have validated the existence of such an optimal position. In addition, the number of the expansion terms and other parameters, namely, the number of measurement points, microphone spacing, standoff distances, measurement aperture sizes versus source surface areas, SNR, etc. are all important in implementing the HELS-based NAH.

Listed below are the guidelines for implementing the HELS-based NAH [84] to reconstruct the vibro-acoustic responses on the surface of a highly non-spherical surface (see Fig. 5.1).

1. Origin position d_{origin}: The inherent difficulty in HELS is to approximate the vibro-acoustic quantities on a highly non-spherical surface by using the spherical wave functions. If the origin of the coordinate system is placed too close to the surface, errors in reconstruction, though still bounded, can be quite large because the point of the origin is excluded in the region of validity for an exterior problem [85]. On the other hand, if the origin is placed too far from the surface, detailed features in the vibro-acoustic responses associated with the higher-order expansions in the HELS formulations may diminish, leading to discrepancies in reconstruction. Therefore, there exits an optimal position for placing the origin of the coordinate system. Since there are no analytic formulations for selecting the optimal origin of the coordinate system, numerical simulations are employed and results suggest that the optimal position d_{origin} falls within ± 10 % of the characteristic dimension of the plate D,

$$d_{\mathrm{origin}} = (0.9 \sim 1.1)D, \qquad (5.1)$$

where $D = 0.5\sqrt{L^2 + W^2}$, L and W are the length and width of the plate, respectively.
2. The critical spatial wavelength λ_{cr}: The spatial wavelengths of any vibrating structure are usually unknown, so λ_{cr} is a target value. Consider a rectangular plate of dimensions $L \times W$. Suppose that we aim at reconstructing up to the (n, m)th mode of this plate, where n is the modal index in the longitudinal direction with a dimension L and m is that in the transverse direction with a dimension W. Then $\lambda_{\mathrm{cr}}/2$ is the smaller of L/n and W/m,

Fig. 5.1 Schematic of measurement setup for a plate type structure

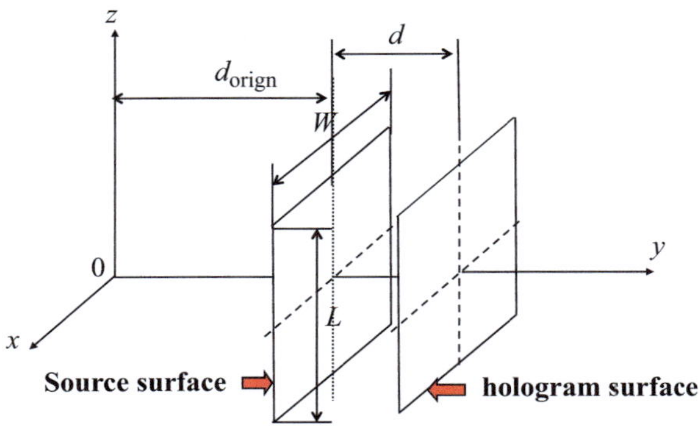

$$\lambda_{\mathrm{cr}}/2 = \min(L/n, W/m). \tag{5.2}$$

For example, consider a rectangular plate of dimensions 0.3×0.2 m^2 and the (4, 3)th mode of this plate is the highest mode to be reconstructed. The critical spatial wavelength is given by Eq. (5.2), where $n = 4$, $m = 3$, $L = 0.3$ m, and $W = 0.2$ m. Substituting these values into Eq. (5.2) yields $\lambda_{\mathrm{cr}}/2 = 0.067$ m, or $\lambda_{\mathrm{cr}} = 0.134$ m.

It is important to remember that the test setup in NAH is gauged with respect to the spatial wavelength or spatial frequency, not the acoustic wavelength or temporal frequency. If we can reconstruct structural waves up to the critical spatial wavelength λ_{cr}, the mechanical energies of all components of the structural waves up to the spatial wavelength λ_{cr} are captured. Since the vibration energy of a structural wave decays with spatial wavelength, we only miss a small portion of the total vibration energy that is the sum of structural waves whose wavelengths are shorter than λ_{cr}.

It is emphasized that the acoustic wave, regardless of its wavelength, is of no concern in NAH reconstruction. So long as the structural waves of wavelengths up to λ_{cr} are reconstructed, all acoustic waves can be reconstructed. For example, consider the plate as cited above. Suppose that structural waves of wavelengths up to $\lambda_{\mathrm{cr}} = 0.134$ m are reconstructed. Then all acoustic waves whose wavelengths are shorter than 0.134 m or equivalently, the acoustic frequencies higher than $f > c/\lambda_{\mathrm{cr}} = 340/0.134 = 2{,}537$ Hz can be reconstructed. The acoustic waves whose frequencies are lower than 2,537 Hz can always be reconstructed because the longest acoustic wavelength of the audible sound wave is always shorter than the longest spatial wavelength, i.e., the rigid body motion.

Note that if test setup is gauged with respect to the acoustic wavelength or temporal frequency, but not the spatial wavelength or spatial frequency, it will be acoustical holography, not NAH. Accordingly, one will only be able to reconstruct the radiated acoustic pressure and nothing else.

3. Number of measurement points M: This parameter is critical in practice. In theory, the more the measurement points are taken, the more information is

collected, and the more accurate the reconstruction becomes. However, an excessive number of measurement points may not be acceptable in practice. A compromise is to link the number of measurement points M to the required reconstruction surface area S through a target structural wavelength λ_{cr},

$$M > 4S/\lambda_{cr}^2 \quad \text{or} \quad M_{min} = 44. \tag{5.3}$$

Note that we have imposed a minimal number of measurement points $M_{min} = 44$ [84]. This is because for a small plate and a lower order of vibration mode, the value of M can be very small, leading to an omission of the critical information in the input data and large errors in a reconstruction. Once again, there is no analytic solution to determine M_{min}. The value of $M_{min} = 44$ is to ensure that the HELS expansion includes at least the fifth-order spherical Hankel function, namely, $n = 0$–5, in reconstruction, guaranteeing certain levels of details in the reconstructed vibro-acoustic responses.

4. Microphone spacing δ: Unlike the Fourier acoustics-based NAH, HELS does not require a uniform microphone spacing on the hologram surface. However, it is a good practice to set the microphone spacing to be less than one-half of the target structural wavelength λ_{cr} [86],

$$\delta < \lambda_{cr}/2. \tag{5.4}$$

5. Standoff distances d: The goal of NAH is to reconstruct vibro-acoustic quantities without the wavelength resolution limit in theory. This is possible when all the near-field effects are collected, which may be accomplished by placing microphones infinitely close to the target vibrating surface and infinitely close to each other. Such a scenario is unrealistic and unattainable in reality. Practical considerations such as the working condition, temperature, and accessory component attached to a structure require that microphones be placed at certain distances away from the structure. Thus, there is an upper limit in the spatial resolution in a reconstruction. To strike a balance between the theoretical goal and practical consideration, we recommend that the standoff distance d be less than one-eighth of the value of λ_{cr},

$$d < \lambda_{cr}/8. \tag{5.5}$$

Notice that there is an important distinction between the standoff distances for the Fourier acoustics-based NAH and those for the HELS-based NAH. The former utilizes the discrete spatial Fourier transform, and its accuracy in reconstruction is critically dependent on the spatial sampling frequency that is intimately related to the standoff distances. If the spatial sampling frequency is so low that the microphone spacing becomes larger than the standoff distance, "undersampling" may happen, causing spatial aliasing in a reconstruction. Hence the standoff distances in the Fourier acoustics-based NAH are kept at least one microphone spacing to avoid "undersampling" in data acquisition. The situations are quite different in the HELS-based NAH, where the acoustic quantities are reconstructed by superimposing the spherical wave functions. There is no direct correlation between the spatial resolution and spatial sampling frequency. In fact, the spatial resolution in HELS is directly related to the number of the spherical wave functions employed. In order for the high-order spherical wave functions to function the way they are supposed to, the standoff distances should be as close to the target surface as possible in order to collect enough near-field information. Experimental results have confirmed that the smaller the value d is, the more accurate the reconstruction is, regardless of the microphone spacing δ. This property of decoupling the measurement distance from microphone spacing is unique to the HELS method.

6. Target source surface S: Because the spherical wave functions are used in the HELS-based NAH to approximate the acoustic fields generated by non-spherical vibrating structures, it is a good idea to limit the overall size of a target source surface S so that reconstruction can be done all at once. Consider a plate of dimensions $S = L \times W$. We recommend that the length and width be no more than twice the target structural wavelength, namely,

$$L, W \leq 2\lambda_{\mathrm{cr}}. \tag{5.6}$$

This imposes some restriction on the overall dimensions of the structure that HELS may be attempted at once, but nevertheless leads to satisfactory reconstruction on a target surface. For surfaces whose overall lengths or widths are larger than $2\lambda_{\mathrm{cr}}$, patch reconstruction may be utilized. In performing patch reconstructions, the origin of the coordinate system should move with each patch, and the rest remains the same. The measurement aperture A_m must be at least one row and one column larger than a target reconstruction surface area S. Note that there is a difference between a patch measurement and patch reconstruction. The former refers reconstruction of the acoustic quantities on a portion of a large surface, whereas the latter indicates a specific measurement setup, which is often the case in practice for a finite number of microphones. For example, a specific reconstruction requires 100 measurement points, but only 20 microphones are available. Then reconstruction may be done by taking five patches of measurements sequentially.

7. Aspect ratio: For a planar surface, aspect ratio refers the ratio of its overall length to width. To reconstruct acoustic quantities on a planar surface of dimensions $S = L \times W$, its aspect ratio should be limited to the following range per reconstruction,

$$(1:1) \leq (L:W) \leq (2:1). \tag{5.7}$$

This is because the spherical wave functions and spherical harmonic functions are used in HELS to approximate the vibro-acoustic quantities on a planar surface. It may be difficult to ensure a satisfactory reconstruction over the entire surface area when the aspect ratio is larger than 2:1. For a planar surface with aspect ratio larger than 2:1, patch reconstruction should be used.

8. SNR: This parameter is universal to most measurement methods,

$$\mathrm{SNR} > 10(\mathrm{dB}) \tag{5.8}$$

Physically, this means that the energy or mean-squared acoustic pressure amplitude of the signal is at least ten times higher than that of background noise.

9. Number of reconstruction points N: There is no restriction on the number of reconstruction points, either on the source surface or in the field. However, for engineering applications it is recommended that vibro-acoustic quantities be reconstructed at four points per critical spatial wavelength λ_{cr} on the source surface to produce a smooth ODS. Excessive number of reconstruction points will not provide further information and should be avoided.

10. Number of the expansion functions J: Eq. (3.60) offers an effective way to estimate the optimal number of expansion functions $J_{op,MTR}$ for reconstructing the acoustic pressure and normal velocity on the source surface. When $J_{op,MTR} < M$, we have an overdetermined system. When $J_{op,MTR} > M$, we have an under-determined system. Either way, the system of equations can be solved by SVD. However, an under-determined system tends to yield less satisfactory reconstruction results than an overdetermined system does. Therefore the maximal number of expansion terms is set to be equal to that of measurement points M.

Note that the above guidelines have accounted for the needs to simplify the measurement setup and data acquisition processes in engineering applications. In conducting research projects for which the accuracy of reconstruction is of primary concern, whereas time and effort are of no concern, some of the above guidelines may be tightened as needed.

For example, in our study the area of the square plate is $S = 0.22 \times 0.22 = 0.0484$ m^2, and the (4, 4)th natural mode of this plate is selected as the highest mode to be reconstructed. By using Eq. (5.2), we get $\lambda_{cr}/2 = \min(0.22/4, 0.22/4) = 0.055$ m, so $\lambda_{cr} = 0.11$ m. Next, we use Eq. (5.3) to determine the number of measurement points as $M > 0.0484/0.055^2 = 16$, which is smaller than 44, so we take $M_{min} = 44$. By using Eqs. (5.4) and (5.5), we can set the microphone spacing at $\delta < \lambda_{cr}/2 = 0.05$ m, which is less than 0.055 m as suggested by Eq. (5.4), and the standoff distance of $d = 0.0125$ m, which is shorter than 0.01375 m as suggested by Eq. (5.5).

These parameters would suffice to produce a quick reconstruction with a decent accuracy. However, we want to establish the accuracy in reconstructing the normal surface velocity by using the HELS-based NAH, and do not mind spending extra time and efforts in collecting input data. Therefore, we take three patches of measurements using a 12×4 microphone array, resulting in a total $M = 144$ measurement points over the measurement aperture A_m, which is one row and one column larger the surface area of the square plate S. The corresponding microphone spacing is set to be $\delta = 0.03$ m, which is less than the suggested microphone spacing ($\delta = 0.05$ m). Moreover, we set the standoff distance to be $d = 0.01$ m, which is less than the suggested standoff distance ($d = 0.0125$ m). This fine measurement grid and close measurement distance ensures the desired accuracy in reconstruction. The origin of the coordinate system is placed at $d_{\mathrm{origin}} = 0.155$ m behind the plate as dictated by Eq. (5.1). The reconstruction results are exhibited in Figs. 3.6, 3.7, 3.8, 3.9, 3.10, and 3.11.

These results demonstrate that satisfactory reconstruction of the vibro-acoustic quantities on the surface of a highly non-spherical vibrating structure can be obtained by using the HELS method. In particular, the target (4, 4)th natural mode can be satisfactorily reconstructed.

5.2 Practical Considerations in Implementing the HELS Method

Noise and vibration abatement have always been one of the primary challenges facing the manufacturing industry, for example, the automobile, aircraft, and appliance manufacturers. The first step toward noise and vibration abatement is to identify their root causes, their interrelationships, and the key components that play the critical roles in generating undesirable noise and vibration.

The traditional technologies such as EMA [87] and operational modal analysis (OMA) [88] can provide an insight into the integrity of a vibrating structure by extracting its modal parameters that include the natural frequency, the natural mode, and the damping ratio. The knowledge acquired from EMA and OMA, however, may not be employed directly in noise abatement because these modal parameters are not related to sound radiation.

Traditional measurement devices such as microphones, intensity probes, and accelerometers enable one to measure the acoustic pressure, acoustic intensity, and normal surface velocity on specific locations that are very important to understanding the interrelationships between acoustic radiation and structural vibrations. However, the information captured in measurements is usually isolated and uncorrelated to each other. In other words, one can obtain a local and direct view of sound or vibration at a specific location, but not the global view of how sound is generated by a vibrating structure, and how it is correlated to structural vibrations.

Invention of the NAH technology has fundamentally changed the diagnostics and analyses of noise and vibration problems in that it enables one to visualize all acoustic quantities, including the acoustic pressure, particle velocity vector, acoustic intensity vector, and out-of-plane structural vibration distributions on the surface of a structure by taking acoustic pressure measurements at a very close distance. The insights acquired from NAH to the characteristics of vibrating structures and resultant acoustic radiation cannot be matched by traditional measurement methods.

Because NAH requires taking measurements at very close distances to a target structure to capture the near-field information, a conformal array of microphones is usually required. This conformal measurement will ensure that the standoff distance is uniform and the accuracy in the input data is consistent. The time and effort involved in setting up a conformal array can be quite intensive. Moreover, the measurement environment in engineering practice is usually not echo free, which means there are sound reflections and reverberation inside a test chamber. Accordingly, it is important to take these effects into consideration in order to get the desired reconstruction.

In addition to the general guidelines presented in Sect. 5.1, we offer further suggestions to optimize the measurement setups for various test configurations and test environments that are often encountered in engineering applications.

5.3 Test Configuration

Whenever possible, a conformal array should be utilized instead of a flat array. There is no doubt that a conformal microphone array will take time to make and set up. However, this is well worth the effort because the accuracy in reconstruction will be directly related to that in the input data. Accurate and consistent input data will lead to accurate and consistent reconstruction of all acoustic quantities in three-dimensional space, including the source surfaces. Figures 5.2 and 5.3 show schematics of conformal microphone arrays for reconstructing the acoustic fields generated by arbitrarily shaped vibrating structures in exterior and interior regions, respectively.

Oftentimes the source surfaces may be larger than the measurement aperture. Hence patch measurements are required. The number of measurement points as suggested in the guidelines is minimal for data acquisition. The number of reconstruction points, however, may be higher than that of measurement points. The suggested number of reconstruction points in HELS is up to but no more than four times that of measurement points.

Figures 5.4 and 5.5 illustrate, respectively, examples of using the HELS method to analyze the acoustic fields generated by an automobile transaxle in the exterior region, and by an aircraft inside its cabin while the aircraft was cruising at 0.8 Mach number 30,000 ft above the ground. The microphone array was mounted on a track so that it could travel along the longitudinal direction to measure the near-field acoustic pressure inside the cabin.

It must be pointed out that in engineering applications the demand for easy-of-use often overrides everything else. As a result, a planar array of microphones is used to collect the input data, even though a target source surface is nonplanar (see Fig. 5.6). As a result, the near-field information is completely lost because the measurement distances are varying and too large, and the reconstructed acoustic

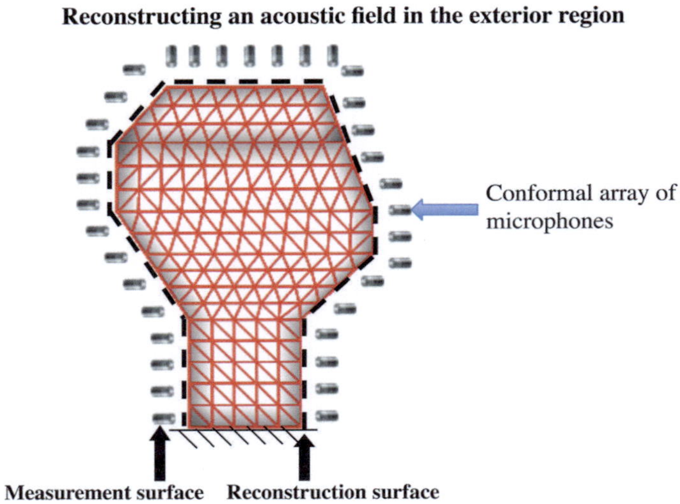

Fig. 5.2 Schematic of a conformal microphone array for reconstructing the acoustic field produced by an arbitrarily shaped source in the exterior region

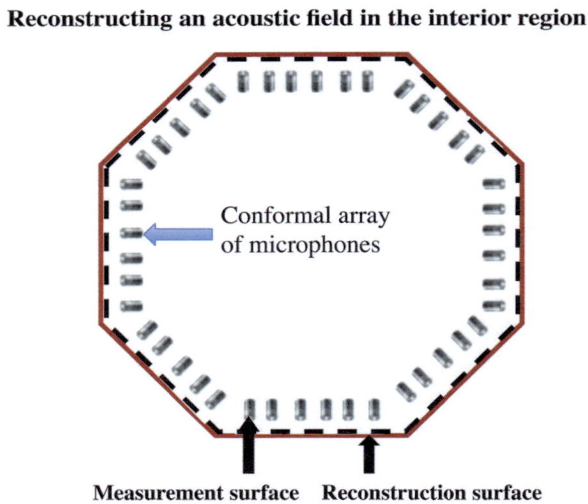

Fig. 5.3 Schematic of a conformal microphone array for reconstructing the acoustic field produced by an arbitrarily shaped source in the interior region

quantities are useless or even misleading. Therefore, even though this approach seems to save time and effort in data acquisition, it actually wastes all of them including those in post processing. The well-known statement in the field of computer science "Garbage in, garbage out" holds exactly true in this case.

Fig. 5.4 A conformal array of measurement microphones around an automobile transaxle at very close distances to the target source surface

Fig. 5.5 A conformal array of measurement microphones around along the circumference of the interior space of an aircraft while it was cruising at 0.8 Mach number 30,000 ft above the ground. The microphone array was mounted on a track so that it could travel in the longitudinal direction to measure the nearfield acoustic pressure inside the cabin

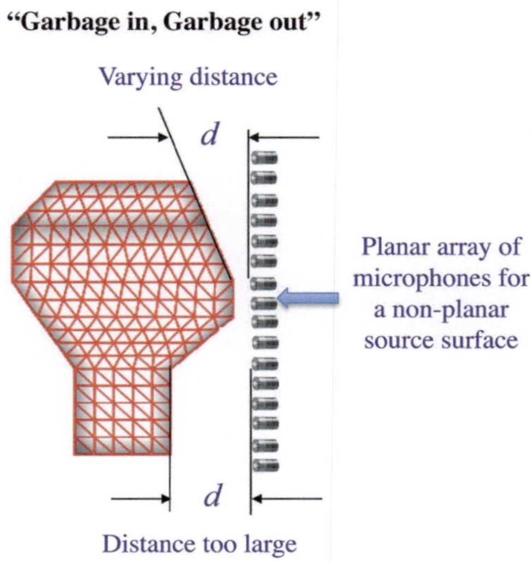

Fig. 5.6 Schematic of using a planar array of microphones to collect input data for an arbitrarily shaped vibrating structure. The near-field information is all lost in this case because measurement distances are varying and too large. As a result the reconstruction results are useless and even misleading

5.4 Test Environment

The HELS method is valid in both exterior and interior regions. In either case it is critical for the measurement surface to cover the entire source surface. If the source surface is relatively large, patch measurements should be taken. Covering a portion of the source surface may produce satisfactory reconstruction locally, but not acceptable globally.

Figure 5.7 displays a scenario of reconstructing the acoustic field using the HELS method in the exterior region. In this case the measurement surface A_m is on one side at close range. Hence, reconstruction may be acceptable on the covered source surface area, but not elsewhere.

Similarly, if a conformal array of microphones covers only a portion of the interior surface of a vibrating structure (see Fig. 5.8), $S = S_1 + S_2 + S_3$, the reconstructed acoustic quantities may be acceptable on the covered surfaces, but not on other surfaces, nor in the interior region.

Oftentimes we are dealing with a vibrating structure inside a large room (see Fig. 5.9), where the total acoustic pressure is the sum of the direct and reflected sound waves.

Under this condition, it will be critical to ensure that there is enough space between the source of interest and reflective walls, and SNR is at least 10 dB or higher in order to minimize the effects of reflected sound waves in data acquisition. If these conditions are met, the reconstructed results might be acceptable on the source surfaces covered by a conformal array of microphones. If these conditions cannot be met, reconstruction should not be carried out because the input data will be severely contaminated by the interfering sound signals.

This is because a reflecting surface behaves like an image source. Consider the case in which a source is situated on two infinitely large reflecting surfaces as shown in Fig. 5.10. Then the acoustic pressure measured in this confined space consists of the direct sound radiated from the source (ray 1) and those reflected from walls (rays 2–4). This is equivalent to the case where the source and its three images lie in free space, and the measured sound pressure will consist of the contributions

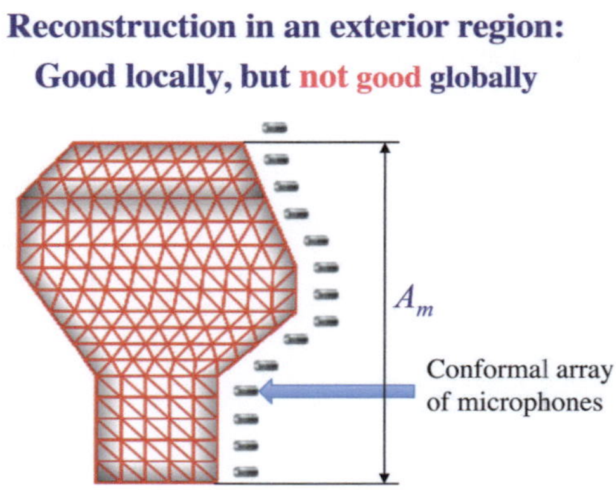

Fig. 5.7 Schematic of a conformal microphone array covering only a portion of the target surface S in the exterior region. Accordingly, the reconstructed acoustic quantities may be acceptable locally, but not globally

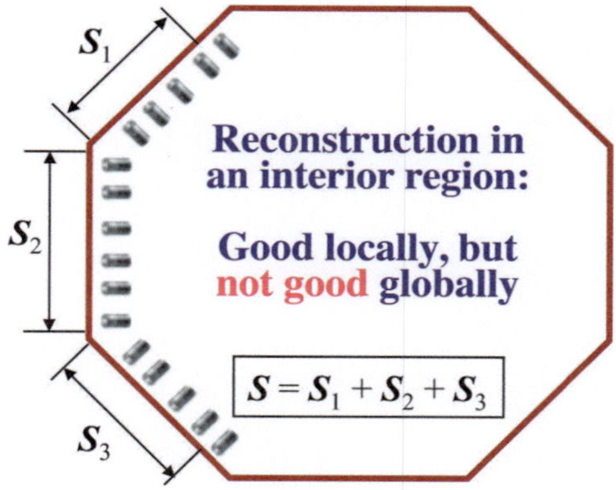

Fig. 5.8 Schematic of a conformal microphone array covering only a portion of the interior surface. The resultant reconstruction may be satisfactory on the cover surface area and immediately adjacent to it, but not satisfactory on other surfaces as well as in the interior region

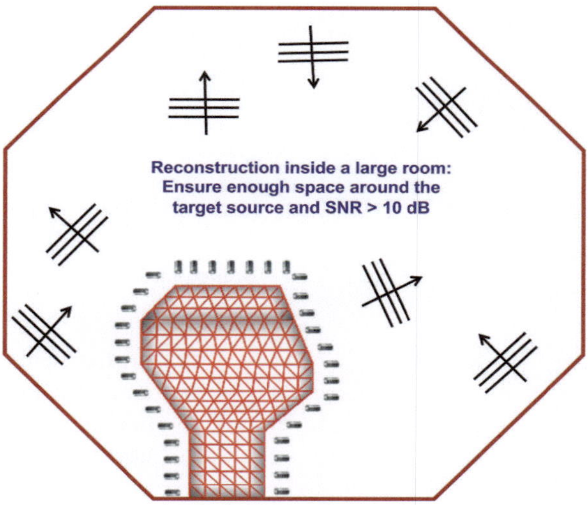

Fig. 5.9 Schematic of reconstructing the acoustic quantities generated by a vibrating structure inside a large room with reflecting surfaces. The total acoustic pressure is the sum of the direct and reflected sound waves

from all these sources. If a source is inside a room consisting four walls, one ceiling and one floor, then the measured acoustic pressure will consist of one direct sound wave emitted from the source and an infinite number of reflected sounds. Mathematically, this is expressible as

$$\hat{p}^{\,2}_{\mathrm{rms}}(r, \theta, \phi; \omega) = \rho_0 c P(\omega) \left[\frac{Q_\theta}{4\pi r^2} + \frac{4}{R_{\mathrm{rc}}(\omega)} \right], \tag{5.9}$$

where $\hat{p}^{\,2}_{\mathrm{rms}}(r, \theta, \phi; \omega)$ indicates the measured mean-squared acoustic pressure inside a large room; $P(\omega)$ depicts the acoustic power, which is a function of frequency but independent of measurement location; Q_θ is the directivity factor given by

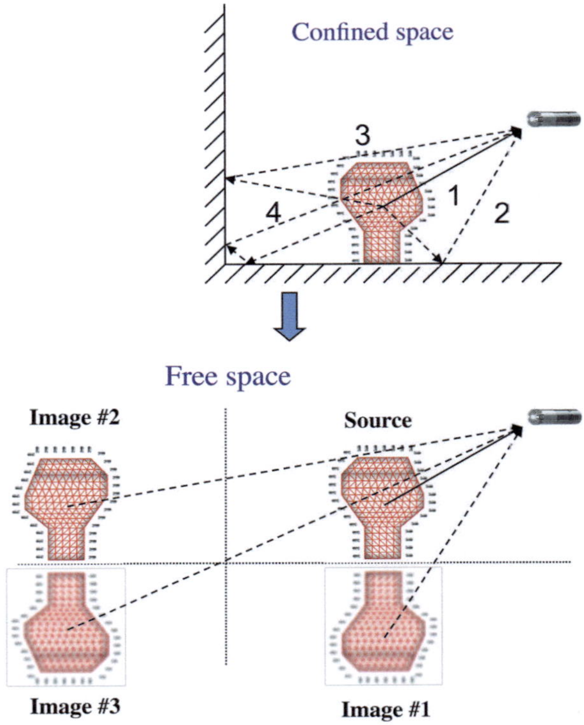

Fig. 5.10 Schematic of the effects of reflecting surfaces on the measured acoustic pressure

$$Q_\theta = \begin{cases} 1, & \text{in a free field;} \\ 2, & \text{on a large floor;} \\ 4, & \text{near any edge;} \\ 8, & \text{near any corner;} \end{cases} \tag{5.10}$$

The parameter $R_{rc}(\omega)$ in Eq. (5.9) represents the room constant defined as

$$R_{rc}(\omega) = \frac{S_{total}\overline{\alpha}(\omega)}{1 - \overline{\alpha}(\omega)}, \tag{5.11}$$

where S_{total} stands for the total reflecting surface area of the room and $\overline{\alpha}(\omega)$ represents the spatial averaged acoustic pressure absorption coefficient given by

$$\overline{\alpha}(\omega) = \frac{\sum_i S_i \alpha_i(\omega)}{S_{total}}. \tag{5.12}$$

The first term on the right side of Eq. (5.9) describes the direct sound wave emitted from the source, and the second term depicts the effect of reverberation of sounds inside the room. The smaller the room constant $R_{rc}(\omega)$ is, the higher the effect of reverberation of the room becomes and the worse the measurement condition is for reconstruction. Since in general the values of $\alpha_i(\omega)$ of all reflecting surfaces inside a room are unknown a priori, there is no way of determining the value of $R_{rc}(\omega)$. Hence, the measured mean-squared acoustic pressure does not reflect the true acoustic pressure emitted from the source, but includes all reverberation effects of the room. Accordingly, the input data to reconstruction will be severely contaminated.

There are several methods that can be used to determine the reverberation effect of a room, for example, reference source method and double-concentric-surface method [89]. All these methods require taking two sets of measurements and therefore are known as the indirect methods.

In the reference source method, a reference source with a known power spectrum is placed at the position of or close to a target source inside a room. Next the acoustic power radiated by this reference source is calculated based on the mean-squared acoustic pressure measured on a surface enclosing this reference source. Using Eq. (5.9), $R_{rc}(\omega)$ at any frequency can be written as

$$\frac{4}{R_{rc}(\omega)} = \frac{\hat{p}^2_{rms}(r_{ref}, \theta_{ref}, \phi_{ref}; \omega)}{\rho_0 c P_{ref}(\omega)} - \frac{Q_\theta}{4\pi r^2_{ref}}, \tag{5.13}$$

where $\hat{p}^2_{rms}(r_{ref}, \theta_{ref}, \phi_{ref}; \omega)$ represent the measured mean-squared acoustic pressures radiated by the reference source, and $P_{ref}(\omega)$ is the known acoustic power of this reference source. Once the value of room constant $R_{rc}(\omega)$ is specified, the true mean-squared acoustic pressure is given by

$$\hat{p}^2_{rms}(r, \theta, \phi; \omega) = \rho_0 c P(\omega) \left[\frac{\hat{p}^2_{rms}(r_{ref}, \theta_{ref}, \phi_{ref}; \omega)}{\rho_0 c P_{ref}(\omega)} + \frac{Q_\theta}{4\pi r^2} - \frac{Q_\theta}{4\pi r^2_{ref}} \right]. \tag{5.14}$$

In the double-concentric-surface method, the room constant $R_{rc}(\omega)$ is determined by taking the acoustic pressures over two concentric surfaces S_1 and S_2, respectively,

$$\hat{p}^2_{rms,1}(r_1, \theta_1, \phi_1; \omega) = \rho_0 c P(\omega) \left[\frac{Q_\theta}{4\pi r_1^2} + \frac{4}{R_{rc}(\omega)} \right], \tag{5.15}$$

$$\hat{p}^2_{rms,2}(r_2, \theta_2, \phi_2; \omega) = \rho_0 c P(\omega) \left[\frac{Q_\theta}{4\pi r_2^2} + \frac{4}{R_{rc}(\omega)} \right]. \tag{5.16}$$

Assume that $S_2 > S_1$, then $\hat{p}^2_{rms,2}(r_2, \theta_2, \phi_2; \omega) < \hat{p}^2_{rms,1}(r_1, \theta_1, \phi_1; \omega)$ because S_1 is closer to the source, and the amplitude of the acoustic pressure decays from S_1 to S_2, even though the decay rate is not known. Meanwhile, the acoustic power radiated from the target source $P(\omega)$ and directivity factor Q_θ remain the same. Therefore, combining Eqs. (5.15) and (5.16) we can express the room constant $R_{rc}(\omega)$ as

$$\frac{4}{R_{\rm rc}(\omega)} = \frac{\frac{Q_\theta}{4\pi r_1^2}\hat{p}^{\,2}_{\rm rms,2}(r_2,\theta_2,\phi_2;\omega) - \frac{Q_\theta}{4\pi r_2^2}\hat{p}^{\,2}_{\rm rms,1}(r_1,\theta_1,\phi_1;\omega)}{\hat{p}^{\,2}_{\rm rms,1}(r_1,\theta_1,\phi_1;\omega) - \hat{p}^{\,2}_{\rm rms,2}(r_2,\theta_2,\phi_2;\omega)}. \quad (5.17)$$

The acoustic power of the target source $P(\omega)$ can then be given by

$$\rho_0 c P(\omega) = \frac{\hat{p}^{\,2}_{\rm rms,1}(r_1,\theta_1,\phi_1;\omega) - \hat{p}^{\,2}_{\rm rms,2}(r_2,\theta_2,\phi_2;\omega)}{\frac{Q_\theta}{4\pi r_1^2} - \frac{Q_\theta}{4\pi r_2^2}}. \quad (5.18)$$

Once $R_{\rm rc}(\omega)$ and $P(\omega)$ are determined, the true mean-squared acoustic pressure emitted by the target source inside a large room is given by

$$\hat{p}^{\,2}_{\rm rms}(r,\theta,\phi;\omega) = \frac{\left[\hat{p}^{\,2}_{\rm rms,1}(r_1,\theta_1,\phi_1;\omega) - \hat{p}^{\,2}_{\rm rms,2}(r_2,\theta_2,\phi_2;\omega)\right]}{\left(\frac{Q_\theta}{4\pi r_1^2} - \frac{Q_\theta}{4\pi r_2^2}\right)}$$
$$\times \left[\frac{Q_\theta}{4\pi r^2} + \frac{\frac{Q_\theta}{4\pi r_1^2}\hat{p}^{\,2}_{\rm rms,2}(r_2,\theta_2,\phi_2;\omega) - \frac{Q_\theta}{4\pi r_2^2}\hat{p}^{\,2}_{\rm rms,1}(r_1,\theta_1,\phi_1;\omega)}{\hat{p}^{\,2}_{\rm rms,1}(r_1,\theta_1,\phi_1;\omega) - \hat{p}^{\,2}_{\rm rms,2}(r_2,\theta_2,\phi_2;\omega)}\right], \quad (5.19)$$

where $\hat{p}^{\,2}_{\rm rms}(r,\theta,\phi;\omega)$ depicts the true mean-squared acoustic pressure radiated from a target source inside a large room; $\hat{p}^{\,2}_{\rm rms,1}(r_1,\theta_1,\phi_1;\omega)$ and $\hat{p}^{\,2}_{\rm rms,2}(r_2,\theta_2,\phi_2;\omega)$ indicate the mean-squared acoustic pressures measured on two concentric surfaces. Note that this method works most effectively when the difference between two concentric surfaces is large enough so that the sound pressure level L_1 measured on the first surface S_1 is at least 3 dB higher than L_2 measured on the second surface S_2. In practice, hemispherical surfaces or rectangular parallelepiped surfaces are often selected for S_1 and S_2.

5.5 Clarifications

In implementing the Fourier acoustics-based NAH, measurement distance and microphone spacing should always be gauged with respect to the spatial frequency or spatial wavenumber, but not with respect to the acoustic frequency or acoustic wavenumber. That point was made clear in one of the original NAH papers [7], which stated "The minimum resolvable distance is on the order of $R = \pi/k_{\max}$, where $k_{\max}$ is the highest spatial frequency for a measurable Fourier component $\tilde{\psi}(k_x, k_y, z_H)$. In conventional optical and acoustical holography no evenescent waves are used in the field reconstructions so that $k_{\max} = k$ and $R = \lambda/2$, where λ is the acoustic wavelength."

Here the highest spatial frequency or the spatial wavenumber k_{max} is linked to the shortest spatial wavelength λ_{min} ($=2\pi/k_{max}$) that contains significant vibration energy [90]. Note that vibration energy of any structure decays with the spatial wavelength. The shorter the spatial wavelength is, the less the vibration energy it contains, whereas the longer the spatial wavelength is, the more the vibration energy it contains. Therefore, if measurement setup is gauged with respect to the highest spatial frequency k_{max} or the shortest spatial wavelength λ_{min}, structural vibrations that contain the vibration energy up to the shortest spatial wavelength can be reconstructed. Once this is done, the entire acoustic field, including the surface acoustic pressure, the normal surface acoustic intensity, and radiated acoustic power, can be reconstructed, and the correlation between structural vibration and acoustic radiation can be established. This is the advantage of NAH that cannot be matched by other methodologies. All that is lost during this process are the components of structural vibrations whose spatial wavelengths are shorter than λ_{min}, whose vibration energies are insignificant.

For example, in setting up the measurement microphone array, one can gauge the microphone spacing with respect to a target spatial resolution in reconstruction of structural vibration in, say, the z-axis direction. If the target spatial resolution is defined as R, then $R=\lambda_{cr}/2$, where λ_{cr} is the smallest axial wavelength corresponding to the maximum value of k_{max} [90]. This criterion agrees perfectly with Eq. (5.4). Note that the spatial sampling must be high enough to avoid spatial aliasing. In other words, the highest spatial wavenumber k_{max} "containing significant energy must be sampled at least at the rate of two samples per wavelength to prevent spatial aliasing which causes high wavenumbers to be converted to low wavenumbers [5]."

There is a huge difference between gauging microphone spacing with respect to the spatial wavelength and that with respect to the acoustic wavelength. This is because vibration of any structure can be expressed as a superposition of an infinite number of spatial waves, each of which has a specific spatial wavelength. This vibrating structure, however, can only produce a finite number of acoustic waves, each of which has a specific acoustic wavelength, which radiate into the surrounding fluid medium. The number of the acoustic waves generated by any vibrating elastic structure is always much less than that of the spatial waves. This is why we say, "Although sound can be generated by vibrations, not all vibrations can produce sound." In fact, only a small portion of structural vibrations can produce sound. The majority of the mechanical energy of an elastic structure stays close to the structure to maintain its vibration without emitting much sound into the surrounding fluid medium at all.

This phenomenon can be best illustrated by placing our ears next to a large window, where we can clearly sense the rumbly sound due to structural vibrations of the window, but nothing at all when we step back a little from the window. Another example is to put our ears near a railway track to find out if a train is coming. If a train is approaching, we will hear rumbling sound due to vibrations of the railway track excited by a train, even though we cannot see it. However, when we stand up, we hear nothing! This is because what we have sensed is the structural vibration whose amplitude decays exponentially with respect to the distance away from the structure. These exponentially decaying waves are known as the evanescent waves that are insignificant in acoustic radiation, but critically important for structural vibrations.

The primary objective of NAH is to reconstruct the evanescent waves so as to acquire a better understanding of structural vibrations, and how they are correlated to sound radiation. The insight into the interrelationships between structural vibrations and sound radiation will enable us to device the most cost-effective measures to tackle undesirable noise emission problems.

To illustrate this point, we consider an acoustic wave at 1,000 Hz or any frequency for that matter emitted by a vibrating panel of infinite dimensions in longitudinal and transverse directions. The acoustic wavelength for this 1,000 Hz sound wave is 0.343 m, given that the speed of sound is 343 m/s. If the microphone spacing is set with respect to the acoustic wavelength, we have $\delta = 0.171 \text{ m} < \lambda/2$. Meanwhile, the measurement distance cannot be less than microphone spacing, so we can select $d = 0.172 > \lambda/2$, which is slightly more than one-half the acoustic wavelength.

To capture the critical evanescent component k_c that carries significant amount of energy, the dynamic range D (SNR of the measurement system) must satisfy the following condition:

$$10^{D/20} > \mathrm{e}^{k_c(z_h - z_S)}, \tag{5.20}$$

where $(z_h - z_S) = d$ is the measurement distance.

Since the amplitudes of the evanescent waves decay exponentially as $\mathrm{e}^{-k_c d}$, SNR will drop by 27.2 dB or 95.7 % at a measurement distance of $d = 0.172 \approx \lambda/2$! This will make it impossible for the evanescent components to be captured. Therefore, if we gauge the microphone array with respect to the acoustic frequency, we will not be able to reconstruct structural vibrations at all.

This example demonstrates that in order to capture the evanescent waves, the measurement setup must be gauged with respect to the target spatial, not the acoustic, wavelength. If the measurement spacing is gauged with respect to the acoustic frequency, we are in fact conducting acoustical holography, which will produce an image of the far-field component of the acoustic pressure radiated from a vibrating structure but nothing else! It cannot tell us anything about the acoustic pressure distribution on the surface of a vibrating structure, the normal surface velocity distribution of the structure, and the normal component of the time-averaged acoustic intensity or acoustic energy flow out of the structure. Moreover, the acoustic pressure reconstructed by using acoustical holography cannot be compared with respect to measured acoustic pressures because the measured data contain both near- and far-field components of the acoustic pressure. Further, the spatial resolution of acoustical holography is no better than one wavelength of the acoustic wave radiated from a target source. In contrast, NAH can produce, theoretically, an infinitely high spatial resolution if all near-field effects are captured, plus all the acoustic quantities anywhere in three-dimensional space.

Once the intent of NAH is understood, it becomes obvious that all documents, regardless what their titles claim, are in fact performing acoustic holography but not NAH, if the measurement setup is gauged with respect to the acoustic frequency or acoustic wavelength.

Problems

5.1. In implementing the HELS-based NAH or any NAH technologies, what is the single most important parameter that we should target in designing our measurement setup? Why?

5.2. Should we consider the frequency of the emitted sound wave in setting up our microphone array in conducting NAH measurements? Why?

5.3. Is the microphone spacing related to the standoff distance in the HELS-based NAH setup the same way as that in the Fourier transform-based NAH? Why?

5.4. How should input data be taken in general when reconstruction of the acoustic quantities on the surface of a vibrating structure is desired by using the HELS method in a nonideal environment such as inside a large room with unspecified reflecting objects and surfaces?

5.5. A prefixed and planar microphone array is easy to use and requires no setup time. Is it a good idea to use such a planar array to reconstruct the acoustic quantities generated by an arbitrarily shaped vibrating structure? Why?

5.6. Sometimes measurements can only be taken on one side of a vibrating structure in practice. In fact, this type of scenario happens almost all the time in engineering applications. What the impacts of this restriction may have on the resultant reconstruction? What should we expect under this condition?

5.7. When reconstruction must be conducted inside a large room in which the sound reflection and reverberation effects are not necessarily negligible, what should we do to minimize the impacts of sound reflection and reverberation?

5.8. What is wrong to target the measurement setup with respect to the frequency of the sound wave generated by a vibrating structure?

Chapter 6
Combined Helmholtz Equation Least-Squares (CHELS) Method

Although the HELS method has exhibited a great promise in reconstructing the acoustic fields in both exterior and interior regions, the accuracy in reconstruction for an arbitrarily shaped structure can be unsatisfactory. This is because the expansion based on the spherical waves for an acoustic field generated by on an arbitrarily shaped surface is incomplete.

An alternative for reconstructing acoustic radiation from an arbitrary structure is to use the Helmholtz integral theory. In implementing this integral theory, BEM is used and the surface is discretized into segments and the acoustic field is specified on the nodes of these segments using a particular interpolation scheme. This BEM-based NAH has been used to reconstructing acoustic radiation from structures in the exterior and interior regions.

The main advantage of the BEM-based NAH is its generality, allowing users to tackle an arbitrarily shaped structure. The disadvantage is that it may fail to yield a unique solution for the exterior problem when the excitation frequencies are close to one of the eigenfrequencies of the boundary value problem in the corresponding interior region. While this nonuniqueness difficulty may be overcome by the CHIEF method, the efficiency and accuracy of its reconstruction can be significantly affected.

The main drawback of the BEM-based NAH, however, is due to the fact that the acoustic field is reconstructed via spatial discretization. In other words, we must have a minimum number of nodes per wavelength in order to achieve the desired resolution in reconstruction. Accordingly, one must take enough measurements of the radiated acoustic pressures to determine the acoustic quantities specified on discrete nodes. For complex structures vibrating at mid-to-high frequencies, the number of nodes necessary to describe the surface acoustic quantities can be large. Hence the number of measurements is large, which makes the reconstruction process very time consuming. Although there are techniques developed recently to avoid the singularity problem inherent in the Helmholtz integral equation and methodologies to optimize the measurement locations by using an effective

S.F. Wu, *The Helmholtz Equation Least Squares Method*, Modern Acoustics and Signal Processing, DOI 10.1007/978-1-4939-1640-5_6

independence [27] implementation of the BEM-based NAH is not straightforward and reconstruction of the acoustic field is extremely slow and inefficient.

It is emphasized that in using the BEM-based NAH it is not necessary to have exactly the same measurements as discrete nodes. This is because using SVD and regularization procedures, one can have either an over- or under-determined systems of equations. Accordingly, one can use fewer measurements than the discrete nodes. However, the accuracy of reconstruction cannot be guaranteed if the measurements are substantially fewer than the discrete nodes. This is because the measured data are not error free and background noises are always present. In order to obtain a convergent solution, the equation must be truncated to filter out the evanescent waves that fall under the background noises. If measurements are too few, the equivalent cutoff wavenumber is forced to be very low. As a result, the high spatial wavenumber contents are filtered out and aliasing occurs in reconstruction.

In this chapter we show that by combining the HELS- and BEM-based NAH, the efficiency of reconstruction can be significantly enhanced and satisfactory reconstruction be obtained by using relatively few measurements [91]. First, we present a brief account of the Helmholtz integral theory.

6.1 The Helmholtz Integral Theory

The key to the acoustic radiation problems is to solve the wave equation subject to certain boundary conditions, which for a harmonic excitation reduces to the Helmholtz equation

$$\nabla^2\hat{p}\left(\vec{x};\omega\right)+k^2\hat{p}\left(\vec{x};\omega\right)=0, \tag{6.1}$$

where $\hat{p}\left(\vec{x};\omega\right)$ is the complex amplitude of the acoustic pressure at any field point $\vec{x}$ and satisfies the Sommerfeld radiation condition at infinity,

$$\lim_{\left|\vec{x}\right|\to\infty}\left|\vec{x}\right|\left(\frac{\partial\hat{p}}{\partial\left|\vec{x}\right|}-ik\hat{p}\right)=0\quad\text{as}\ \left|\vec{x}\right|\to\infty. \tag{6.2}$$

The Helmholtz equation (6.1) subject to the boundary condition can be solved for source surfaces that are expressible as one of 11 coordinate systems [92]. For arbitrary geometry, there is no analytic solution; hence, numerical solutions are sought. However, the efforts involved may be significant because one must discretize the entire three-dimensional space.

To enhance the efficiency in numerical computations, we can utilize the Helmholtz integral theory, which can be derived by making use of the free-space Green's function. First, we consider the inhomogeneous Helmholtz equation for the free-space Green's function,

$$\nabla^2 G\left(\vec{x}\middle|\vec{x}_s;\omega\right) + k^2 G\left(\vec{x}\middle|\vec{x}_s;\omega\right) = -4\pi\delta\left(\vec{x} - \vec{x}_s\right), \tag{6.3}$$

where $G = e^{ikR}/R$, where $R = \left|\vec{x} - \vec{x}_s\right|$ is the distance between the source at $\vec{x}_s$ and a receiver at $\vec{x}$ in free space, and $\delta\left(\vec{x} - \vec{x}_s\right)$ is the Dirac delta function [93], which can be considered as a function that is 0 everywhere except at the origin, where it is infinite,

$$\delta(x - x_0) = \begin{cases} 0, & x \neq x_0 \\ \infty, & x = x_0 \end{cases}. \tag{6.4}$$

The Dirac delta function has the sifting property,

$$\int_{-\infty}^{\infty} \delta(x - x_0) f(x) dx = f(x_0). \tag{6.5}$$

Therefore, for $f(x) \equiv 1$, the integration of the Dirac delta function is identically unity,

$$\int_{-\infty}^{\infty} \delta(x - x_0) dx = 1 \tag{6.6}$$

Multiply Eq. (6.1) by G and Eq. (6.3) by $\hat{p}\left(\vec{x};\omega\right)$, and use the chain rule to rewrite the Laplacian operator ∇^2 as

$$\begin{aligned} \nabla\cdot\left[G\left(\vec{x}\middle|\vec{x}_s;\omega\right)\nabla\hat{p}\left(\vec{x};\omega\right)\right] - \nabla G\left(\vec{x}\middle|\vec{x}_s;\omega\right)\cdot\nabla\hat{p}\left(\vec{x};\omega\right) \\ + k^2 G\left(\vec{x}\middle|\vec{x}_s;\omega\right)\hat{p}\left(\vec{x};\omega\right) = 0, \end{aligned} \tag{6.7}$$

$$\begin{aligned} \nabla\cdot\left[\hat{p}\left(\vec{x};\omega\right)\nabla G\left(\vec{x}\middle|\vec{x}_s;\omega\right)\right] - \nabla G\left(\vec{x}\middle|\vec{x}_s;\omega\right)\cdot\nabla\hat{p}\left(\vec{x};\omega\right) \\ + k^2 G\left(\vec{x}\middle|\vec{x}_s;\omega\right)\hat{p}\left(\vec{x};\omega\right) = -4\pi\hat{p}\left(\vec{x};\omega\right)\delta\left(\vec{x} - \vec{x}_s\right). \end{aligned} \tag{6.8}$$

Subtracting Eq. (6.8) from (6.7) and integrating both sides over the volume enclosed by the source surface and that at infinity leads to

$$\iiint\limits_V \nabla \cdot \left[G\left(\vec{x}\middle|\vec{x}_s;\omega\right) \nabla \hat{p}\left(\vec{x};\omega\right) - \hat{p}\left(\vec{x};\omega\right) \nabla G\left(\vec{x}\middle|\vec{x}_s;\omega\right) \right] dV$$

$$= 4\pi \iiint\limits_V \hat{p}\left(\vec{x};\omega\right) \delta\left(\vec{x} - \vec{x}_s\right) dV. \tag{6.9}$$

The volume integral on the right side of Eq. (6.9) leads to the acoustic pressure itself due to the sifting property of the Dirac delta function Eq. (6.5). The volume integral on the left side can be expressed as a surface integral by using the Gauss theorem or divergence theorem [94],

$$\hat{p}\left(\vec{x};\omega\right) = \frac{1}{4\pi} \iint\limits_{S'} \left[\hat{p}\left(\vec{x}_{s'};\omega\right) \frac{\partial G\left(\vec{x}\middle|\vec{x}_{s'};\omega\right)}{\partial \mathrm{n}\left(\vec{x}_{s'}\right)} - \frac{\partial \hat{p}\left(\vec{x}_{s'};\omega\right)}{\partial \mathrm{n}\left(\vec{x}_{s'}\right)} G\left(\vec{x}\middle|\vec{x}_{s'};\omega\right) \right] dS', \tag{6.10}$$

where $\partial/\partial \mathrm{n}$ represents a normal derivative with respect to the outward unit vector on the surface S, and $\partial\hat{p}/\partial \mathrm{n}$ is related to the normal surface velocity through the Euler's equation,

$$\frac{\partial \hat{p}\left(\vec{x}_s;\omega\right)}{\partial \mathrm{n}\left(\vec{x}_s\right)} = i\omega\rho_0 \hat{v}_{\mathrm{n}}\left(\vec{x}_s;\omega\right). \tag{6.11}$$

Note that there is a change in sign on the right side of Eq. (6.10) because the unit normal vector on the source surface S, which points to the interior region enclosed by S as required by the Gauss theorem, should point into the region external to S in the surface integral.

Equation (6.10) is known as the Helmholtz integral formulation, which states that $\hat{p}\left(\vec{x};\omega\right)$ anywhere in free space may be specified by integrating the surface acoustic $\hat{p}\left(\vec{x}_s;\omega\right)$ and normal surface velocity $\hat{v}_{\mathrm{n}}\left(\vec{x}_s;\omega\right)$ through the free-space Green's function $G\left(\vec{x}\middle|\vec{x}_s;\omega\right)$. Accordingly, the dimensionality of the original problem given by Eq. (6.1) is reduced by 1.

Note that the surface acoustic pressure $\hat{p}\left(\vec{x}_s;\omega\right)$ and normal surface velocity $\hat{v}_{\mathrm{n}}\left(\vec{x}_s;\omega\right)$ are interrelated together and should not be specified simultaneously in practice. For acoustic radiation problems, the normal surface velocity $\hat{v}_{\mathrm{n}}\left(\vec{x}_s;\omega\right)$ is usually specified in the boundary condition. So the first step in predicting the radiated acoustic pressure $\hat{p}\left(\vec{x};\omega\right)$ is to specify the surface acoustic pressure

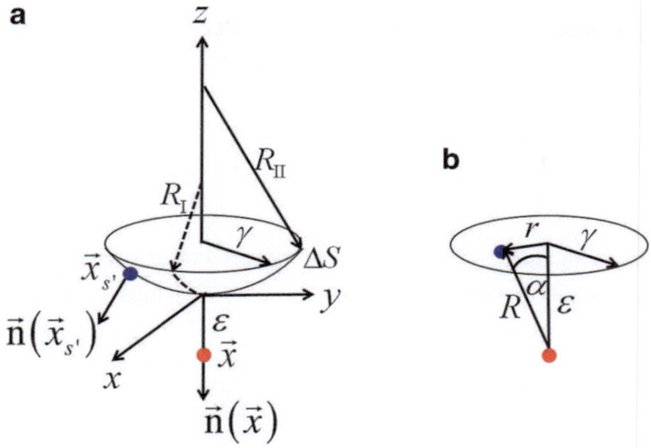

Fig. 6.1 Schematic of taking the limit as the field point approaches the surface $\vec{x} \to \vec{x}_s$. (**a**) The field point is located at $\vec{x} = \vec{x}_s + \varepsilon\vec{n}\left(\vec{x}_s\right)$, the surface point is at $\vec{x}_{s'}$ with the unit normal $\vec{n}\left(\vec{x}_{s'}\right)$, and the distance between these two points is $R = \left|\vec{x} - \vec{x}_{s'}\right|$. (**b**) Close-up view of taking the limit as $R \to 0$. The correct sequence is to take $\varepsilon \to 0$ first, and then $\gamma \to 0$

$\hat{p}\left(\vec{x}_s;\omega\right)$. This may be done by solving an integral equation obtained by taking the limit as the field point approaches the surface $\vec{x} \to \vec{x}_s$ in Eq. (6.10). Note that all integrands in Eq. (6.10) become singular because $R \to 0$ as $\vec{x} \to \vec{x}_s$. This difficulty can be overcome by taking the Cauchy principal value [95]. It is emphasized that the sequence in which this limit is taken is critical. A wrong sequence in taking the limit $\vec{x} \to \vec{x}_s$ leads to a wrong result.

Figure 6.1 depicts schematic of taking the limit $\vec{x} \to \vec{x}_s$. For simplicity yet without loss of generality, we consider the case in which both $\vec{x}_{s'}$ and $\vec{x}_s$ are located inside ΔS, which is a circular segment of the source surface S, centered at $\vec{x}_s$ of a radius γ. The segment ΔS is shaped like a bowl with radii R_I and R_{II} in the perpendicular directions, respectively. The remainder of the surface is denoted as $(S - \Delta S)$. The field point is at $\vec{x} = \vec{x}_s + \varepsilon\vec{n}\left(\vec{x}_s\right)$, which is a small distance ε away from the surface point $\vec{x}_s$ along the unit normal direction $\vec{n}\left(\vec{x}_s\right)$. Therefore the distance between these two points is $R = \left|\vec{x} - \vec{x}_{s'}\right|$ (see Fig. 6.1a). As $\vec{x} \to \vec{x}_s$, $R \to 0$ and all integrands in Eq. (6.10) become singular. So care must be taken in taking the limit of $\vec{x} \to \vec{x}_s$.

Note that if $\vec{x}_{s'}$ is in $(S - \Delta S)$, R will not be 0 as $\vec{x} \to \vec{x}_s$, and all integrands in Eq. (6.10) will be finite. So we only need to concentrate on the situation in which $\vec{x}_{s'}$ falls inside ΔS.

Given that ΔS is small, the z component of a generic surface point $\vec{x}_{s'}$ can be expanded into a power series of x, y, R_{I}, and R_{II}, where $x/R_{\mathrm{I}} \ll 1$ and $y/R_{\mathrm{II}} \ll 1$,

$$z = C\sqrt{1 - \left(\frac{x^2}{R_{\mathrm{I}}^2} + \frac{y^2}{R_{\mathrm{II}}^2}\right)} \approx C\left(1 - \frac{x^2}{2R_{\mathrm{I}}^2} - \frac{y^2}{2R_{\mathrm{II}}^2}\right). \tag{6.12}$$

Equation (6.12) is derived from the ellipsoid equation $(x^2/R_{\mathrm{I}}^2) + (y^2/R_{\mathrm{II}}^2) + (z^2/C^2) = 1$, where C is a constant. Therefore for $x/R_{\mathrm{I}} \ll 1$ and $y/R_{\mathrm{II}} \ll 1$, we have

$$z - C \approx -\frac{Cx^2}{2R_{\mathrm{I}}^2} - \frac{Cy^2}{2R_{\mathrm{II}}^2}. \tag{6.13}$$

By using $(z - C)$, we have moved the origin of the coordinate system describing the ellipsoid to the surface point $\vec{x}_s$. The unit outward normal vector $\vec{\mathrm{n}}\left(\vec{x}_{s'}\right)$ at $\vec{x}_{s'}$ can now be approximated by

$$\vec{\mathrm{n}}\left(\vec{x}_{s'}\right) \approx \left(\frac{x}{R_{\mathrm{I}}}\right)\vec{e}_x + \left(\frac{y}{R_{\mathrm{II}}}\right)\vec{e}_y + \vec{e}_z, \tag{6.14}$$

because $\vec{x}_{s'} = x\vec{e}_x + y\vec{e}_y + (z - C)\vec{e}_z$. Similarly, $\vec{x} = \vec{x}_s + \varepsilon\vec{\mathrm{n}}\left(\vec{x}_s\right) = \varepsilon\vec{e}_z$. Thus $\vec{R} = \vec{x} - \vec{x}_{s'}$ is given by

$$\begin{aligned} \vec{R} &= \varepsilon\vec{e}_z - x\vec{e}_x - y\vec{e}_y - (z - C)\vec{e}_z \\ &\approx \left(\varepsilon - \frac{Cx^2}{2R_{\mathrm{I}}^2} - \frac{Cy^2}{2R_{\mathrm{II}}^2}\right)\vec{e}_z - x\vec{e}_x - y\vec{e}_y, \end{aligned} \tag{6.15}$$

and the normal derivative of the free-space Green's function $\partial G\left(\vec{x}|\vec{x}_s;\omega\right)/\partial \mathrm{n}$ in Eq. (6.10) is

$$\begin{aligned} \frac{\partial G\left(\vec{x}|\vec{x}_{s'};\omega\right)}{\partial \mathrm{n}\left(\vec{x}_{s'}\right)} &= \vec{\mathrm{n}}\left(\vec{x}_{s'}\right) \cdot \nabla_{s'} G\left(\vec{x}|\vec{x}_{s'};\omega\right) \\ &\approx -\left(\frac{x}{R_{\mathrm{I}}}\vec{e}_x + \frac{y}{R_{\mathrm{II}}}\vec{e}_y + \vec{e}_z\right) \cdot \left(\frac{ikR - 1}{R^2}\right)\left(\frac{\vec{x} - \vec{x}_{s'}}{R}\right)\mathrm{e}^{ikR} \\ &= \left(\frac{x}{R_{\mathrm{I}}}\vec{e}_x + \frac{y}{R_{\mathrm{II}}}\vec{e}_y + \vec{e}_z\right) \cdot \left[\frac{x\vec{e}_x + y\vec{e}_y - \left(\varepsilon + \frac{Cx^2}{2R_{\mathrm{I}}^2} + \frac{Cy^2}{2R_{\mathrm{II}}^2}\right)\vec{e}_z}{R^3}\right](ikR - 1)\mathrm{e}^{ikR} \\ &\approx \left(\frac{x^2}{R_{\mathrm{I}}} + \frac{y^2}{R_{\mathrm{II}}} - \varepsilon - \frac{Cx^2}{2R_{\mathrm{I}}^2} - \frac{Cy^2}{2R_{\mathrm{II}}^2}\right)(ikR - 1)(1 + ikR) = \left(\varepsilon - \frac{Cx^2}{2R_{\mathrm{I}}^2} - \frac{Cy^2}{2R_{\mathrm{II}}^2}\right)\frac{\left(1 + k^2R^2\right)}{R^3} \end{aligned}, \tag{6.16}$$

where we have expanded e^{ikR} into the Taylor series for a small R as $R \to 0$, and R is written as

$$R = \sqrt{x^2 + y^2 + (\varepsilon - z)^2} \approx \sqrt{\varepsilon^2 + x^2 + y^2} = \sqrt{\varepsilon^2 + r^2}. \tag{6.17}$$

Meanwhile, Fig. 6.1b implies that for a small area ΔS, $dS \approx rdr$, $x = r\cos\phi$, and $y = r\sin\phi$, where r varies from 0 to γ and ϕ from 0 to 2π. With Eqs. (6.16) and (6.17) we can evaluate the integrals in Eq. (6.10) as $\vec{x} \to \vec{x}_s$. Let us consider the first integral on the right side of Eq. (6.10). In particular, we divide the surface into ΔS and $(S - \Delta S)$ in taking the limit as $\Delta S \to 0$,

$$\begin{aligned}\iint\limits_S \hat{p}\left(\vec{x}_{s'};\omega\right)\frac{\partial G\left(\vec{x}_s\middle|\vec{x}_{s'};\omega\right)}{\partial n\left(\vec{x}_{s'}\right)}dS' &= \lim_{\Delta S\to 0}\iint\limits_{\Delta S}\hat{p}\left(\vec{x}_{s'};\omega\right)\frac{\partial G\left(\vec{x}_s\middle|\vec{x}_{s'};\omega\right)}{\partial n\left(\vec{x}_{s'}\right)}dS' \\ &+ \lim_{\Delta S\to 0}\iint\limits_{S-\Delta S}\hat{p}\left(\vec{x}_{s'};\omega\right)\frac{\partial G\left(\vec{x}_s\middle|\vec{x}_{s'};\omega\right)}{\partial n\left(\vec{x}_{s'}\right)}dS',\end{aligned} \tag{6.18}$$

where all surface points in the first integral fall inside ΔS so that as $\vec{x} \to \vec{x}_s$, $R \to 0$. For the second integral, $\vec{x}_s$ is in $(S - \Delta S)$, while $\vec{x}_{s'}$ is in ΔS. Hence as $\vec{x} \to \vec{x}_s$, $R \neq 0$. So we only need to focus on the first integral on the right side of Eq. (6.18) because the second integral is regular as the field point approaches the surface $\vec{x} \to \vec{x}_s$.

The required limit $\Delta S \to 0$ in Eq. (6.18) can be accomplished by taking $\varepsilon \to 0$ with γ fixed, followed by $\gamma \to 0$. The order in which these limits are taken is very important.

Substituting Eqs. (6.16) and (6.17) into the first integral on the right side of Eq. (6.18) yields

$$\begin{aligned}\lim_{\Delta S\to 0}\iint\limits_{\Delta S}\hat{p}\left(\vec{x}_{s'};\omega\right)\frac{\partial G\left(\vec{x}_s\middle|\vec{x}_{s'};\omega\right)}{\partial n\left(\vec{x}_{s'}\right)}dS' &= \lim_{\substack{\varepsilon\to 0\\ \gamma\to 0}}\iint\limits_{S-\Delta S}\hat{p}\left(\vec{x}_{s'};\omega\right)\frac{\varepsilon}{R^3}rdrd\phi \\ &+ \lim_{\substack{\varepsilon\to 0\\ \gamma\to 0}}\iint\limits_{S-\Delta S}\hat{p}\left(\vec{x}_{s'};\omega\right)\left[\frac{\varepsilon k^2 r}{R} - \frac{k^2 r^3}{2R}\left(\frac{\cos^2\phi}{R_{\mathrm{I}}} + \frac{\sin^2\phi}{R_{\mathrm{II}}}\right)\right]drd\phi \\ &- \lim_{\substack{\varepsilon\to 0\\ \gamma\to 0}}\iint\limits_{S-\Delta S}\frac{\hat{p}\left(\vec{x}_{s'};\omega\right)r^3}{2R^3}\left(\frac{\cos^2\phi}{R_{\mathrm{I}}} + \frac{\sin^2\phi}{R_{\mathrm{II}}}\right)drd\phi.\end{aligned} \tag{6.19}$$

From Fig. 6.1b we see that we can use trigonometric properties to rewrite $r = \varepsilon \tan\alpha$ and $R = \varepsilon \sec\alpha$. Accordingly, we have $dr = \varepsilon \sec^2\alpha d\alpha$, $rdr = \varepsilon^2 \tan\alpha \sec^2\alpha d\alpha$, and $(\varepsilon/R^3) rdr = \sin\alpha d\alpha$. As $\varepsilon \to 0$, $\alpha \to \pi/2$. Therefore, the first integral on the right side of Eq. (6.19) leads to

$$\begin{aligned}
\lim_{\substack{\varepsilon \to 0 \\ \gamma \to 0}} \iint_{S-\Delta S} \hat{p}\left(\vec{x}_{s'}; \omega\right) \frac{\varepsilon}{R^3} r dr d\phi &= \lim_{\substack{\varepsilon \to 0 \\ \gamma \to 0}} \int_0^{2\pi} \int_0^{\pi/2} \hat{p}\left(\vec{x}_{s'}; \omega\right) \frac{\varepsilon^3 \tan\alpha \sec^2\alpha}{\varepsilon^3 \sec^3\alpha} d\alpha d\phi \\
&= \lim_{\substack{\varepsilon \to 0 \\ \gamma \to 0}} \int_0^{2\pi} \int_0^{\pi/2} \hat{p}\left(\vec{x}_{s'}; \omega\right) \sin\alpha d\alpha d\phi \\
&= -2\pi \hat{p}\left(\vec{x}_s; \omega\right) \cos\alpha \Big|_0^{2\pi} = 2\pi \hat{p}\left(\vec{x}_s; \omega\right). \qquad (6.20)
\end{aligned}$$

Note that if we let $\gamma \to 0$ first in Eq. (6.19), the integration limits for α would be all 0, making the integral to vanish, which will be obviously wrong.

The second and third integrals on the right side of Eq. (6.19) are given by

$$\begin{aligned}
&\lim_{\substack{\varepsilon \to 0 \\ \gamma \to 0}} \iint_{S-\Delta S} \hat{p}\left(\vec{x}_{s'}; \omega\right) \left[\frac{\varepsilon k^2 r}{R} - \frac{k^2 r^3}{2R}\left(\frac{\cos^2\phi}{R_{\mathrm{I}}} + \frac{\sin^2\phi}{R_{\mathrm{II}}}\right)\right] dr d\phi \\
&= \lim_{\substack{\varepsilon \to 0 \\ \gamma \to 0}} \int_0^{2\pi} \int_0^{\pi/2} \hat{p}\left(\vec{x}_{s'}; \omega\right) \left[\frac{k^2 \varepsilon^2}{\sec\alpha} - \frac{k^2 \varepsilon^3 \tan^2\alpha}{2\sec\alpha}\left(\frac{\cos^2\phi}{R_{\mathrm{I}}} + \frac{\sin^2\phi}{R_{\mathrm{II}}}\right)\right] \tan\alpha \sec^2\alpha d\alpha d\phi \equiv 0.
\end{aligned} \qquad (6.21)$$

$$\begin{aligned}
&\lim_{\substack{\varepsilon \to 0 \\ \gamma \to 0}} \iint_{S-\Delta S} \frac{\hat{p}\left(\vec{x}_{s'}; \omega\right) r^3}{2R^3} \left(\frac{\cos^2\phi}{R_{\mathrm{I}}} + \frac{\sin^2\phi}{R_{\mathrm{II}}}\right) dr d\phi \\
&= \lim_{\substack{\varepsilon \to 0 \\ \gamma \to 0}} \int_0^{2\pi} \int_0^{\pi/2} \frac{\hat{p}\left(\vec{x}_{s'}; \omega\right) \varepsilon^2 \tan^3\alpha \sec^2\alpha}{2\varepsilon \sec^3\alpha} \left(\frac{\cos^2\phi}{R_{\mathrm{I}}} + \frac{\sin^2\phi}{R_{\mathrm{II}}}\right) d\alpha d\phi \equiv 0.
\end{aligned} \qquad (6.22)$$

Meanwhile, the second integral on the right side of Eq. (6.18) is regular as $\Delta S \to 0$. So we have

$$\begin{aligned}
&\lim_{\Delta S \to 0} \iint_{S-\Delta S} \hat{p}\left(\vec{x}_{s'}; \omega\right) \frac{\partial G\left(\vec{x}_s | \vec{x}_{s'}; \omega\right)}{\partial \mathrm{n}\left(\vec{x}_{s'}\right)} dS' \\
&= \iint_{S'} \hat{p}\left(\vec{x}_{s'}; \omega\right) \frac{\partial G\left(\vec{x}_s | \vec{x}_{s'}; \omega\right)}{\partial \mathrm{n}\left(\vec{x}_{s'}\right)} dS'. \qquad (6.23)
\end{aligned}$$

Following the same procedures as outlined above, we can show that the second integral on the right side of Eq. (6.10) is regular as the field point approaches the surface,

$$\iint_{S'} \frac{\partial \hat{p}\left(\vec{x}_{s'};\omega\right)}{\partial n\left(\vec{x}_{s'}\right)} G\left(\vec{x}\middle|\vec{x}_{s'};\omega\right) dS' = \lim_{\Delta S \to 0} \iint_{\Delta S} \frac{\partial \hat{p}\left(\vec{x}_{s'};\omega\right)}{\partial n\left(\vec{x}_{s'}\right)} G\left(\vec{x}\middle|\vec{x}_{s'};\omega\right) dS' + \lim_{\Delta S \to 0} \iint_{S-\Delta S} \frac{\partial \hat{p}\left(\vec{x}_{s'};\omega\right)}{\partial n\left(\vec{x}_{s'}\right)} G\left(\vec{x}\middle|\vec{x}_{s'};\omega\right) dS' = \iint_{S'} \frac{\partial \hat{p}\left(\vec{x}_{s'};\omega\right)}{\partial n\left(\vec{x}_{s'}\right)} G\left(\vec{x}\middle|\vec{x}_{s'};\omega\right) dS'. \tag{6.24}$$

Substituting Eqs. (6.23) and (6.24) into (6.10) with $\vec{x} \to \vec{x}_s$ then leads to

$$\hat{p}\left(\vec{x}_s;\omega\right) = \frac{1}{2\pi} \iint_{S'} \left[\hat{p}\left(\vec{x}_{s'};\omega\right) \frac{\partial G\left(\vec{x}_s\middle|\vec{x}_{s'};\omega\right)}{\partial n\left(\vec{x}_{s'}\right)} - i\omega\rho_0 \hat{v}_n\left(\vec{x}_{s'};\omega\right) G\left(\vec{x}_s\middle|\vec{x}_{s'};\omega\right) \right] dS', \tag{6.25}$$

where both $\vec{x}_s$ and $\vec{x}_{s'}$ are on the surface S.

Equation (6.25) is known as the surface Helmholtz integral equation because there is an unknown variable under the integral sign. Once the surface acoustic pressure $\hat{p}\left(\vec{x}_s;\omega\right)$ and normal surface velocity $\hat{v}_n\left(\vec{x}_s;\omega\right)$ are specified, the acoustic pressure $\hat{p}\left(\vec{x};\omega\right)$ in free space is completely determined by Eq. (6.10). The complexities of the problem are significantly reduced because one only deals with discretization of a two-dimensional source surface. The trouble is that Eq. (6.25) may fail to produce a unique solution whenever the frequency is equal to 1 of the characteristic frequencies of the corresponding boundary value problem in the interior region.

6.2 Nonuniqueness Difficulties

The nonuniqueness difficulties of the surface Helmholtz integral equation (6.25) can be examined by looking at a general Fredholm integral equation of the second kind [96],

$$u(\varsigma) - \Lambda \iint_S K(\varsigma, \xi) u(\xi) dS(\xi) = F(\varsigma), \tag{6.26}$$

where $u(\varsigma)$ is unknown, $K(\varsigma, \xi)$ is called "L^2 kernels," meaning that they are square integrable over S in the Lebesgue sense, Λ is some value whose meaning will be specified shortly, and $F(\varsigma)$ is a known function. The associate homogeneous Fredholm integral equation is given by

$$u_0(\varsigma) - \Lambda \iint_S K(\varsigma, \xi) u_0(\xi) dS(\xi) = 0. \tag{6.27}$$

The Sturm-Liouville theory [97] states that if the associated homogeneous equation (6.27) has a nontrivial solution $u_0(\varsigma)$, then Λ is a characteristic value or eigenvalue of the kernel $K(\varsigma, \xi)$ and $u_0(\varsigma)$ is a characteristic function of $K(\varsigma, \xi)$ belonging to Λ. Otherwise Λ is a regular value.

The adjoint inhomogeneous Fredholm integral equation of the second kind is expressible as

$$y(\varsigma) - \Lambda^* \iint_S K(\varsigma, \xi)^* y(\xi) dS(\xi) = H(\varsigma), \tag{6.28}$$

where $K(\varsigma, \xi)^*$ is the adjoint kernel of $K(\varsigma, \xi)$ and $H(\varsigma)$ is given. The adjoint homogeneous equation for Eq. (6.28) is

$$y_0(\varsigma) - \Lambda^* \iint_S K(\varsigma, \xi)^* y_0(\xi) dS(\xi) = 0. \tag{6.29}$$

The following theorems have been proven by Smithies [97]

Theorem 6.1 *If Λ is a regular value of $K(\varsigma, \xi)$, then Λ^* is a regular value of $K(\varsigma, \xi)^*$, the homogeneous Eqs.* (6.27) *and* (6.29) *have only trivial solutions, and Eqs.* (6.26) *and* (6.28) *have unique solutions for any L^2 functions $F(\varsigma)$ and $H(\varsigma)$.*

Theorem 6.2 *If Λ is a characteristic value of $K(\varsigma, \xi)$, then Λ^* is a characteristic value of $K(\varsigma, \xi)^*$, and the homogeneous Eqs.* (6.27) *and* (6.29) *have nontrivial solutions.*

Theorem 6.3 *If Λ is a characteristic value of $K(\varsigma, \xi)$, then the inhomogeneous equation* (6.27) *has an L^2 solution if and only if $F(\varsigma)$ is orthogonal to every L^2 solution of the adjoint homogeneous equation* (6.29), *i.e., if $F(\varsigma)$ satisfies*

$$\iint_S y_0(\xi)^* F(\xi) dS(\xi) = 0. \tag{6.30}$$

Furthermore, even if the compatibility condition (6.30) *is satisfied, the solution to Eq.* (6.26) *is not determined uniquely since any multiple of $u_0(\varsigma)$ can be added to a particular solution of Eq.* (6.26).

Now let us apply these theorems to examining the solution of the surface Helmholtz integral equation (6.25). Let

$$u(\varsigma) = \hat{p}\left(\vec{x}_s;\omega\right), K(\varsigma,\xi) = \frac{1}{2\pi}\frac{\partial G\left(\vec{x}_s|\vec{x}_{s'};\omega\right)}{\partial \mathrm{n}\left(\vec{x}_{s'}\right)}, \Lambda = 1, \quad \text{and}$$

$$F(\varsigma) = -\frac{i\omega\rho_0}{2\pi}\iint\limits_{S}\hat{v}_{\mathrm{n}}\left(\vec{x}_{s'};\omega\right)G\left(\vec{x}_s|\vec{x}_{s'};\omega\right)dS'. \tag{6.31}$$

Substituting Eq. (6.31) into Eq. (6.26), we obtain

$$\hat{p}\left(\vec{x}_s;\omega\right) - \frac{1}{2\pi}\iint\limits_{S'}\hat{p}\left(\vec{x}_{s'};\omega\right)\frac{\partial G\left(\vec{x}_s|\vec{x}_{s'};\omega\right)}{\partial \mathrm{n}\left(\vec{x}_{s'}\right)}dS'$$
$$= -\frac{i\omega\rho_0}{2\pi}\iint\limits_{S}\hat{v}_{\mathrm{n}}\left(\vec{x}_{s'};\omega\right)G\left(\vec{x}_s|\vec{x}_{s'};\omega\right)dS', \tag{6.32}$$

where $\hat{v}_{\mathrm{n}}\left(\vec{x}_s;\omega\right)$ is specified, and $\hat{p}\left(\vec{x}_S;\omega\right)$ is to be determined. The corresponding homogeneous equation is given by

$$\hat{p}\left(\vec{x}_s;\omega\right) - \frac{1}{2\pi}\iint\limits_{S'}\hat{p}\left(\vec{x}_{s'};\omega\right)\frac{\partial G\left(\vec{x}_s|\vec{x}_{s'};\omega\right)}{\partial \mathrm{n}\left(\vec{x}_{s'}\right)}dS' = 0. \tag{6.33}$$

From Theorems 6.1 and 6.2 we know that Eq. (6.32) has a unique solution, except at some characteristic frequencies for which Eq. (6.33) has nontrivial solutions.

From Theorem 6.3 we further learn that at these characteristic frequencies, Eq. (6.33) has no solution unless the compatibility condition [see Eq. (6.30)],

$$-\frac{i\omega\rho_0}{2\pi}\iint\limits_{S}\hat{p}_0\left(\vec{x}_s;\omega\right)^*\left[\iint\limits_{S'}\hat{v}_{\mathrm{n}}\left(\vec{x}_{s'};\omega\right)G\left(\vec{x}_s|\vec{x}_{s'};\omega\right)dS'\right]dS = 0, \tag{6.34}$$

holds for all $\hat{p}_0\left(\vec{x}_s;\omega\right)^*$, which satisfies the adjoint homogeneous equation (6.29),

$$\hat{p}_0\left(\vec{x}_s;\omega\right)^* - \frac{1}{2\pi}\iint\limits_{S'}\hat{p}_0\left(\vec{x}_{s'};\omega\right)^*\frac{\partial G\left(\vec{x}_s|\vec{x}_{s'};\omega\right)}{\partial \mathrm{n}\left(\vec{x}_{s'}\right)}dS' = 0. \tag{6.35}$$

To show that compatibility condition (6.30) is satisfied, we consider the interior problem, for which the Helmholtz integral formulation can be written as

$$\hat{p}^{\mathrm{I}}\left(\vec{x};\omega\right) = -\frac{1}{4\pi}\iint_{S'} \hat{p}^{\mathrm{I}}\left(\vec{x}_{s'};\omega\right)\frac{\partial G\left(\vec{x}|\vec{x}_{s'};\omega\right)}{\partial \mathrm{n}\left(\vec{x}_{s'}\right)}dS' + \frac{i\omega\rho_0}{4\pi}\iint_{S'} \hat{v}_{\mathrm{n}}^{\mathrm{I}}\left(\vec{x}_{s'};\omega\right)G\left(\vec{x}|\vec{x}_{s'};\omega\right)dS. \tag{6.36}$$

Note that Eq. (6.36) can be derived in the same way as that of Eq. (6.10), except that the sign of the unit normal on the surface should be reversed for the interior problem.

Taking the limit as the field point approaches the surface from the inside, we obtain

$$\hat{p}^{\mathrm{I}}\left(\vec{x}_{\mathrm{s}};\omega\right) + \frac{1}{2\pi}\iint_{S'} \hat{p}^{\mathrm{I}}\left(\vec{x}_{s'};\omega\right)\frac{\partial G\left(\vec{x}_{\mathrm{s}}|\vec{x}_{s'};\omega\right)}{\partial \mathrm{n}\left(\vec{x}_{s'}\right)}dS' = \frac{i\omega\rho_0}{2\pi}\iint_{S'} \hat{v}_{\mathrm{n}}^{\mathrm{I}}\left(\vec{x}_{s'};\omega\right)G\left(\vec{x}_{\mathrm{s}}|\vec{x}_{s'};\omega\right)dS'. \tag{6.37}$$

On the other hand, taking the normal derivative of the Helmholtz integral formulation for the interior region, Eq. (6.36), we obtain

$$i\omega\rho_0\hat{v}_{\mathrm{n}}^{\mathrm{I}}\left(\vec{x};\omega\right) = -\frac{1}{4\pi}\frac{\partial}{\partial \mathrm{n}}\iint_{S'} \hat{p}^{\mathrm{I}}\left(\vec{x}_{s'};\omega\right)\frac{\partial G\left(\vec{x}|\vec{x}_{s'};\omega\right)}{\partial \mathrm{n}\left(\vec{x}_{s'}\right)}dS' + \frac{i\omega\rho_0}{4\pi}\iint_{S'} \hat{v}_{\mathrm{n}}^{\mathrm{I}}\left(\vec{x}_{s'};\omega\right)\frac{\partial G\left(\vec{x}|\vec{x}_{s'};\omega\right)}{\partial \mathrm{n}\left(\vec{x}_{s'}\right)}dS'. \tag{6.38}$$

Taking the limit as $\vec{x} \to \vec{x}_s$ from the inside yields

$$i\omega\rho_0\hat{v}_{\mathrm{n}}^{\mathrm{I}}\left(\vec{x}_s;\omega\right) - \frac{i\omega\rho_0}{2\pi}\iint_{S'} \hat{v}_{\mathrm{n}}^{\mathrm{I}}\left(\vec{x}_{s'};\omega\right)\frac{\partial G\left(\vec{x}_s|\vec{x}_{s'};\omega\right)}{\partial \mathrm{n}\left(\vec{x}_{s'}\right)}dS' = -\frac{1}{2\pi}\frac{\partial}{\partial \mathrm{n}}\iint_{S'} \hat{p}^{\mathrm{I}}\left(\vec{x}_{s'};\omega\right)\frac{\partial G\left(\vec{x}_s|\vec{x}_{s'};\omega\right)}{\partial \mathrm{n}\left(\vec{x}_{s'}\right)}dS'. \tag{6.39}$$

For the Dirichlet problem for which $\hat{p}^{\mathrm{I}}\left(\vec{x}_s;\omega\right) = 0$ on the source surface, Eq. (6.39) reduces to

$$i\omega\rho_0\hat{v}_{\mathrm{n}}^{\mathrm{I}}\left(\vec{x}_s;\omega\right)-\frac{i\omega\rho_0}{2\pi}\iint_{S'}\hat{v}_{\mathrm{n}}^{\mathrm{I}}\left(\vec{x}_{s'};\omega\right)\frac{\partial G\left(\vec{x}_s|\vec{x}_{s'};\omega\right)}{\partial\mathrm{n}\left(\vec{x}_{s'}\right)}dS'=0. \tag{6.40}$$

The roots of this homogeneous equation are called the Dirichlet eigenfrequencies, $k=k_{\mathrm{D}}$. At these characteristic frequencies Eq. (6.40) has nontrivial solutions that are known as the characteristic functions belonging to the characteristic frequencies k_{D}.

Note that Eq. (6.40) has the same form as the homogeneous surface Helmholtz integral equation (6.33). Accordingly, they share the same characteristic frequencies k_{D}. In other words, the set of solutions $\hat{v}_{\mathrm{n}}^{\mathrm{I}}\left(\vec{x}_s;\omega\right)$ for the interior Dirichlet problem are the same as those of $\hat{p}\left(\vec{x}_s;\omega\right)$ for the exterior Neumann problem for which $\hat{v}_{\mathrm{n}}\left(\vec{x}_s;\omega\right)=0$ on the surface. In fact, applying this Neumann boundary condition in Eq. (6.32), we obtain Eq. (6.33), which has the same form as that of Eq. (6.40).

Now applying the Dirichlet boundary condition to the interior Helmholtz integral equation (6.37), we obtain

$$\iint_{S'}\hat{v}_{\mathrm{n}}^{\mathrm{I}}\left(\vec{x}_{s'};\omega\right)G\left(\vec{x}_s|\vec{x}_{s'};\omega\right)dS'=0. \tag{6.41}$$

Because of the equivalence of solutions sets $\hat{v}_{\mathrm{n}}^{\mathrm{I}}\left(\vec{x}_s;\omega\right)$ for the interior region and $\hat{p}\left(\vec{x}_s;\omega\right)$ for the exterior region, we can interchange these two sets and rewrite Eq. (6.41) as

$$\iint_{S'}\hat{p}\left(\vec{x}_{s'};\omega\right)G\left(\vec{x}_s|\vec{x}_{s'};\omega\right)dS'=0 \quad \text{or} \quad \iint_{S'}\hat{p}\left(\vec{x}_{s'};\omega\right)^{*}G\left(\vec{x}_s|\vec{x}_{s'};\omega\right)dS'=0. \tag{6.42}$$

Substituting Eq. (6.42) into Eq. (6.34) and interchanging the order of integrations, we obtain

$$\begin{aligned}&-\frac{i\omega\rho_0}{2\pi}\iint_{S}\hat{p}_0\left(\vec{x}_s;\omega\right)^{*}\left[\iint_{S'}\hat{v}_{\mathrm{n}}\left(\vec{x}_{s'};\omega\right)G\left(\vec{x}_s|\vec{x}_{s'};\omega\right)dS'\right]dS\\&=-\frac{i\omega\rho_0}{2\pi}\iint_{S}\hat{v}_{\mathrm{n}}\left(\vec{x}_s;\omega\right)\left[\iint_{S'}\hat{p}_0\left(\vec{x}_{s'};\omega\right)^{*}G\left(\vec{x}_s|\vec{x}_{s'};\omega\right)dS'\right]dS\equiv 0.\end{aligned} \tag{6.43}$$

Equation (6.43) shows that the compatibility condition is perfectly satisfied and Eq. (6.33) has nontrivial solution. However, the solution to the surface Helmholtz integral equation (6.32) may be nonunique because any multiple of $\hat{v}_{\mathrm{n}}\left(\vec{x}_s;\omega\right)$ may be added to the particular solution and the compatibility condition (6.43) is still satisfied.

Therefore the surface Helmholtz integral equation (6.25) fails to yield a unique solution whenever the frequency coincides with one of the characteristic values for the interior Dirichlet boundary value problem. However, among all these characteristic frequencies there is only one set that also satisfies the interior Helmholtz integral formulation simultaneously. This is the basis for the Combined Helmholtz Integral Equation Formulation or CHIEF for short that provides unique solutions for acoustic radiation problems at any frequency.

6.3 Discrete Helmholtz Integral Formulations

For arbitrarily shaped surfaces, the Helmholtz integral formulation (6.10) and the surface Helmholtz integral equation (6.25) cannot be solved analytically. Hence numerical solutions are sought. Suppose that the surface is discretized into segments with N nodes, then Eqs. (6.10) and (6.25) can be rewritten as

$$\hat{p}\left(\vec{x};\omega\right) = \left\{T_p\left(\vec{x}\middle|\vec{x}_s;\omega\right)\right\}_{1\times N}\left\{\hat{p}\left(\vec{x}_s;\omega\right)\right\}_{N\times 1}, \tag{6.44}$$

$$\hat{p}\left(\vec{x};\omega\right) = \left\{T_v\left(\vec{x}\middle|\vec{x}_s;\omega\right)\right\}_{1\times N}\left\{\hat{v}_{\mathrm{n}}\left(\vec{x}_s;\omega\right)\right\}_{N\times 1}, \tag{6.45}$$

where $\left\{T_p\left(\vec{x}\middle|\vec{x}_s;\omega\right)\right\}_{1\times N}$ and $\left\{T_v\left(\vec{x}\middle|\vec{x}_s;\omega\right)\right\}_{1\times N}$ represent, respectively, the transfer functions that correlate the surface acoustic pressure and normal surface velocity to the field acoustic pressure, and are given by

$$\left\{T_p\left(\vec{x}\middle|\vec{x}_s;\omega\right)\right\}_{1\times N} = (4\pi)^{-1}\left(\{D\}_{1\times N} + \{M\}_{1\times N}[M_s]_{N\times N}^{-1}\left(2\pi[\mathrm{I}]_{N\times N} - [D_s]_{N\times N}\right)\right), \tag{6.46}$$

$$\left\{T_v\left(\vec{x}\middle|\vec{x}_s;\omega\right)\right\}_{1\times N} = (4\pi)^{-1}\left(\{D\}_{1\times N}\left(2\pi[\mathrm{I}]_{N\times N} - [D_s]_{N\times N}\right)^{-1}[M_s]_{N\times N} + \{M\}_{1\times N}\right), \tag{6.47}$$

where $[\mathrm{I}]_{N\times N}$ is a unitary matrix, $[M_s]_{N\times N}$ and $[D_s]_{N\times N}$ depict the effects of monopoles and dipoles on a surface point, respectively, and $[M]_{N\times N}$ and $[D]_{N\times N}$ describe those of monopoles and dipoles on a field point, respectively. The $\mu\nu$th elements of these matrices are given by

$$M_{\mu\nu} = G\left(\vec{x}_{\mu}\middle|\vec{x}_{s',\nu};\omega\right)J_{\mu\nu} \quad \text{and} \quad D_{\mu\nu} = \frac{\partial G\left(\vec{x}_{\mu}\middle|\vec{x}_{s',\nu};\omega\right)}{\partial \mathrm{n}\left(\vec{x}_{s'}\right)}J_{\mu\nu}, \tag{6.48}$$

$$M_{s,\mu\nu} = G\left(\vec{x}_{s,\mu}\middle|\vec{x}_{s',\nu};\omega\right)J_{\mu\nu} \quad \text{and} \quad D_{s,\mu\nu} = \frac{\partial G\left(\vec{x}_{s,\mu}\middle|\vec{x}_{s',\nu};\omega\right)}{\partial \mathrm{n}\left(\vec{x}_{s'}\right)}J_{\mu\nu}, \tag{6.49}$$

where $J_{\mu\nu}$ indicates the Jacobian of the surface integration in Eqs. (6.46) and (6.47) [98].

In the BEM-based NAH, the goal is to reconstruct the surface acoustic quantities based on discrete N nodes. So we need to take at least N measurement points of the field acoustic pressures to form a square matrix in Eqs. (6.44) and (6.45). Accordingly, we can rewrite Eqs. (6.44) and (6.45) as

$$\left\{\hat{p}\left(\vec{x}_m;\omega\right)\right\}_{N\times 1} = \left[T_p\left(\vec{x}\middle|\vec{x}_s;\omega\right)\right]_{N\times N}\left\{\hat{p}\left(\vec{x}_s;\omega\right)\right\}_{N\times 1}, \tag{6.50}$$

$$\left\{\hat{p}\left(\vec{x}_m;\omega\right)\right\}_{N\times 1} = \left[T_v\left(\vec{x}\middle|\vec{x}_s;\omega\right)\right]_{N\times N}\left\{\hat{v}_{\mathrm{n}}\left(\vec{x}_s,\omega\right)\right\}_{N\times 1}, \tag{6.51}$$

where $\hat{p}\left(\vec{x}_m;\omega\right)$, $m=1$ to N, represent the measured acoustic pressures.

Equations (6.50) and (6.51) enables one to reconstruct surface acoustic pressure $\hat{p}\left(\vec{x}_s;\omega\right)$ and normal surface velocity $\hat{v}_{\mathrm{n}}\left(\vec{x}_s;\omega\right)$ through inversion of matrices $\left[T_p\left(\vec{x}\middle|\vec{x}_s;\omega\right)\right]_{N\times N}$ and $\left[T_v\left(\vec{x}\middle|\vec{x}_s;\omega\right)\right]_{N\times N}$, respectively. In practice, the measured acoustic pressures $\hat{p}\left(\vec{x}_m;\omega\right)$ are not error free. As a result, the matrix equations (6.50) and (6.51) may be ill conditioned. To overcome this difficulty, regularization can be employed, the simplest one being a TSVD to eliminate the evanescent waves that fall below the background noise level. Accordingly, $\hat{p}\left(\vec{x}_s;\omega\right)$ and $\hat{v}_{\mathrm{n}}\left(\vec{x}_s;\omega\right)$ can be written as

$$\left\{\hat{p}\left(\vec{x}_s;\omega\right)\right\}_{N\times 1} = \left[V_p\right]_{N\times N}\left[\Sigma_p^{-1}\right]_{N\times N}\left[U_p\right]_{N\times N}^{\mathrm{T}}\left\{\hat{p}\left(\vec{x}_m,\omega\right)\right\}_{N\times 1}, \tag{6.52}$$

$$\left\{\hat{v}_{\mathrm{n}}\left(\vec{x}_s;\omega\right)\right\}_{N\times 1} = \left[V_v\right]_{N\times N}\left[\Sigma_v^{-1}\right]_{N\times N}\left[U_v\right]_{N\times N}^{\mathrm{T}}\left\{\hat{p}\left(\vec{x}_m,\omega\right)\right\}_{N\times 1}, \tag{6.53}$$

where $[V_p]_{N\times N}$ and $[U_p]_{N\times N}$, respectively, are the right and left unitary matrices of the transfer matrix $\left[T_p\left(\vec{x}\middle|\vec{x}_s;\omega\right)\right]_{N\times N}$, namely, they satisfy $[V_p]_{N\times N}[V_p]_{N\times N}^{\mathrm{T}} = [\mathrm{I}]_{N\times N}$ and $[U_p]_{N\times N}[U_p]_{N\times N}^{\mathrm{T}} = [\mathrm{I}]_{N\times N}$, and $[\Sigma_p]^{-1} = \mathrm{diag}[\ldots, 1/\sigma_{p,n}, \ldots]$ stands for the diagonal matrix containing inversions of the non-zero singularities $\sigma_{p,n}$ of the matrix $\left[T_p\left(\vec{x}\middle|\vec{x}_s;\omega\right)\right]_{N\times N}$. Similarly, $[V_v]_{N\times N}$ and $[U_v]_{N\times N}$ are the right and left unitary matrices of the transfer matrix $\left[T_v\left(\vec{x}|\vec{x}_s;\omega\right)\right]_{N\times N}$, respectively, $[V_v]_{N\times N}[V_v]_{N\times N}^{\mathrm{T}}=[\mathrm{I}]_{N\times N}$ and $[U_v]_{N\times N}[U_v]_{N\times N}^{\mathrm{T}}=[\mathrm{I}]_{N\times N}$, and $[\Sigma_v]^{-1}=\mathrm{diag}[\ldots, 1/\sigma_{v,n},\ldots]$ is the diagonal matrix that contains inversions of the non-zero singularities $\sigma_{v,n}$ of the matrix $\left[T_v\left(\vec{x}|\vec{x}_s;\omega\right)\right]_{N\times N}$.

Equations (6.52) and (6.53) gives the reconstructed acoustic quantities on the surface of any arbitrary structure. A rule of thumb in discretization for the BEM method is to have a minimum of six nodes per wavelength. For a complex structure vibrating at mid-to-high frequencies, the total number of discrete nodes needed to depict the surface acoustic quantities can be extremely large. As a result, the number of measurements required to reconstruct the acoustic quantities can be very high, thus making the reconstruction process unrealistically time consuming.

6.4 The Combined Helmholtz Equation Least-Squares Method

To enhance the efficiency of the BEM-based NAH and improve the accuracy of the HELS method for reconstructing the acoustic field generated by an arbitrary structure, we combine these two methods and describe the procedures as follows:

1. Take the acoustic pressures $\hat{p}\left(\vec{x}_m^{\,\mathrm{meas}};\omega\right)$, $m=1$ to M, on the hypothetical spherical surface that encloses the target source surface.
2. Divide the measurement points into two groups M_1 and M_2, where $M=M_1+M_2$.
3. Use M_1 as the input to establish the HELS formulations to reconstruct the acoustic pressure on M_2 points on the measurement surface,

$$\left\{\hat{p}\left(\vec{x}_m^{\,\mathrm{rec}};\omega\right)\right\}_{M_2\times 1}=\left[G_{pp}\left(\vec{x}_m^{\,\mathrm{rec}}|\vec{x}_m^{\,\mathrm{meas}};\omega\right)\right]_{M_2\times M_1}\left\{\hat{p}\left(\vec{x}_m^{\,\mathrm{meas}};\omega\right)\right\}_{M_1\times 1}, \tag{6.54}$$

where $\left[G_{pp}\left(\vec{x}_m^{\,\mathrm{rec}}|\vec{x}_m^{\,\mathrm{meas}};\omega\right)\right]_{N\times M_1}$ is defined in Eq. (3.16).

Note that since the least-squares method is used, the expansion solution (6.54) with $J=M_1$ will be the best fit at M_1 measurement locations, but it may not be the best approximation for the remaining M_2 values. This is especially true when the measured acoustic pressures contain errors. Consequently, the accuracy in reconstruction at M_2 locations will increase with J first, and then deteriorates thereafter.

4. Find the optimal expansion term J_{op}, which is equivalent to finding a low-pass filter for the spherical harmonics such that the evanescent waves below the background noise level are eliminated. There are many regularization techniques available for solving a set of linear equations. The simplest yet very effective for the HELS method is an iteration scheme,

$$\min_J \sum_{i=1}^{M} \left\| \hat{p}\left(\vec{x}_{m,i}^{\text{rec}}; \omega\right) - \hat{p}\left(\vec{x}_{m,i}^{\text{meas}}; \omega\right) \right\|_2^2 \to J_{\text{op}}, \tag{6.55}$$

where reconstruction is done on the measurement surface to determine the value of J_{op}.

5. Use J_{op} in Eq. (6.55) to regenerate the acoustic pressures at as many points as necessary on the measurement surface, say, the same as that of discrete notes,

$$\left\{\hat{p}\left(\vec{x}_m^{\text{rec}}; \omega\right)\right\}_{N\times 1} = \left[G_{pp}\left(\vec{x}_m^{\text{rec}} \middle| \vec{x}_m^{\text{meas}}; \omega\right)\right]_{N\times M_1} \left\{\hat{p}\left(\vec{x}_m^{\text{meas}}; \omega\right)\right\}_{M_1\times 1}. \tag{6.56}$$

6. Take these regenerated acoustic pressures $\hat{p}\left(\vec{x}_m^{\text{rec}}; \omega\right)$ as input data to the BEM-based NAH formulations to reconstruct the acoustic pressure $\hat{p}\left(\vec{x}_s^{\text{rec}}; \omega\right)$ and normal velocity $\hat{v}_{\text{n}}\left(\vec{x}_s^{\text{rec}}; \omega\right)$ on the source surface,

$$\begin{aligned}&\left\{\hat{p}\left(\vec{x}_s^{\text{rec}}; \omega\right)\right\}_{N\times 1}\\ &= \left[V_p\right]_{N\times N} \left[\Sigma_p^{-1}\right]_{N\times N} \left[U_p\right]_{N\times N}^{\text{T}} \left[G_{pp}\left(\vec{x}_m^{\text{rec}} \middle| \vec{x}_m^{\text{meas}}; \omega\right)\right]_{N\times M_1} \left\{\hat{p}\left(\vec{x}_m^{\text{meas}}; \omega\right)\right\}_{M_1\times 1},\end{aligned} \tag{6.57}$$

$$\begin{aligned}&\left\{\hat{v}_{\text{n}}\left(\vec{x}_s^{\text{rec}}; \omega\right)\right\}_{N\times 1}\\ &= \left[V_v\right]_{N\times N} \left[\Sigma_v^{-1}\right]_{N\times N} \left[U_v\right]_{N\times N}^{\text{T}} \left[G_{pp}\left(\vec{x}_m^{\text{rec}} \middle| \vec{x}_m^{\text{meas}}; \omega\right)\right]_{N\times M_1} \left\{\hat{p}\left(\vec{x}_m^{\text{meas}}; \omega\right)\right\}_{M_1\times 1},\end{aligned} \tag{6.58}$$

where the matrix $\left[G_{pp}\left(\vec{x}_m^{\text{rec}} \middle| \vec{x}_m^{\text{meas}}; \omega\right)\right]_{N\times M_1}$ is given, respectively, by

$$\begin{aligned}&\left[G_{pp}\left(\vec{x}_m^{\text{rec}} \middle| \vec{x}_m^{\text{meas}}; \omega\right)\right]_{N\times M_1}\\ &= \left[\Psi\left(\vec{x}_m^{\text{rec}}; \omega\right)\right]_{N\times J_{\text{op}}} \left(\left[\Psi\left(\vec{x}_m^{\text{meas}}; \omega\right)\right]_{J_{\text{op}}\times M_1}^{H} \left[\Psi\left(\vec{x}_m^{\text{meas}}; \omega\right)\right]_{M_1\times J_{\text{op}}}\right)^{-1}\\ &\quad \left[\Psi\left(\vec{x}_m^{\text{meas}}; \omega\right)\right]_{J_{\text{op}}\times M_1}^{H},\end{aligned} \tag{6.59}$$

where the elements of $\left[\Psi\left(\vec{x}_m^{\text{meas}}; \omega\right)\right]_{M_1\times J_{\text{op}}}$ are given in Eq. (3.2).

The enhancement in the reconstruction efficiency is obvious. Equation (6.59) shows that $\hat{p}\left(\vec{x}_s^{\,\mathrm{rec}}, \omega\right)$ and $\hat{v}_{\mathrm{n}}\left(\vec{x}_s^{\,\mathrm{rec}}, \omega\right)$ on N nodes of the arbitrarily shaped surface can now be reconstructed using M_1 measurements. Since $M << N$, the required measurement time is significantly reduced.

Note that the accuracy of the regenerated field acoustic pressure is consistent with that of measured data for $\vec{x}_m \in \vec{X}$. This is because the acoustic pressure for $\vec{x}_m \in \vec{X}$ can be completely and uniquely described by Eq. (6.54) as $J \to \infty$. The omission of the higher-order terms, namely, the evanescent waves have a negligibly small impact on the resultant field acoustic pressure. Hence, there is no need to take more measurements than necessary. In fact, the accuracy of reconstruction would remain unchanged, even if the regenerated field acoustic pressures were replaced by the real measurements. The trade-off is that the accuracy in reconstruction may be limited because certain evanescent waves have been lost as measurements are taken over a spherical surface rather than a conformal surface at close range.

It is emphasized that one cannot extend the processes discussed above to the region inside the minimum sphere, either by taking measurements or regenerating the acoustic pressures. This is because the acoustic pressure there cannot be represented adequately by the spherical waves.

6.5 Applications of the CHELS Method

In this section we examine the performance of the CHELS method and compare its results with that of the BEM-based NAH. In particular, we want to check if CHELS can yield satisfactory reconstruction of an acoustic field accurately and efficiently based on greatly reduced input data.

Example 6.1 Consider a partially vibrating sphere of radius $a = 0.1$ m. The reason for selecting this example is because it contains very rich evanescent waves and yet the analytic solution is readily available. The normal surface velocity distribution $\hat{v}_{\mathrm{n}}$ can be written as

$$\hat{v}_{\mathrm{n}}(a, \theta, \varphi; \omega) = \begin{cases} v_0, & 0 \leq |\theta| \leq \theta_0 \\ 0, & \text{otherwise} \end{cases}, \tag{6.60}$$

where v_0 is a constant and the half vertex angle is, say, $\angle\theta_0 = 36°$.

Since the source is a sphere, the minimum surface is spherical. Following the guidelines as given in Chap. 5, we gauge the measurement distance d and microphone spacing δ with respect to the critical spatial wavelength λ_{cr}. Suppose that as an initial guess, we set $\lambda_{\mathrm{cr}} = a/3$ and $\delta < \lambda_{\mathrm{cr}}/2$. Since $a = 0.1$ m, we find $\lambda_{\mathrm{cr}} \approx 0.033$ m and $\delta = 0.0165$ m. The number of measurement points can be estimated by using Eq. (5.3). Since the source surface $S = 2\pi a^2$, $a = 0.1$ m, and $\lambda_{cr} \approx 0.033$ m, we have $M > 4 \times (2\pi \times 0.1^2)/0.033^2 \approx 278$. All the input acoustic pressures are calculated by using the formulation given by Morse and Ingard [99], of which $M_1 = 76$ are used in Eq. (6.56) to regenerate the acoustic pressures on the spherical measurement surface and the rest $M_2 = 202$ points are used to optimize the number of expansion functions J_{op}. In this case $J_{op} = 26$ is found to be acceptable for $0 < ka \leq 10$. The regenerated field acoustic pressures are taken as input to Eqs. (6.57) and (6.58).

Note that for any given set of measurements in engineering practice, a larger value of J_{op} indicates an inclusion of more evanescent waves and the higher the accuracy of reconstruction can be. A smaller value of J_{op} implies a lower SNR and less evanescent waves included in input data. As a result, the reconstructed acoustic field may be unsatisfactory.

For comparison, we use the BEM-based NAH to reconstruct the surface acoustic quantities. To ensure the accuracy in reconstruction, we use six discrete nodes per structural wavelength to depict the surface acoustic quantities. Suppose that we take six discrete nodes per critical spatial wavelength, $\delta = \lambda_{cr}/6$. Since $\lambda_{cr} \approx 0.033$ m, we have $\delta = 0.0055$ m, which leads to a total number of $N = 602$ discrete nodes. Accordingly, we need to take $M = 602$ measurement points of the acoustic pressures, which are obtained by using the formulation given by Morse and Ingard [99] and taken as input to the BEM-based NAH Eqs. (6.52) and (6.53) to reconstruct the surface acoustic pressure and normal component of the surface velocity.

In this example, we show the reconstruction results based on a coarse mesh with an average distance between neighboring discrete nodes $\delta = 0.032$ m, which is twice the value of $\delta = 0.0165$ m as suggested for CHELS. Accordingly, the number of nodes by using a triangular element and the first-order interpolation is reduced to $N = 152$. The number of measurement points is the same as that of the discrete nodes, i.e., $M = 152$.

Figure 6.2 shows the comparison of the reconstructed acoustic pressures at $ka = 1.46$ on the generator of the sphere. Results show that the surface acoustic pressures reconstructed by CHELS with $N = 152$ agree very well with those of the BEM codes with $N = 602$ and the analytic solutions.

While a fine mesh does not make much difference in reconstructing the acoustic pressure, it does have a significant impact on reconstructing the normal surface velocity. Figure 6.3 shows that a coarse mesh with $N = 152$ nodes only enables one to capture the main characteristics of the normal surface velocity distribution. By using a fine mesh of $N = 602$ nodes and the same number of the input data points regenerated by Eq. (6.56), we can significantly improve the reconstruction accuracy. This is because the surface normal velocity distribution has a sharp edge that contains higher wavenumber contents than the surface acoustic pressure does.

The fact that the CHELS method can yield satisfactory reconstruction with relatively few measurements is of a great significance. It indicates that the fidelity of the input data regenerated by Eq. (6.56) is preserved. Hence, one does not need to take more measurements than necessary. Moreover, it shows that one can improve

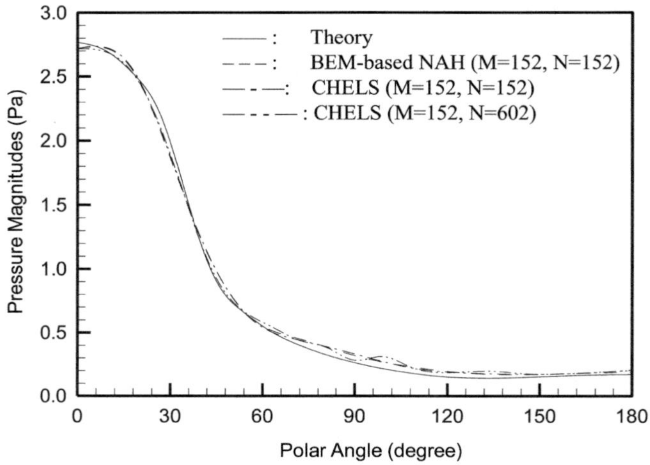

Fig. 6.2 Comparison of the reconstructed acoustic pressure distributions on the surface of a partially vibrating sphere at $ka = 1.46$ based on measurements taken at $r = 0.105$ m

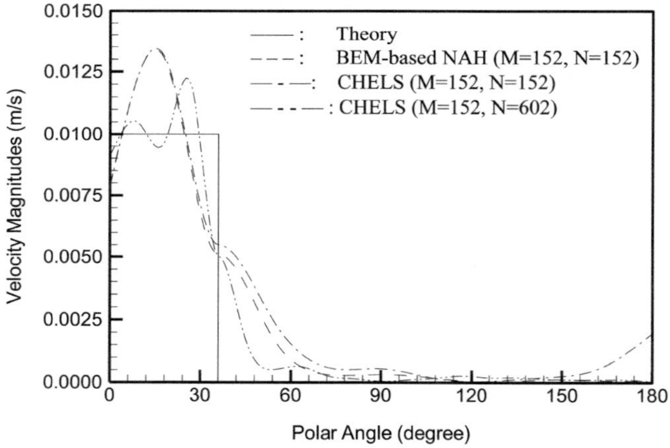

Fig. 6.3 Comparison of reconstructed normal component of velocity distributions on the surface of a partially vibrating sphere at $ka = 1.46$ based on measurements taken at $r = 0.105$ m

the accuracy by increasing the input data. Because these data are calculated but not measured, the efficiency of reconstruction is significantly enhanced.

However, one should not expect the normal surface velocity to converge to the true value even as the number of regenerated input data approaches infinity. This is because the accuracy of reconstruction is controlled by the amount of evanescent waves captured in the measured data. The closer the measurements are to the source surface, the more the evanescent waves are captured, and the more accurate the

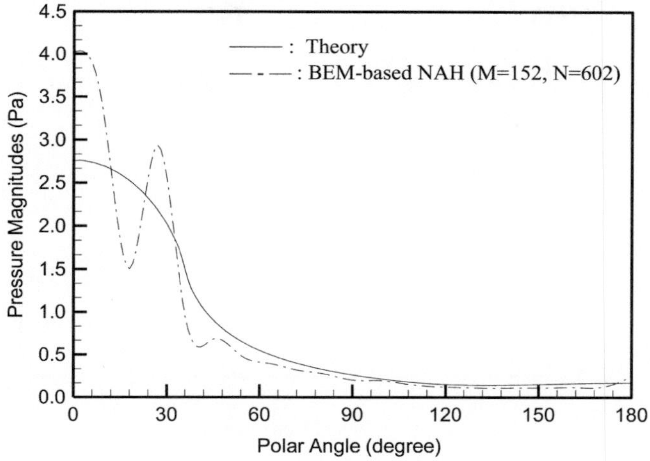

Fig. 6.4 Reconstructed acoustic pressure distribution on the surface of a partially vibrating sphere at $ka = 1.46$ using the BEM-based NAH with 602 discrete nodes and 152 measurements taken at $r = 0.105$ m

reconstruction is. Once the measurement distance is fixed, so is the amount of the evanescent waves that can be captured. Thus the improvement in the reconstruction accuracy through increasing the number of regenerated input data is limited.

Also, it is emphasized that we do not need to have exactly the same measurement number as the discrete nodes. This is because using SVD and regularization, we can have either an over- or under-determined system of equations, or equivalently, take more or fewer measurements than the discrete nodes. However, if the measurements are too few, a spatial aliasing may occur and the resulting reconstruction can be greatly distorted. Figure 6.4 displays that when 152 field acoustic pressures are taken as the input to the BEM-based NAH for a surface with 602 discrete nodes, the resulting reconstruction of the surface acoustic pressure is severely distorted.

To show the effect of measurement distances on the reconstruction accuracy, we present the reconstructed surface acoustic quantities based on conformal measurements taken at different radial distances $r = 0.105$, 0.110, 0.125, and 0.150 m under $ka = 1.46$. Figure 6.5 depicts that as measurement distances increase, more and more evanescent waves are lost in the input data. As a result, the reconstructed normal surface velocity becomes more and more distorted. However the accuracy in reconstruction of the surface acoustic pressure remains essentially unchanged (results omitted for brevity). This is because the normal surface velocity contains more near-field effects than the surface acoustic pressure does. These results demonstrate the importance of keeping the measurements very close to the target source surface.

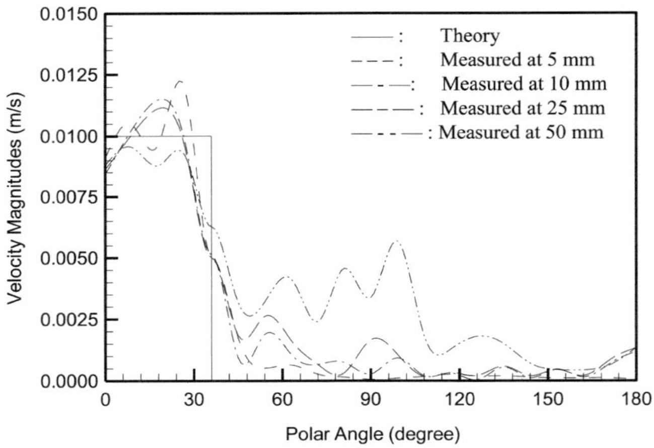

Fig. 6.5 Comparison of the reconstructed normal component of velocity on the surface of a partially vibrating sphere $ka = 1.46$ using the CHELS method based on measurements taken at different radial distances

Example 6.2 Consider a simplified engine block with an overall length of 0.460 m, overall width of 0.435 m, and overall height of 0.630 m. To test the effectiveness of the CHELS method, sharp edges and corners and abrupt changes in surface contours are built in this model. For such arbitrary geometry, analytic solutions do not exist and numerical solutions must be sought.

To simulate acoustic radiation from this engine block in a free field, harmonic excitations of different amplitudes are assumed on three arbitrarily selected surfaces: 5×10^5 N/m^2 on the top and 2×10^5 N/m^2 on part of the front and back surfaces at various frequencies (see Fig. 6.6). The bottom of the engine block is clamped with zero displacement and slope, and the rest surfaces are left unconstrained. The normal surface velocity distributions are obtained using the standard FEM codes and the surface acoustic pressures are specified using the BEM codes with 1,548 triangular elements and 776 discrete nodes. Once the surface acoustic quantities are specified, field acoustic pressures are calculated.

To reconstruct the surface acoustic quantities using the CHELS method, we take $M = 277$ measurement points over an imaginary sphere of radius $r = 0.408$ m that encloses the engine block, which is much fewer than 776 points. In particular, we select $M_1 = 56$ as input to set up the HELS formulations and $M_2 = 221$ to optimize the number of expansion functions. For the frequency range considered, this optimal value is found to be approximately $J_{op} = 22$. Once this is done, the acoustic pressures on the measurement surface are regenerated by Eq. (6.56). The results are taken as input data to Eqs. (6.57) and (6.58) to reconstruct the surface acoustic quantities.

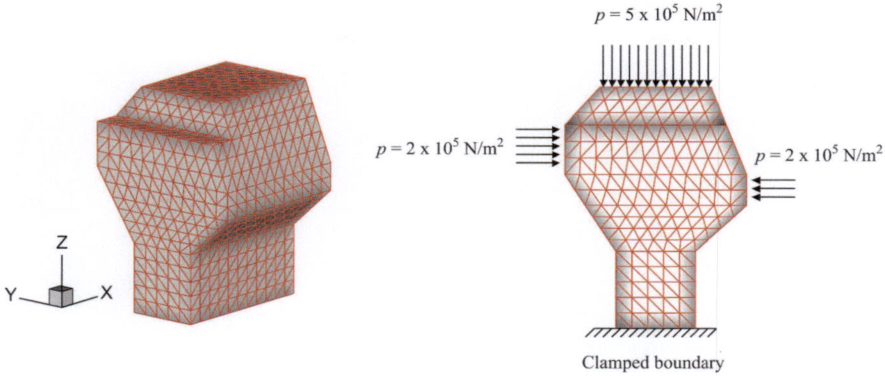

Fig. 6.6 Schematic of an engine block subject to distributed harmonic force excitations

Note that one can divide the measurements into two groups of any sizes. However, for an arbitrarily shaped surface, the clearance between the measurement and source surfaces may vary from one point to another. As a result, the amounts of the evanescent waves captured in the input data may be different. To minimize the impact of the loss of evanescent waves in reconstruction, a low-pass filter must be used to eliminate the evanescent waves that drop below the background noise level. This is equivalent to specifying the optimal expansion number J_{op}. Experiment results indicate that in most cases it is better to select a smaller value for M_1 and a larger value for M_2 to achieve the desired reconstruction.

It is emphasized that even if all steps as suggested above are followed, it will be unrealistic to expect ideal reconstruction. This is because: (1) measurements are taken on a spherical surface, not a conformal surface, and (2) the number of measurement points is greatly reduced in CHELS to alleviate the complexities involved taking an excessive number of measurements demanded by the BEM-based NAH.

Also noted is the fact that the accuracy of reconstruction varies with frequency. To ensure that measurements are taken in the near field, we gauge the standoff distance d with respect to the critical spatial wavelength λ_{cr}, which is set at $\lambda_{cr} = \pi a/8$, and require that the following conditions be satisfied: (1) $d \ll a$, and (2) $d \ll \lambda_{cr}$. Because in the CHELS method the standoff distance d is nonuniform, we take the maximum clearance between the measurement and source surfaces d_{max}. Therefore, these conditions are rewritten as $d_{max} \ll a$ and $d_{max} \ll \lambda_{cr} = \pi a/8$.

In this case the characteristic dimension of the engine is $a = (0.435 + 0.460 + 0.630)/3 = 0.508$ m and the maximum clearance between measurement and source surface is $d_{max} = 0.19$ m < 0.508 m, so the first condition is satisfied. However, the second condition is not because $d_{max} = 0.19$ m and $\lambda_{cr} = \pi a/8 = 0.197$ m. As a result, some near-field information is lost in the input data.

In what follows, we present the reconstructed acoustic fields on the engine block surfaces based on $M = 277$ measurement points. For validation purposes, we use

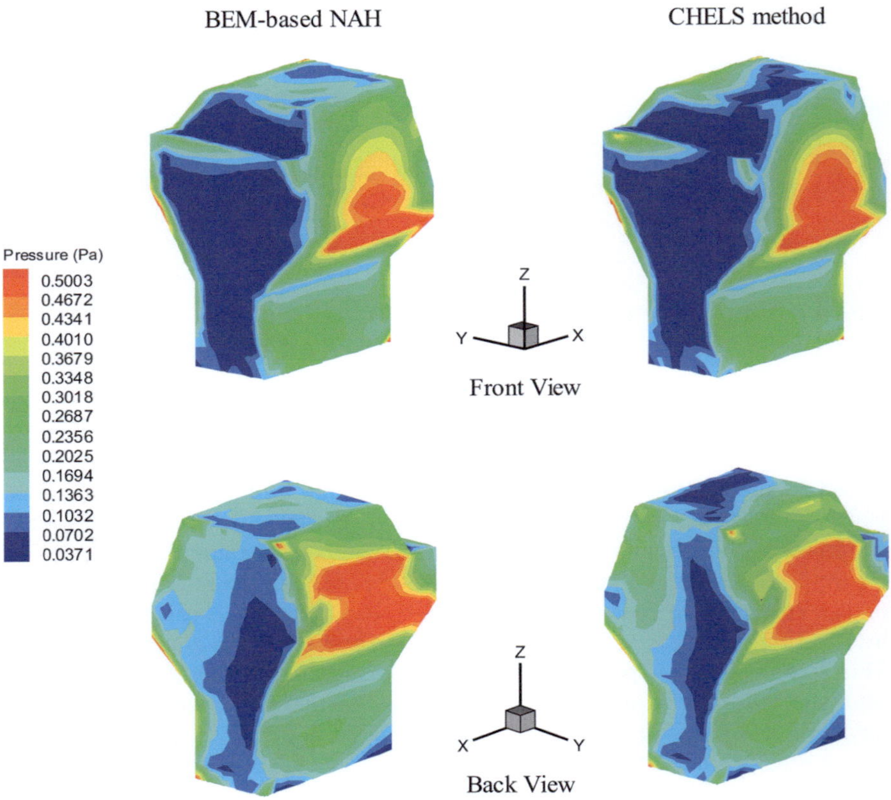

Fig. 6.7 Comparison of the reconstructed acoustic pressure distributions over the surfaces of engine block using the CHELS method with $M = 277$ data points (*right column*) and benchmark values with $M = 776$ data points (*left column*) at $ka = 1$

Eq. (6.3-6) to reconstruct the surface acoustic quantities based on $M = 776$ measurement points, which are the same as the discrete nodes. Figures 6.7 and 6.8 display the comparisons of the reconstructed acoustic pressure and normal velocity distributions on the surfaces of the engine block by using the CHELS method and benchmark values, respectively, at $ka = 1$.

It is emphasized that this engine block represents a fairly complex structure that contains sharp edges, corners, and abrupt changes along the surface contours. Yet satisfactory reconstruction is obtained by using the CHELS method with a reduction in input data points by more than 63 %. In contrast, when data points are reduced to one-half, $M = 386$, aliasing occurs in the reconstructed acoustic quantities obtained by using the BEM-based NAH because input data are severely under sampled (see Fig. 6.9) [100].

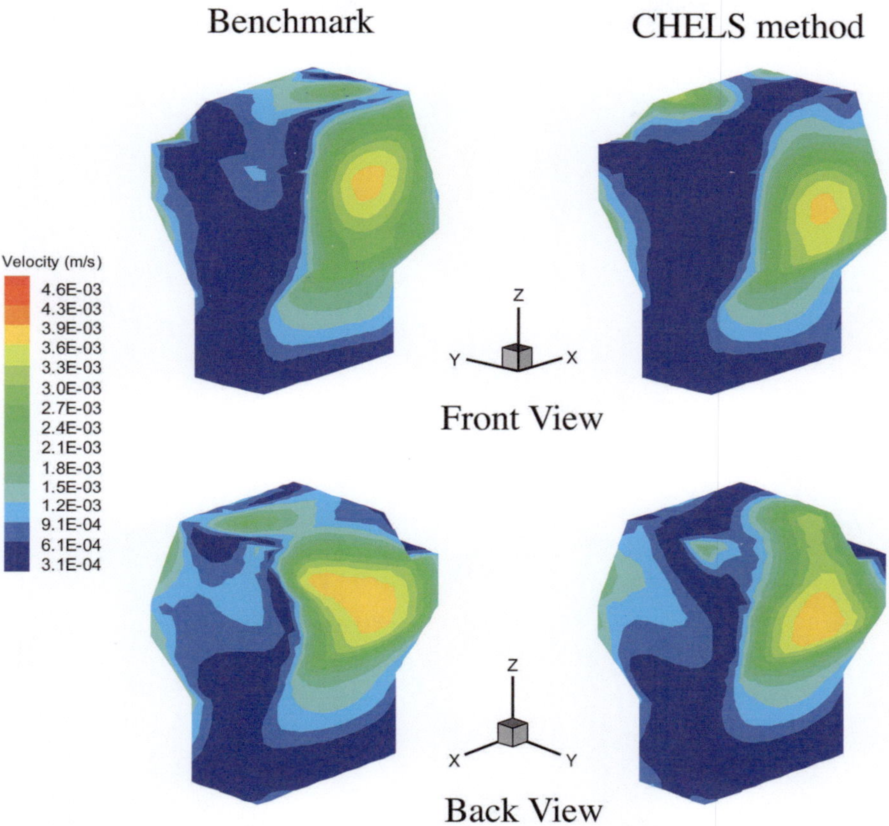

Fig. 6.8 Comparison of the reconstructed normal surface velocity distributions over the surfaces of engine block using the CHELS method with $M = 277$ data points (*right column*) and benchmark values with $M = 776$ data points (*left column*) at $ka = 1$

Equations (6.57) and (6.58) can be used to reconstruct acoustic radiation in the far field by setting the reconstruction point at $\vec{x}_s^{\,rec} = \vec{x}^{\,rec}$. This is straightforward because all field points are now outside the minimum sphere, so the acoustic field can be adequately represented by the outgoing spherical waves. Moreover, the loss of the evanescent waves has a negligible impact on reconstruction [101].

Figure 6.10 demonstrates comparisons of the reconstructed normal component of the time-averaged acoustic intensity by using the CHELS method on two planes (2.6×2.6 m^2) at $y = \pm 3$ m measured from the center of the engine block versus the BEM results. Note that the peak amplitude of the time-averaged intensity on the front plane is slightly lower than that of the back plane. This is because more forces are acting on the backside than on the front side of the engine block.

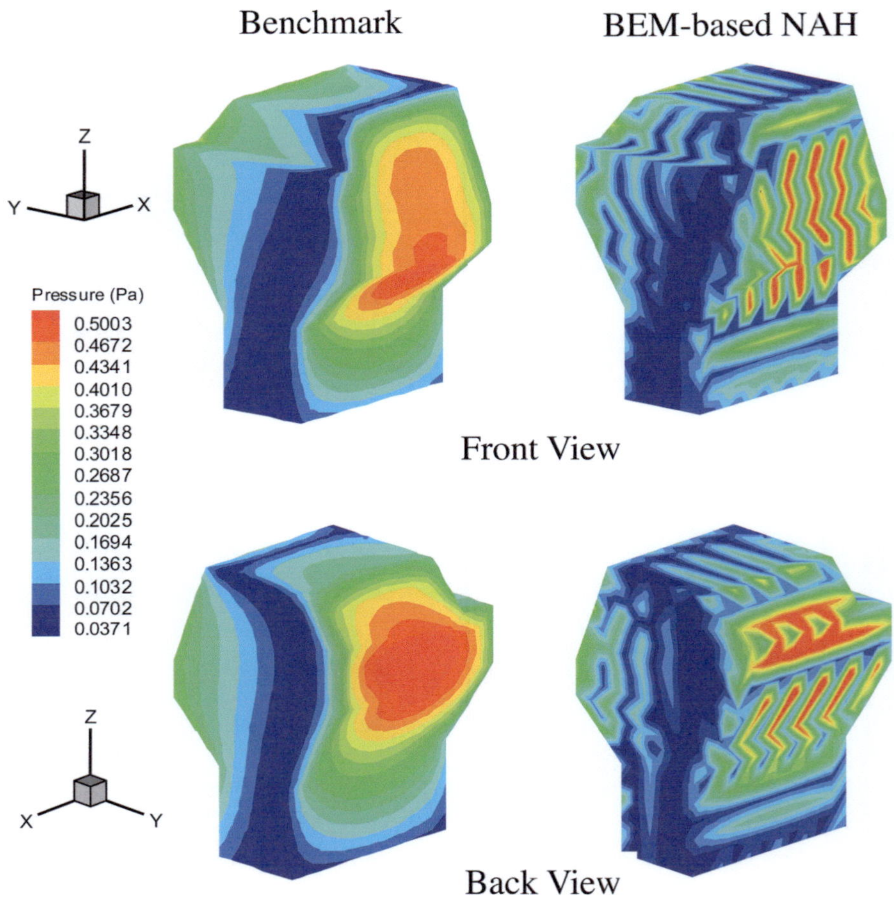

Fig. 6.9 Comparison of the reconstructed acoustic pressure distributions on the engine block surfaces by using the BEM-based NAH with $M = 388$ input data point and the benchmark results. In this case, aliasing occurs because the input data are severely under sampled

These results demonstrate that the CHELS method can be used to enhance the efficiency in reconstruction by taking relatively fewer measurement points on a minimum sphere enclosing the target source, yet still allowing for a relatively accurate reconstruction of the acoustic quantities. This is done by setting up the HELS formulations using a finite number measurement points, and regenerating as many acoustic pressures as those required by the BEM-based NAH to reconstruct the acoustic quantities on the source surface as well as in the field.

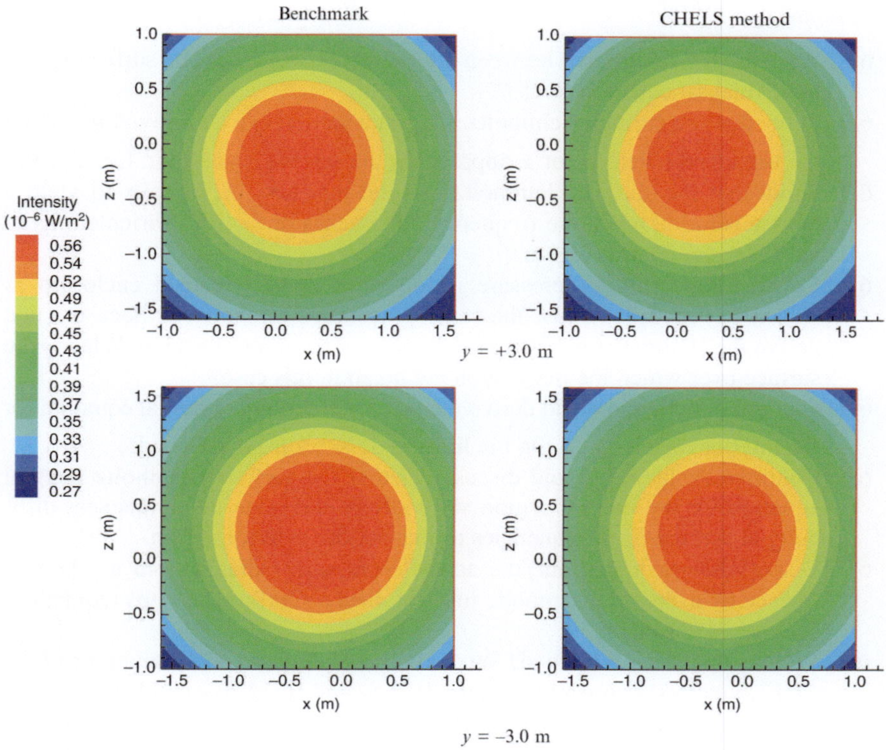

Fig. 6.10 Comparison of the reconstructed normal component of the time-averaged acoustic intensity distribution over two planes at $y = \pm 3.0$ m measured with respect to the center of the engine block using the CHELS method (*right column*) and BEM-based NAH (*left column*) at $ka = 1$

Problems

6.1. What are the differences between the Helmholtz equation and surface Helmholtz integral equation (6.25)?

6.2. Show that the surface Helmholtz integral equation can be derived by taking the limit as the field point $\vec{x}$ approaches the surface point $\vec{x}_s$ in Eq. (6.10).

6.3. Show that the surface Helmholtz integral equation (6.25) fails to yield a unique solution when the frequency approaches one of the Dirichlet eigenvalue for the interior region.

6.4. Consider the acoustic pressure inside an arbitrarily shaped enclosure as shown in Fig. 5.9. Follow the same procedures as shown in Sect. 6.1 and derive the Helmholtz integral formulation for the interior region. What is the surface over which the integration are taken in this case?

6.5. Continue Problem 6.4 and derive the surface Helmholtz integral equation for the interior region by taking the limit as $\vec{x} \to \vec{x}_s$ from the inside.

6.6. Continue Problem 6.5 and discuss whether the surface Helmholtz integral equation for the interior region suffers from the same nonuniqueness difficulty at certain eigenfrequencies as that for the exterior region.

6.7. Discuss how to determine the acoustic pressure radiated from a vibrating surface by using the Helmholtz integral theory. Outline the steps required in solving this problem.

6.8. What is the CHELS method? What are the advantages and limitations of the CHELS method compared with the HELS- and BEM-based NAH?

6.9. Discuss the implementation of the CHELS method and compare it to that of HELS method.

6.10. Discuss the implementation of the CHELS method and compare it to that of BEM-based NAH.

Chapter 7
Hybrid NAH

All traditional NAH techniques are limited to handle cases where sound sources are only on one side of an array of microphones. The reality is much more complicated however. A typical example is the analysis of noise radiation from a vehicle stationed on the chassis dynamometers inside a semi-anechoic chamber. For safety and durability concerns, the surfaces of the chamber cannot be made as acoustically absorptive as they should be. Consequently, the measured acoustic pressures consist of both direct and reflected waves. Also, the dynamometer is generating its own noise, making NAH application and analysis very difficult. To date, vehicle noise is still analyzed by measuring transfer functions between a source and receiver, or by sweeping an intensity probe over a target source surface at close range. The information obtained is often isolated and valid at the measurement locations. These traditional noise diagnosis and analysis processes cannot reveal much insightful information of the root causes of noise and structural vibrations.

In this chapter we present hybrid NAH to reconstruct acoustic radiation from an arbitrary object in confined or free space in a cost-effective manner [102]. This hybrid NAH can be derived from a modified HELS, which expands the acoustic pressures in terms of both outgoing and incoming spherical waves, and combines it with the BEM-based NAH and regularization. Since this hybrid NAH allows for regeneration of the acoustic pressure on a measurement surface, both the accuracy and efficiency of reconstruction are enhanced.

7.1 Modified HELS

In order to enhance the efficiency of reconstruction of acoustic radiation from an arbitrarily shaped source in confined space, we rewrite the HELS expansion (3.2) as

S.F. Wu, *The Helmholtz Equation Least Squares Method*, Modern Acoustics and Signal Processing, DOI 10.1007/978-1-4939-1640-5_7

$$p\left(\vec{x};\omega\right) = \sum_{j=1}^{J} \Psi_j^{(1)}\left(\vec{x};\omega\right) C_j(\omega) + \sum_{j=1}^{J} \Psi_j^{(2)}\left(\vec{x};\omega\right) D_j(\omega), \quad (7.1)$$

where $\Psi_j^{(1)}\left(\vec{x};\omega\right)$ is defined in Eq. (3.2) and $\Psi_j^{(2)}\left(\vec{x};\omega\right) \equiv \Psi_{nl}^{(2)}(r,\theta,\varphi) = h_n^{(2)}(kr)Y_n^l(\theta,\varphi)$, where $h_n^{(2)}(kr)$ is the second kind spherical Hankel functions of order n.

Physically, the terms on the right side of Eq. (7.1) represent the outgoing and incoming spherical waves, respectively. To facilitate the derivations of hybrid NAH formulations, we rewrite the expansion functions as a matrix and determine the expansion coefficients by solving an overdetermined linear system of equations obtained by matching the assumed-form solution (7.1) to the acoustic pressures measured on Γ through least squares.

$$\left\{\hat{p}\left(\vec{x}_m^{\Gamma};\omega\right)\right\}_{M\times 1} = \left[\widetilde{\Psi}\left(\vec{x}_m^{\Gamma};\omega\right)\right]_{M\times 2J}\left\{\widetilde{C}(\omega)\right\}_{2J\times 1}, \quad \vec{x}_m^{\Gamma}\in\Gamma, \ m = 1 \text{ to } M, \quad (7.2)$$

where $\left\{\hat{p}\left(\vec{x}_m^{\Gamma};\omega\right)\right\}_{M\times 1}$ are the acoustic pressures measured on Γ; $\left[\widetilde{\Psi}\left(\vec{x}_m^{\Gamma};\omega\right)\right]_{M\times 2J}$ and $\left\{\widetilde{C}(\omega)\right\}_{2J\times 1}$ are defined as

$$\left[\widetilde{\Psi}\left(\vec{x}_m^{\Gamma};\omega\right)\right]_{M\times 2J} = \begin{bmatrix} \Psi_{11}^{(1)} & \Psi_{11}^{(2)} & \Psi_{12}^{(1)} & \Psi_{12}^{(2)} & \cdots & \Psi_{1J}^{(1)} & \Psi_{1J}^{(2)} \\ \Psi_{21}^{(1)} & \Psi_{21}^{(2)} & \Psi_{22}^{(1)} & \Psi_{22}^{(2)} & \cdots & \Psi_{2J}^{(1)} & \Psi_{2J}^{(2)} \\ \vdots & \vdots & \vdots & \vdots & \ddots & \vdots & \vdots \\ \Psi_{J1}^{(1)} & \Psi_{J1}^{(2)} & \Psi_{J2}^{(1)} & \Psi_{J2}^{(2)} & \cdots & \Psi_{JJ}^{(1)} & \Psi_{JJ}^{(2)} \\ \vdots & \vdots & \vdots & \vdots & \ddots & \vdots & \vdots \\ \Psi_{\widetilde{M1}}^{(1)} & \Psi_{\widetilde{M1}}^{(2)} & \Psi_{\widetilde{M2}}^{(1)} & \Psi_{\widetilde{M2}}^{(2)} & \cdots & \Psi_{\widetilde{MJ}}^{(1)} & \Psi_{\widetilde{MJ}}^{(2)} \end{bmatrix}_{M\times 2J}, \quad (7.3)$$

$$\left\{\widetilde{C}(\omega)\right\}_{2J\times 1} = \begin{Bmatrix} C_1 \\ D_1 \\ C_2 \\ D_2 \\ \vdots \\ C_J \\ D_J \end{Bmatrix}_{2J\times 1}, \quad M > 2J. \quad (7.4)$$

Because reconstruction is an ill-posed problem, $\left[\widetilde{\Psi}\left(\vec{x}_m^{\Gamma};\omega\right)\right]_{M\times 2J}$ may be ill conditioned. So regularization must be utilized to ensure that the reconstructed acoustic quantities are bounded. The situation here is worse because the source is not in a free field, and input data may be contaminated either through sound reflections from unspecified boundary surfaces or through interferences due background noise. Such a scenario is often encountered in engineering applications.

To overcome these difficulties, we take measurements over two conformal surfaces Γ_1 and Γ_2 around a target source at close distances, with Γ_2 being inside Γ_1 by a small separation Δ apart. In particular, we take M_1 measurements on Γ_1 to establish the expansion coefficients $\left\{\widetilde{C}(\omega)\right\}_{2J\times 1}$, and optimize the number of expansion terms J_{op} by minimizing reconstruction errors with respect to additional M_2 measurements taken on Γ_2.

$$\min_{\widetilde{C}_j} \left|\left|\hat{p}_j^{\,\text{rec}}\left(\vec{x}_m^{\Gamma_1};\omega\right) - \hat{p}_j^{\,\text{meas}}\left(\vec{x}_m^{\Gamma_1};\omega\right)\right|\right|_2^2, \quad \vec{x}_m^{\Gamma_1} \in \Gamma_1,\ j = 1 \text{ to } 2J, \tag{7.5}$$

$$\min_{2J} \left|\left|\hat{p}_j^{\,\text{rec}}\left(\vec{x}_m^{\Gamma_2};\omega\right) - \hat{p}_j^{\,\text{meas}}\left(\vec{x}_m^{\Gamma_2};\omega\right)\right|\right|_2^2, \quad \vec{x}_m^{\Gamma_2} \in \Gamma_2,\ m = 1 \text{ to } M_2. \tag{7.6}$$

The reason for taking measurements over two concentric conformal surfaces for a source in confined space is to ensure that we acquire not only acoustic pressures but also their gradients so as to discern the directions of wave propagations. This is especially important inside a reactive acoustic field, namely, the boundary surfaces are highly reflective. Examples of such are seen in visualizing acoustic fields bounded by hard surfaces, for which reconstruction cannot possibly be done correctly without the use of double layers of measurements [103].

It is emphasized that the modified HELS formulation (7.2) can also be used to reconstruct acoustic radiation from arbitrarily shaped structures at constant frequencies in a free field. Under this condition, it is sufficient to take measurements over a single conformal surface at a very close distance to the target source. Taking measurements on two conformal surfaces only prolongs the reconstruction process with no apparent benefits of improving the reconstruction accuracy.

7.2 Hybrid NAH

Equation (6.52) has provided the BEM solutions to reconstructing the acoustic quantities generated by an arbitrary source in a free field. This BEM-based NAH formulation can be utilized in a non-free field with some modifications.

Suppose that the source surface is discretized into elements with N discrete nodes. Because both the acoustic pressure and normal velocity on the source surface need to be specified, we solve Eq. (6.53) simultaneously. Accordingly, we take M measurement points, which are at least equal to N discrete nodes, as input to Eq. (6.53) to determine these surface acoustic quantities. Since BEM requires a minimum of six discrete nodes per spatial wavelength to avoid distortions, the required number of measurement points M may be very high. This is especially true when the frequency is relatively high. The CHELS method offers an effective way to overcome this difficulty by using a finite number of measurement points to establish the HELS formulations first, and regenerating as many input points as necessary for the BEM formulations.

The procedures for hybrid NAH are described as follows:

1. Take M_1 measurements of the acoustic pressure on a conformal surface Γ_1 around a target source at close range, and M_2 measurements of the acoustic pressure on another conformal surface Γ_2, which is inside Γ_1. Both M_1 and M_2 are finite.
2. Use Eq. (7.5) to determine the expansion coefficients $\widetilde{C}_j$, $j=1$ to $2J$, and Eq. (7.6) to specify the value of J_{op} by minimizing reconstruction errors with respect to additional M_2 measurements on Γ_2. This guarantees accurate reconstruction of acoustic pressures on Γ_2.
3. Use Eq. (7.2) to regenerate as many acoustic pressures as needed on $\vec{x}_m^{\Gamma} \in \Gamma_2$, $m=1$ to N, and take them as input data to the BEM formulations to reconstruct the vibro-acoustic quantities on the source surface,

$$\left\{\hat{p}\left(\vec{x}_s^{\text{rec}};\omega\right)\right\}_{N\times 1} = \left[\widetilde{T}_p\left(\vec{x}_s|\vec{x}_m^{\text{meas}};\omega\right)\right]_{N\times M_1}\left\{\hat{p}\left(\vec{x}_m^{\text{meas}};\omega\right)\right\}_{M_1\times 1}, \quad (7.7)$$

$$\left\{\hat{v}_{\text{n}}\left(\vec{x}_s^{\text{rec}};\omega\right)\right\}_{N\times 1} = \left[\widetilde{T}_v\left(\vec{x}_s|\vec{x}_m^{\text{meas}};\omega\right)\right]_{N\times M_1}\left\{\hat{p}\left(\vec{x}_m^{\text{meas}};\omega\right)\right\}_{M_1\times 1}, \quad (7.8)$$

where $\left[\widetilde{T}_p\left(\vec{x}_s|\vec{x}_m^{\text{meas}};\omega\right)\right]_{N\times M_1}$ and $\left[\widetilde{T}_v\left(\vec{x}_s|\vec{x}_m^{\text{meas}};\omega\right)\right]_{N\times M_1}$ are defined, respectively, as

$$\left[\widetilde{T}_p\left(\vec{x}_s|\vec{x}_m^{\text{meas}};\omega\right)\right]_{N\times M_1} = \left[T_p\left(\vec{x}_s|\vec{x}_m^{\text{meas}};\omega\right)\right]_{N\times N}^{-1}\left[\widetilde{G}_{pp}\left(\vec{x}_m^{\text{rec}}|\vec{x}_m^{\text{meas}};\omega\right)\right]_{N\times M_1}, \quad (7.9)$$

$$\left[\widetilde{T}_v\left(\vec{x}_s|\vec{x}_m^{\text{meas}};\omega\right)\right]_{N\times M_1} = \left[T_v\left(\vec{x}_s|\vec{x}_m^{\text{meas}};\omega\right)\right]_{N\times N}^{-1}\left[\widetilde{G}_{pp}\left(\vec{x}_m^{\text{rec}}|\vec{x}_m^{\text{meas}};\omega\right)\right]_{N\times M_1}, \quad (7.10)$$

where $\left[T_p\left(\vec{x}_s|\vec{x}_m^{\text{meas}};\omega\right)\right]_{N\times N}^{-1}$ and $\left[T_v\left(\vec{x}_s|\vec{x}_m^{\text{meas}};\omega\right)\right]_{N\times N}^{-1}$ stand for the inversions of matrices of $\left[T_p\left(\vec{x}_s|\vec{x}_m^{\Gamma};\omega\right)\right]_{N\times N}$ and $\left[T_v\left(\vec{x}_s|\vec{x}_m^{\Gamma};\omega\right)\right]_{N\times N}$, respectively, which are defined in Eqs. (6.46) and (6.47), and the matrix $\left[\widetilde{G}_{pp}\left(\vec{x}_m^{\text{rec}}|\vec{x}_m^{\text{meas}};\omega\right)\right]_{N\times M_1}$ is given by

$$\left[\widetilde{G}_{pp}\left(\overrightarrow{x}_m^{\text{rec}}\middle|\overrightarrow{x}_m^{\text{meas}};\omega\right)\right]_{N\times M_1}$$
$$=\left[\widetilde{\Psi}\left(\overrightarrow{x}_m^{\text{rec}};\omega\right)\right]_{N\times J_{\text{op}}}\left(\left[\widetilde{\Psi}\left(\overrightarrow{x}_m^{\text{meas}};\omega\right)\right]^H_{J_{\text{op}}\times M_1}\left[\widetilde{\Psi}\left(\overrightarrow{x}_m^{\text{meas}};\omega\right)\right]_{M_1\times J_{\text{op}}}\right)^{-1} \quad (7.11)$$
$$\left[\widetilde{\Psi}\left(\overrightarrow{x}_m^{\text{meas}};\omega\right)\right]^H_{J_{\text{op}}\times M_1},$$

where $\left[\widetilde{\Psi}\left(\overrightarrow{x}_m^{\Gamma};\omega\right)\right]_{M_1\times J_{op}}$ is defined in Eq. (7.3).

Note that the hybrid NAH solution (7.7) and (7.8) is different from the CHELS solution (6.57) and (6.58) in twofolds: (1) the expansion functions in the hybrid NAH consist of both outgoing and incoming spherical waves, whereas those in the CHELS method involve only the outgoing spherical waves, and (2) the transfer functions in hybrid NAH include transfer functions for both BEM and HELS formulations, while those in CHELS contain the transfer function for the BEM formulations.

As mentioned earlier, all reconstruction problems are mathematically ill posed. Therefore reconstruction formulations must be regularized to smooth the dependence of the solution on input data and minimize the impacts of the errors embedded in input data on reconstruction.

Here MTR is used for regularization (see Sect. 3.5). The first term on the right side of Eq. (3.61) represents an ultra rough least-squares solution for which $\alpha = 0$, and the second term implies an ultra smooth solution for which $\alpha \to \infty$. Thus the choice of α allows one to decide how far to go to achieve certain smoothness. One way of specifying α that requires no prior knowledge of noise variance is to employ GCV [50, 104]. The basic idea of GCV is to leave a particular measured acoustic pressure out of calculations of the cost functions first, and then evaluate the effectiveness of the reconstructed source fields in predicting the values of the omitted data. This process should be repeated for all data points and the resulting regularization parameter α can ensure a best fit for the predicted acoustic pressures at all the data points.

Note that one can use other regularization techniques such as standard TR or Landweber iterations [105] together with Morozov discrepancy principle (MDP) [106] and L-curve to reconstruct the acoustic quantities. Numerical tests demonstrate that GCV may fail to yield a value of α when it is coupled with standard TR. While MDP can always yield a value for α, its accuracy may not be very high when it is coupled with standard TR. An optimal combination is an MTR and GCV, which are adopted here.

To summarize, we reconstruct the acoustic pressure and normal velocity on an arbitrarily shaped source surface using Eqs. (7.7) and (7.8) derived from the BEM-based integral formulation (6.3-5), which is regularized by using an MTR and GCV and implemented through SVD. The input data to Eq. (7.2-8) are regenerated by the modified HELS formulations (7.2). Since the number of measurement points in hybrid NAH are much fewer than the discrete nodes needed required by the BEM codes to describe the acoustic quantities on a source surface, the efficiency of numerical computations is significantly enhanced. Meanwhile, the accuracy of reconstruction is ensured by the Helmholtz integral formulations and an optimal combination of an MTR and GCV.

Note that when hybrid NAH is used to reconstruct acoustic radiation from a source in free space, we must set $D_j = 0$ for there is no incoming wave. Under this condition, the hybrid NAH is the same as the CHELS method.

7.3 Reconstructing Acoustic Fields Using the Hybrid NAH

In this section we use hybrid NAH to reconstruct the acoustic field generated by vibrating objects in half space (see Fig. 7.1). The acoustic pressure anywhere in this half space consists of the direct and reflected sound waves. Because the baffle is infinite, this is equivalent to two sources (the original source and its mirror image) in free space, and the acoustic pressure anywhere is the superposition of the direct sound waves from both of them.

The reason for replacing this one-source in half space scenario by two-source in free space is to facilitate the calculations of acoustic pressures in numerical simulations, including benchmark acoustic pressures in the field and on the source, as well as the measured acoustic pressures on Γ_1 and Γ_2 (see Fig. 7.1) by using the half-space Helmholtz integral formulation,

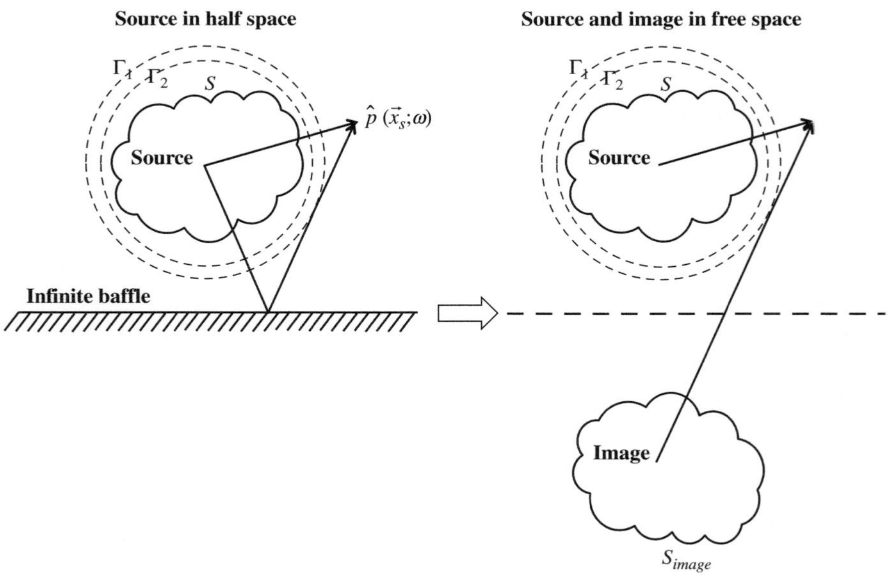

Fig. 7.1 Schematic of equivalence between a source in free space and source-and-image in half space

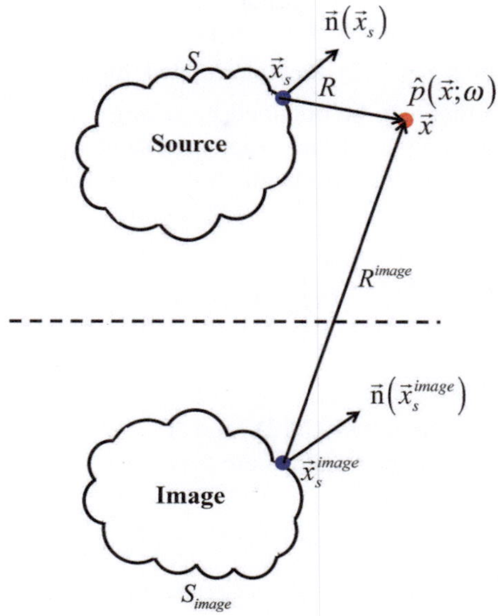

Fig. 7.2 Schematic of calculating the field acoustic pressure radiated from a source and its image using the half-space Helmholtz integral formulation

$$\hat{p}\left(\vec{x};\omega\right) = \frac{1}{4\pi}\iint\limits_{S'}\left[\hat{p}\left(\vec{x}_{s'};\omega\right)\frac{\partial G_{\mathrm{H}}\left(\vec{x}\middle|\vec{x}_{s'};\omega\right)}{\partial \mathrm{n}\left(\vec{x}_{s'}\right)} - i\omega\rho_0\hat{v}_{\mathrm{n}}\left(\vec{x}_{s'};\omega\right)G_{\mathrm{H}}\left(\vec{x}\middle|\vec{x}_{s'};\omega\right)\right]dS', \tag{7.12}$$

where $G_{\mathrm{H}}\left(\vec{x}\middle|\vec{x}_{s'};\omega\right)$ is known as the half-space Green's function defined as

$$G_{\mathrm{H}}\left(\vec{x}_{s'}\right) = \frac{\mathrm{e}^{ikR}}{R} + \beta\frac{\mathrm{e}^{ikR^{\mathrm{image}}}}{R^{\mathrm{image}}}, \tag{7.13}$$

where β represents the acoustic pressure reflection coefficient of the infinite baffle, which is equal to +1 when the baffle is perfectly rigid and the acoustic pressure is doubled on the baffle surface, and to -1 when the baffle is a pressure-release surface on which the acoustic pressure is 0.

For a baffle with finite acoustic impedance, there is no closed-form solution for the half-space Green's function and Eq. (7.12) becomes a good approximation when the field point $\vec{x}$ is at least one-half wavelength away from the baffle. The distance between $\vec{x}$ and source surface $\vec{x}_s$ is $R = \left|\vec{x} - \vec{x}_s\right|$ and that between any field point $\vec{x}$ and surface point of the image source $\vec{x}_s^{\,\mathrm{image}}$ is $R^{\mathrm{image}} = \left|\vec{x} - \vec{x}_s^{\,\mathrm{image}}\right|$ (see Fig. 7.2).

The surface acoustic pressure $\hat{p}\left(\vec{x}_s;\omega\right)$ and normal surface velocity $\hat{v}_{\mathrm{n}}\left(\vec{x}_s;\omega\right)$ in Eq. (7.12) are interrelated through the surface Helmholtz integral equation, which can be obtained by taking a limit as $\vec{x}\rightarrow\vec{x}_s$. The processes of evaluating these integrals are exactly the same as those described in Sect. 6.1, and the resultant surface Helmholtz integral equation for half space is given by

$$\hat{p}\left(\vec{x}_s;\omega\right)=\frac{1}{2\pi}\iint_{S'}p\left(\vec{x}_{s'};\omega\right)\frac{\partial G_{\mathrm{H}}\left(\vec{x}_s\middle|\vec{x}_{s'};\omega\right)}{\partial\mathrm{n}\left(\vec{x}_{s'}\right)}-i\omega\rho_0\hat{v}_{\mathrm{n}}\left(\vec{x}_{s'};\omega\right)G_{\mathrm{H}}\left(\vec{x}_s\middle|\vec{x}_{s'};\omega\right)dS'. \tag{7.14}$$

Equations (7.12) and (7.14) should be solved simultaneously and their discretized versions are given by

$$\hat{p}\left(\vec{x};\omega\right)=\left\{T_p^H\left(\vec{x}\middle|\vec{x}_s;\omega\right)\right\}_{1\times N}\left\{\hat{p}\left(\vec{x}_s;\omega\right)\right\}_{N\times 1}, \tag{7.15}$$

$$\hat{v}_{\mathrm{n}}\left(\vec{x};\omega\right)=\left\{T_v^H\left(\vec{x}\middle|\vec{x}_s;\omega\right)\right\}_{1\times N}\left\{\hat{v}_{\mathrm{n}}\left(\vec{x}_s;\omega\right)\right\}_{N\times 1}, \tag{7.16}$$

where $\left\{T_p^H\left(\vec{x}\middle|\vec{x}_s;\omega\right)\right\}_{1\times N}$ and $\left\{T_v^H\left(\vec{x}\middle|\vec{x}_s;\omega\right)\right\}_{1\times N}$ represent the transfer functions that correlate the surface acoustic pressure and normal surface velocity to the field acoustic pressure, respectively, and are given by

$$\begin{aligned}&\left\{T_p^H\left(\vec{x}\middle|\vec{x}_s;\omega\right)\right\}_{1\times N}\\&=(4\pi)^{-1}\left(\left\{D^H\right\}_{1\times N}+\left\{M^H\right\}_{1\times N}\left[M_s^H\right]_{N\times N}^{-1}\left(2\pi[\mathrm{I}]_{N\times N}-\left[D_s^H\right]_{N\times N}\right)\right),\end{aligned} \tag{7.17}$$

$$\begin{aligned}&\left\{T_v^H\left(\vec{x}\middle|\vec{x}_s;\omega\right)\right\}_{1\times N}\\&=(4\pi)^{-1}\left(\left\{D^H\right\}_{1\times N}\left(2\pi[\mathrm{I}]_{N\times N}-\left[D_s^H\right]_{N\times N}\right)^{-1}\left[M_s^H\right]_{N\times N}+\left\{M^H\right\}_{1\times N}\right),\end{aligned} \tag{7.18}$$

where the elements of matrices $[M_s^H]_{N\times N}$, $[D_s^H]_{N\times N}$, $[M^H]_{1\times N}$, and $[D^H]_{1\times N}$ are defined as

$$M_{\mu\nu}^H=G_{\mathrm{H}}\left(\vec{x}_\mu\middle|\vec{x}_{s',\nu};\omega\right)J_{\mu\nu}\quad\text{and}\quad D_{\mu\nu}^H=\frac{\partial G_{\mathrm{H}}\left(\vec{x}_\mu\middle|\vec{x}_{s',\nu};\omega\right)}{\partial\mathrm{n}\left(\vec{x}_{s'}\right)}J_{\mu\nu}, \tag{7.19}$$

$$M_{s,\mu\nu}^{H} = G_{\rm H}\left(\vec{x}_{s,\mu}\middle|\vec{x}_{s',\nu};\omega\right)J_{\mu\nu} \quad \text{and} \quad D_{s,\mu\nu}^{H} = \frac{\partial G_{\rm H}\left(\vec{x}_{s,\mu}\middle|\vec{x}_{s',\nu};\omega\right)}{\partial {\rm n}\left(\vec{x}_{s'}\right)}J_{\mu\nu}. \tag{7.20}$$

By specifying the acoustic pressures $\hat{p}\left(\vec{x};\omega\right)$ at N field points on two concentric surfaces Γ_1 and Γ_2, we can invert the matrices $\left\{T_p^H\left(\vec{x}\middle|\vec{x}_s;\omega\right)\right\}_{1\times N}$ and $\left\{T_v^H\left(\vec{x}\middle|\vec{x}_s;\omega\right)\right\}_{1\times N}$ in Eqs. (7.15) and (7.16), and determine the surface acoustic pressure $\hat{p}\left(\vec{x}_s;\omega\right)$ and the normal surface velocity $\hat{v}_{\rm n}\left(\vec{x}_s;\omega\right)$ as

$$\left\{\hat{p}\left(\vec{x}_s;\omega\right)\right\}_{N\times 1} = \left\{T_p^H\left(\vec{x}_s\middle|\vec{x}_m;\omega\right)\right\}_{N\times N}^{-1}\left\{\hat{p}\left(\vec{x}_m;\omega\right)\right\}_{N\times 1}, \tag{7.21}$$

$$\left\{\hat{v}_{\rm n}\left(\vec{x}_s;\omega\right)\right\}_{N\times 1} = \left\{T_v^H\left(\vec{x}_s\middle|\vec{x}_m;\omega\right)\right\}_{N\times N}^{-1}\left\{\hat{p}\left(\vec{x}_m;\omega\right)\right\}_{N\times 1}, \tag{7.22}$$

where $\hat{p}\left(\vec{x}_m;\omega\right)$ at N points are regenerated by the modified HELS formulations (7.7) and (7.8) based on a finite number of the acoustic pressures measured on Γ_1,

$$\begin{aligned}\left\{\hat{p}\left(\vec{x}_m;\omega\right)\right\}_{N\times 1} &\equiv \left\{\hat{p}\left(\vec{x}_m^{\,\rm rec};\omega\right)\right\}_{N\times 1} \\ &= \left[\widetilde{G}_{pp}\left(\vec{x}_m^{\,\rm rec}\middle|\vec{x}_m^{\,\rm meas};\omega\right)\right]_{N\times M_1}\left\{\hat{p}\left(\vec{x}_m^{\,\rm meas};\omega\right)\right\}_{M_1\times 1},\end{aligned} \tag{7.23}$$

where $\left[\widetilde{G}_{pp}\left(\vec{x}_m^{\,\rm rec}\middle|\vec{x}_m^{\,\rm meas};\omega\right)\right]_{N\times M_1}$ is given in Eq. (7.11).

Substituting Eq. (7.23) into (7.21), we obtain

$$\begin{aligned}&\left\{\hat{p}\left(\vec{x}_s;\omega\right)\right\}_{N\times 1} \\ &= \left\{T_p^H\left(\vec{x}_s\middle|\vec{x}_m^{\,\rm rec};\omega\right)\right\}_{N\times N}^{-1}\left[\widetilde{G}_{pp}\left(\vec{x}_m^{\,\rm rec}\middle|\vec{x}_m^{\,\rm meas};\omega\right)\right]_{N\times M_1}\left\{\hat{p}\left(\vec{x}_m^{\,\rm meas};\omega\right)\right\}_{M_1\times 1},\end{aligned} \tag{7.24}$$

$$\begin{aligned}&\left\{\hat{v}_{\rm n}\left(\vec{x}_s;\omega\right)\right\}_{N\times 1} \\ &= \left\{T_v^H\left(\vec{x}_s\middle|\vec{x}_m^{\,\rm rec};\omega\right)\right\}_{N\times N}^{-1}\left[\widetilde{G}_{pp}\left(\vec{x}_m^{\,\rm rec}\middle|\vec{x}_m^{\,\rm meas};\omega\right)\right]_{N\times M_1}\left\{\hat{p}\left(\vec{x}_m^{\,\rm meas};\omega\right)\right\}_{M_1\times 1}.\end{aligned} \tag{7.25}$$

Equation (7.24) describes the reconstructed vibro-acoustic quantities on arbitrarily shaped geometry in half space. Although there is a slight increase in computations to evaluate integrals in Eq. (7.25), this increase is more than offset by not having to integrate over the baffle surface.

Example 7.1 Consider a vibrating sphere of radius $a = 0.1$ m in half space bounded by an infinite baffle at $z = 0$ plane and the distance from the source center to baffle is d. In particular, the acoustic pressures are collected on two concentric measurement surfaces Γ_1 and Γ_2 (see Fig. 7.1) and then used as input data to the hybrid NAH formulations to reconstruct the acoustic quantities in the half space. The reconstructed acoustic quantities are validated against the benchmark values. Note that no symmetry is used in numerical computations and the fluid density and speed of sound are set at $\rho_0 = 1.21$ kg/m^3 and $c = 343$ m/s, respectively [107].

In this example the input acoustic pressures are specified over two conformal surfaces that are separated from the source surface at distances of 2 and 5 mm, respectively. Note that it is not necessary, in theory, to take two layers of measurements since the modified HELS formulations has already accounted for the effects of both outgoing and incoming spherical waves. However, our numerical simulations have demonstrated that by taking the acoustic pressure measurements on two conformal surfaces around a target source in the presence of a reflecting surface, we can get more satisfactory results than using one conformal surface. In all cases, we take half of measured data to determine the expansion coefficients and use the rest to specify an optimal number of expansion functions. For brevity, we only display reconstruction results at $ka = 1$.

First, let the amplitude of the normal surface velocity of the sphere be $V_0 = 0.01$ m/s. A total of $M = 152$ measurement points, 76 each along the generator of two conformal surfaces are taken. The discrete nodes for the BEM codes to describe the acoustic quantities on the spherical surface are $N = 602$. Apparently, the number of input data is substantially fewer than that of the discrete nodes. In reconstruction the procedures described in Sect. 7.2 for hybrid NAH are followed. In particular, Eq. (7.23) is used to regenerate $N = 602$ input data points on surface Γ_2 and then are taken as input data to Eqs. (7.24) and (7.25) to reconstruct the acoustic pressure and normal velocity on the source surface. To examine the impacts of the baffle surface on the acoustic fields, we repeat the reconstruction processes with the source located at different distances d. expansion functions provided by Eqs. (7.5) and (7.6) are found to be $J_{op} = 37$ for $d = 1.5a$ and $d = 2a$, and $J_{op} = 26$ for $d = 3a$ and $d = 50$a. Note that J_{op} implies the number of pairs of the outgoing and incoming waves in the expansion functions.

For comparison purposes, we use the modified HELS formulations (7.2) with the value of J_{op} determined by Eq. (7.5) and (7.6) without the need of BEM-based NAH formulations (7.21) and (7.22) under different distances d. The corresponding number of measurement points is $M = 62$, 31 each along the generator of two conformal surfaces. The optimal numbers of the expansion functions are found to be $J_{op} = 7$ for $d = 1.5a$, $d = 2a$, and $d = 3a$, and $J_{op} = 2$ for $d = 50a$. Once again, J_{op} implies the number of pairs of the outgoing and incoming waves in the expansion functions.

Next we consider a sphere oscillating along the z-axis direction with a velocity amplitude $V_z = 0.01\cos\theta$ (m/s), where θ is the angle between the unit normal vector on the spherical surface and the z-axis. The same numbers of measurement points as

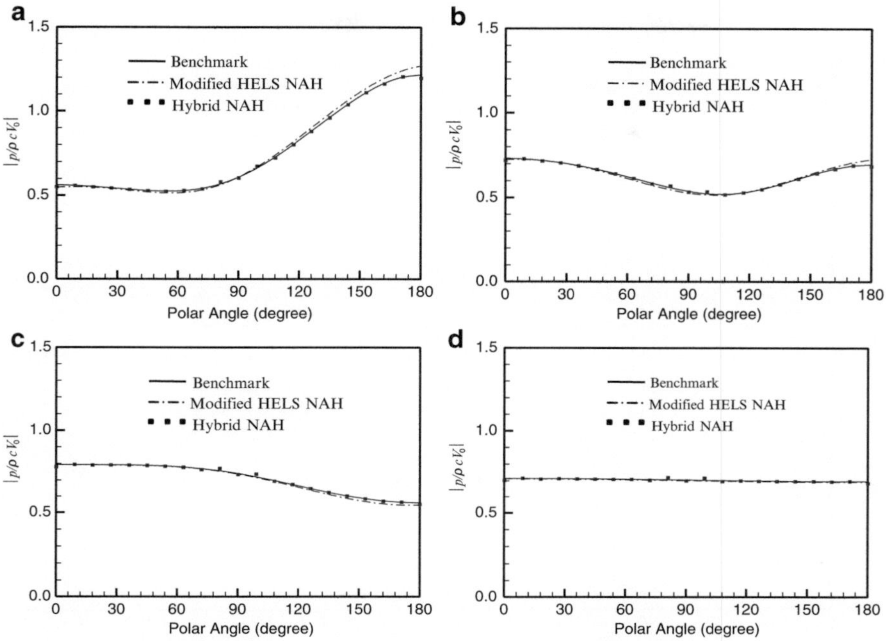

Fig. 7.3 Comparisons of the reconstructed acoustic pressures normalized with respect to the given normal surface velocity distributions on the surface of the dilating sphere at $ka = 1$: (**a**) $d = 1.5a$; (**b**) $d = 2a$; (**c**) $d = 3a$; and (**d**) $d = 50a$

those in a vibrating sphere are used, namely, $M = 152$ for hybrid NAH and $M = 62$ for the modified HELS method. Also, reconstruction is repeated for different distances d. Results show that the optimal pairs of expansion functions for hybrid NAH are $J_{op} = 37$ for $d = 1.5a$; $J_{op} = 26$ for $d = 2a$; and $J_{op} = 27$ for $d = 3a$ and $d = 50a$. On the other hand, $J_{op} = 7$ for $d = 1.5a$, $d = 2a$, and $d = 3a$; and $J_{op} = 4$ for reconstructing surface acoustic pressure and $J_{op} = 2$ for reconstructing surface normal velocity at $d = 50a$.

Figures 7.3 and 7.4 show the comparisons of the reconstructed surface acoustic pressures normalized with respect to the given normal surface velocities versus benchmark values for a dilating sphere and an oscillating sphere, respectively, at different distances d.

In all cases satisfactory agreements are obtained by using hybrid HAN and modified HELS formulations. Results indicate that the presence of an infinite rigid baffle has changed the acoustic pressure distributions on the surfaces of the dilating and oscillating spheres as compared with those of the same spheres in free space (see Figs. 7.3a–c and 7.4a–c). For example, the radiation pattern for a vibrating sphere is no longer omnidirectional as it is in a free field. The differences are due to the interferences between the direct and reflected waves from the baffle. However, these interferences diminish as the sphere moves away from the baffle

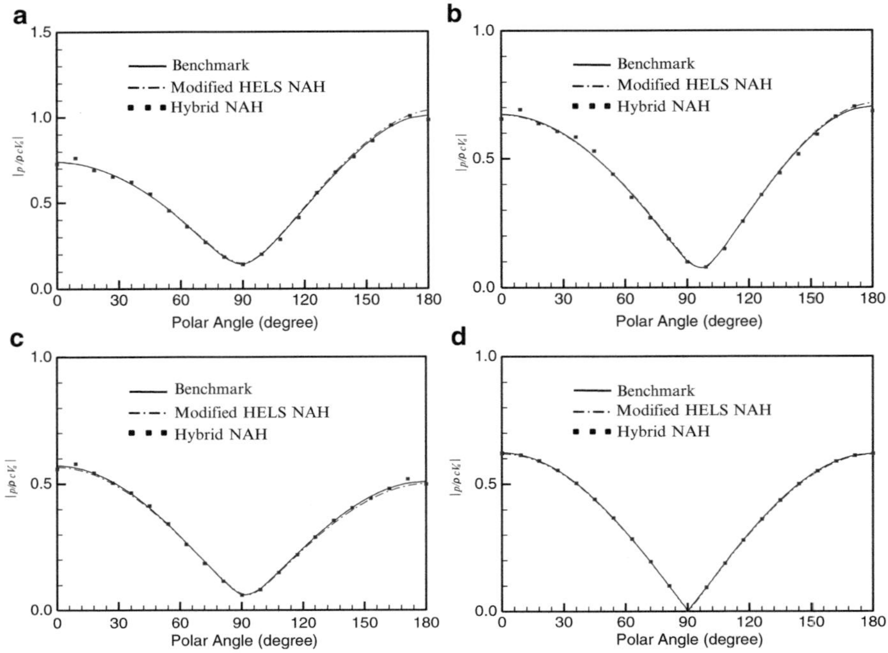

Fig. 7.4 Comparisons of the reconstructed acoustic pressures normalized with respect to the given normal surface velocity on the surface of the oscillating sphere at $ka=1$: (**a**) $d=1.5a$; (**b**) $d=2a$; (**c**) $d=3a$; and (**d**) $d=50a$

(see Figs. 7.3d and 7.4d), and the radiation patterns return to those of a sphere in free space.

Figure 7.5a, b show the comparisons of the reconstructed normal surface velocities normalized with respect to the given normal surface velocity value versus the benchmark values for a dilating and an oscillating sphere, respectively. Again, satisfactory reconstruction is obtained in all cases at different distances d. For brevity, we only present the results of $d=3a$ for the dilating and oscillating spheres.

Example 2 Consider a finite circular cylinder with two spherical endcaps in half space. The radius of this cylinder is a and its half-length is b. Two cases of different half-length to radius ratios are considered: (1) $b/a=0.5$, and (2) $b/a=2$. The distance from the center of the cylinder to baffle is d. For $b/a=0.5$, the cylindrical surface is discretized into 960 triangular elements with 482 discrete nodes, and for $b/a=2$, the cylindrical surface is discretized into 1,536 triangular elements with 770 discrete nodes [107].

Assume that these finite cylinders are either dilating with the amplitude of normal surface velocity of $V_0=0.01$ m/s or oscillating along the z-axis with the normal surface velocity amplitude $V_0=0.01\cos\theta$ m/s. The BEM codes are used to calculate the surface acoustic pressures. Once these surface acoustic quantities are

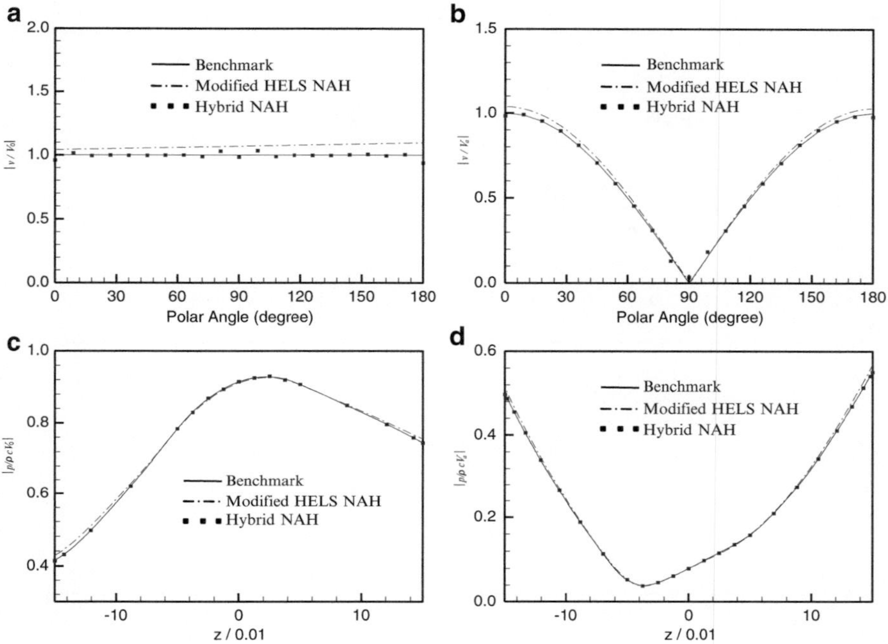

Fig. 7.5 Comparisons of the reconstructed acoustic quantities on the surface of a sphere and cylinder with $b/a = 0.5$ at $ka = 1$ and $d = 3a$. (**a**) Reconstructed normal surface velocity of a dilating sphere; (**b**) Reconstructed normal surface velocity of an oscillating sphere; (**c**) Reconstructed surface acoustic pressure of a dilating cylinder; and (**d**) Reconstructed surface acoustic pressure of an oscillating cylinder

specified, the field acoustic pressures on the measurement surfaces are generated. These field acoustic pressures are taken as input data to hybrid NAH and modified HELS formulations to reconstruct the acoustic fields, and results are compared to the benchmark values obtained by using the BEM codes.

First, we consider the cylinder with an aspect ratio $b/a = 0.5$, here $a = 0.1$ m and $d = 3a$. In using hybrid NAH, we take $M = 121$ measurement points and then regenerate the acoustic pressures at $N = 482$ points on the same measurement surface. These regenerated field acoustic pressures are utilized to reconstruct the acoustic quantities on the surface of the vibrating cylinders. The values of J_{op} are found to be 18 for the dilating cylinder and 27 for the oscillating cylinder.

In using the modified HELS, we take $M = 94$ measurement points and then regenerate the acoustic pressures at $N = 770$ points on the measurement surface. These regenerated field acoustic pressures are used to reconstruct the acoustic quantities on the surface of the vibrating cylinders. The values of J_{op} are 15 and 14, respectively, for the dilating and oscillating cylinders.

Next we consider the cylinder of an aspect ratio $b/a = 2$ with $a = 0.1$ m and $d = 5a$ and follow the same procedures as those described above. In particular, we

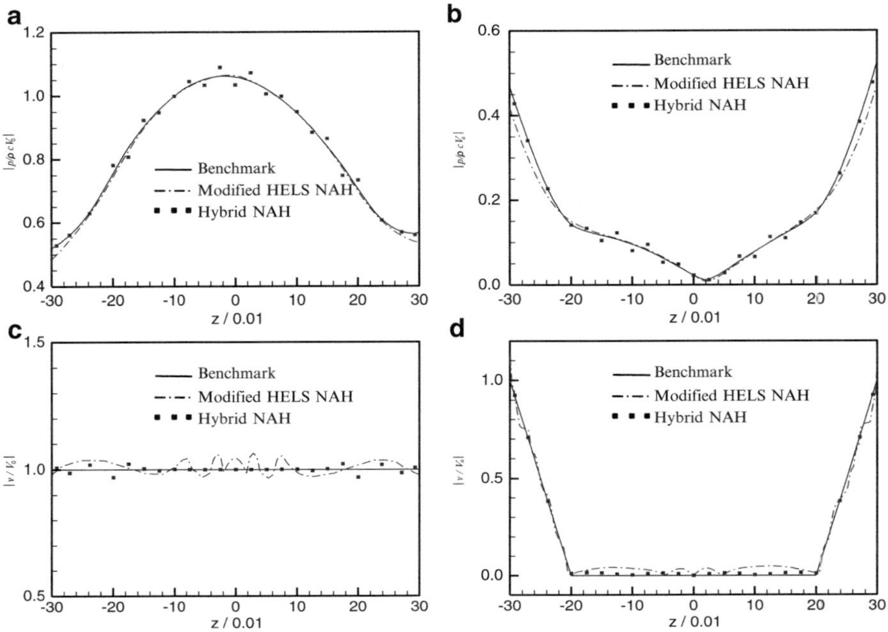

Fig. 7.6 Comparisons of the reconstructed acoustic quantities on the surface of a cylinder with $b/a=2$ at $ka=1$ and $d=5a$. (**a**) Reconstructed surface acoustic pressure for a dilating cylinder; (**b**) Reconstructed surface acoustic pressure for an oscillating cylinder; (**c**) Reconstructed normal surface velocity for a dilating cylinder; and (**d**) Reconstructed normal surface velocity for an oscillating cylinder

take $M=193$ measurement points for hybrid NAH and $M=121$ for modified HELS to reconstruct the surface acoustic quantities at 770 discrete nodes. For hybrid NAH, the optimal pairs of expansion functions are found to be $J_{op}=17$ and 26 for the dilating and oscillating cylinders, respectively. For the modified HELS, the optimal pairs of expansion functions are $J_{op}=9$ and 26 for the dilating cylinder and $J_{op}=10$ and 17 for the oscillating cylinder, respectively, to reconstruct the surface acoustic pressure and normal velocity distributions.

Figure 7.5c, d summarize the comparisons of the normalized reconstructed surface acoustic pressures versus benchmark values for dilating and oscillating cylinders with an aspect ratio $b/a=0.5$, respectively. Similar comparisons for dilating and oscillating cylinders with the aspect ratio $b/a=2$ are given in Figs. 7.6a, b, respectively. Results show that the normal surface velocities are satisfactorily reconstructed for both dilating and oscillating cylinders with different aspect ratios. For brevity, only comparisons of the reconstructed surface normal velocity versus the benchmark values for the dilating and oscillating cylinders with an aspect ratio $b/a=2$ are plotted (see Fig. 7.6c, d).

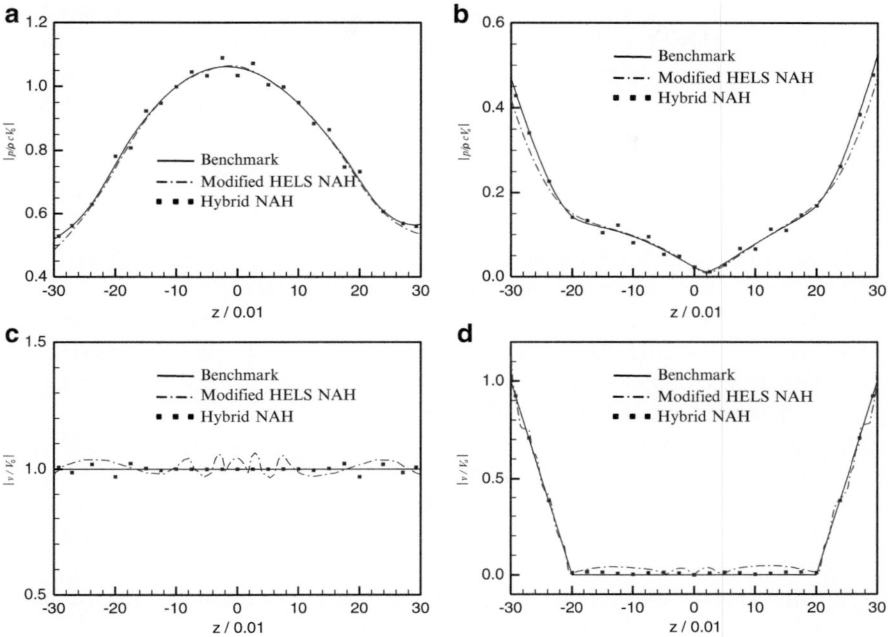

Fig. 7.7 Comparisons of the relative error norms for the reconstructed surface acoustic quantities versus *ka*. (**a**) Error norms for the reconstructed surface acoustic pressures for a dilating sphere at $d=3a$; (**b**) Error norms for the reconstructed normal surface velocity for a dilating sphere at $d=3a$; (**c**) Error norms for the reconstructed surface acoustic pressures for a dilating cylinder with $b/a=2$ and $d=5a$; (**d**) Error norms for the reconstructed normal surface velocity for a dilating cylinder with $b/a=2$ and $d=5a$

To examine the accuracy of using the hybrid NAH and modified HELS to reconstruct the acoustic quantities in half space, we calculate the relative errors defined by

$$\left\|\hat{p}\left(\vec{x}_s^{\,\text{rec}};\omega\right)-\hat{p}\left(\vec{x}_s^{\,\text{bench}};\omega\right)\right\|_2^2/\left\|\hat{p}\left(\vec{x}_s^{\,\text{bench}};\omega\right)\right\|_2^2(\%), \tag{7.26}$$

$$\left\|\hat{v}_{\text{n}}\left(\vec{x}_s^{\,\text{rec}};\omega\right)-\hat{v}_{\text{n}}\left(\vec{x}_s^{\,\text{bench}};\omega\right)\right\|_2^2/\left\|\hat{v}_{\text{n}}\left(\vec{x}_s^{\,\text{bench}};\omega\right)\right\|_2^2(\%), \tag{7.27}$$

for different dimensionless frequency *ka* with sources located at different distances *d*. The symbols $\hat{p}\left(\vec{x}_s^{\,\text{bench}};\omega\right)$ and $\hat{v}_{\text{n}}\left(\vec{x}_s^{\,\text{bench}};\omega\right)$ in Eqs. (7.26) and (7.27) indicate the benchmark values of the surface acoustic pressure and normal surface velocity obtained by using the BEM codes.

Figure 7.7 summarizes the relative errors in reconstruction obtained via hybrid NAH and modified HELS and for a dilating sphere at $d=3a$ and those for a dilating cylinder with an aspect ratio $b/a=2$ at $d=5a$ from the baffle, respectively. Results demonstrate that both hybrid NAH and modified HELS can yield satisfactory reconstruction. However, the errors in reconstructing the surface acoustic pressures are smaller than those in reconstructing the normal surface velocities in general. In particular, hybrid NAH may produce more accurate reconstruction than modified HELS does, but modified HELS requires fewer measurements than hybrid NAH does.

Problems

7.1. What is hybrid NAH? What is the purpose of developing hybrid NAH?
7.2. What are the advantages and limitations of hybrid NAH?
7.3. What are the differences between hybrid NAH and the CHELS method?
7.4. What are the differences between hybrid NAH and the HELS method?
7.5. Discuss the implementation of hybrid NAH and the differences as compared to those of the CHELS and HELS methods.
7.6. Show that the transfer function for reconstructing the acoustic pressure using hybrid NAH is given by Eq. (7.9).
7.7. Continue Problem 7.6. What is the difference between this transfer function and that for reconstructing the acoustic pressure by using the CHELS method?
7.8. Show that the transfer function for reconstructing the normal surface velocity using hybrid NAH is given by Eq. (7.10).
7.9. Continue Problem 7.7. What is the difference between this transfer function and that for reconstructing the normal surface velocity by using the CHELS method?
7.10. Discuss the potential applications of using hybrid NAH for diagnosing noise and vibration problems in engineering applications.

Chapter 8
Equivalent Sources Using HELS

In an effort to reduce the overall measurement points associated with BEM-based NAH, Jeon and Ih [108] explore the use of an equivalent source method where the field acoustic pressures are regenerated by point sources distributed inside the real source surface. To this end, Jeon and Ih reformulate the HELS formulations by expanding the spherical Hankel functions and spherical harmonics with respect to multiple points distributed in the interior region of the source surface. Contributions from all equivalent sources are determined by matching the assumed-form solution to the boundary conditions specified on the source surface [109–111] or to the acoustic pressures on the hologram surface [13, 112–118]. The equivalent sources locations can be optimized by using either the natural algorithm [119] or EfI method [120]. The optimal number of expansion terms is obtained by using a spatial filter and regularization scheme. Once the expansion coefficients are specified, the field acoustic pressures are regenerated and taken as the input data to BEM codes, just like CHELS algorithms. In this way, the overall measurement points are greatly reduced.

All the aforementioned equivalent sources methods rely on a distribution of the monopole sources inside the actual source surface. This chapter presents a more effective equivalent sources by expanding the acoustic pressure field in terms of multipoles [121], which for whatever reasons have escaped the attention of researchers.

We have learned that HELS utilizes an expansion of the spherical waves to approximate the acoustic field generated by an arbitrary source. Similar expansions have been previously utilized to predict acoustic scattering and radiation: the Rayleigh series as discussed in Sect. 4.1, the point-matching method, and least-squares approximation method, which were collectively referred to as Rayleigh methods by some authors [122, 123]. Other expansions that employ outgoing spherical waves that satisfy the Helmholtz equation and Sommerfeld radiation condition to approximate an acoustic field include the localized spherical waves (LSW) [124], distributed spherical waves (DSW) [125], and distributed point sources (DPS) [126]. These expansions are collectively known as the discrete

S.F. Wu, *The Helmholtz Equation Least Squares Method*, Modern Acoustics and Signal Processing, DOI 10.1007/978-1-4939-1640-5_8

sources methods [127]. In Chapter 8 we discuss the discrete sources methods and how to combine them with the HELS formulations to reconstruct the acoustic field generated by an arbitrary source.

8.1 Localized Spherical Waves

LSW was employed to approximate the Green's function included in the Helmholtz integral formulation to predict acoustic radiation [125],

$$\left.\begin{matrix} u\left(\vec{x};\omega\right) \\ 0 \end{matrix}\right\} = \iint_S \left[u\left(\vec{y};\omega\right) \frac{\partial G\left(\vec{x}\,|\,\vec{y};\omega\right)}{\partial \mathrm{n}\left(\vec{y}\right)} - \frac{\partial u\left(\vec{y};\omega\right)}{\partial \mathrm{n}\left(\vec{y}\right)} G\left(\vec{x}\,|\,\vec{y};\omega\right) \right] dS, \quad \begin{cases} \vec{x} \in \Omega_s \\ \vec{x} \in \Omega_i \end{cases}, \tag{8.1}$$

where Ω_s indicates the exterior region bounded by the source surface S and the sphere at infinity, and Ω_i implies the interior region inside the source surface S. In Eq. (8.1) the original format is followed as much as possible.

In an attempt to estimate the acoustic field $u\left(\vec{x};\omega\right)$, Doicu et al. [125] examined the systems of discrete sources as complete systems of functions. They found that there is a close relation between the properties of the acoustic field generated by discrete sources and the structure of their support. For example, a point structure corresponds to the LSW functions. Similarly, a straight line support parallels with the DSW functions, and a surface support is equivalent to the DPS.

Accordingly, if the acoustic field can be approximated by a point structure, LSW functions form a set of characteristic solutions to the Helmholtz equation in the spherical coordinates, which are given by

$$u_{mn}\left(\vec{x};\omega\right) = h_n^{(1)}(kr)P_n^{|m|}(\cos\theta)e^{jm\phi}, \quad n = 0, 1, 2, \ldots, \infty; \quad m = -n \text{ to } +n. \tag{8.2}$$

Using LSW, the Green's function in Eq. (8.1) is expressible as

$$G\left(\vec{x}\,|\,\vec{y};\omega\right) = \frac{jk}{\pi} \sum_{n=0}^{\infty} \sum_{m=-n}^{n} E_{mn} u_{mn}\left(\vec{x}\right) u_{-mn}\left(\vec{y}\right), \tag{8.3}$$

where u_{mn} and u_{-mn} are defined in Eq. (8.2) and the expansion coefficients E_{mn} are given by

$$E_{mn} = \frac{(2n+1)}{4} \frac{(n-|m|)!}{(n+|m|)!}. \tag{8.4}$$

The following theorem has been proven by Doicu et al. [125]

First of all, let us define some terminologies. Let S be the boundary of a bounded domain $D_i \subset \mathbf{R}^3$, namely, a bounded, open, and connected subset of three-dimensional space $\mathbf{R}^3$. We say that the surface S is of class C^2 if for each point $\vec{x} \in S$ there exists a neighborhood $V_{\vec{x}}$ of $\vec{x}$ such that the intersection $V_{\vec{x}} \cap S$ can be mapped bijectively onto a domain $U \subset \mathbf{R}^2$, and this mapping is twice continuously differentiable. We express this property by saying that D_i is of class C^2.

Theorem 8.1 *Let S be a closed surface of class C^2 and $\vec{n}$ denote the unit outward normal to S. Then the system* $u_{mn}\left(\vec{x};\omega\right) = h_n^{(1)}(kr)P_n^{|M|}(\cos\theta)e^{jm\phi}$, $n=0, 1, 2, \ldots, \infty$; $m=-n$ to $+n$, *is complete in* $L^2(S)$.

As discussed in Sect. 4.1, an infinite series is not suitable for our applications, which is especially true for reconstructing the acoustic field generated by an arbitrary source. However, we can adopt the concept of LSW and try instead the following finite expansion:

$$\hat{p}(r,\theta,\phi;\omega) = \sum_{n=0}^{N} \sum_{m=-n}^{n} a_{nm} h_n^{(1)}(kr) P_n^{|m|}(\cos\theta) e^{im\phi}, \tag{8.5}$$

where a_{nm} represent the expansion coefficients, N is the order of expansion, and the total number of expansion terms is $J=(N+1)^2$. Note that there is a subtle difference between LSW and HELS in that the former uses $P_n^{|m|}(\cos\theta)e^{im\phi}$ in the expansion, whereas the latter uses the standard spherical harmonics $P_n^m(\cos\theta)e^{im\phi}$ in the expansion. From Eq. (2.16) we see that $P_n^{|m|}(\cos\theta)e^{im\phi}$ differs from $P_n^m(\cos\theta)e^{im\phi}$ by a constant $(-1)^m$, which may be absorbed by the expansion coefficients a_{nm}. Thus Eq. (8.5) is in effect the same as the HELS expansion. Since LSW corresponds to a point source, the corresponding auxiliary source is located at the origin of the coordinate system.

8.2 Distributed Spherical Waves

In [125] Doicu et al. considered the system of DSW functions, which form a set of radiating solutions to the Helmholtz equation (4.6). These DSW functions are given by

$$u_{m|m|}\left(\vec{x} - z_n\vec{e}_z\right) = h_{|m|}^{(1)}(kr_n)P_{|m|}^{|m|}(\cos\theta)e^{jm\phi}, \tag{8.6}$$

where the discrete sources are distributed along a segment of the z-axis at a radius with respect to the origin, which are expressible in the spherical coordinates as (r_n, θ_n, ϕ_n), $r_n = \sqrt{x^2 + y^2 + (z - z_n)^2}$, and $n = 1, 2, \ldots, \infty$, and $m \in \Xi$, where Ξ is the support of the discrete sources that consist of the origin of the coordinate system.

The following theorem has been proven by Doicu et al. [125]

Theorem 8.2 *Consider the bounded sequence* $(z_n) \subset \Xi$, *where* Ξ *is a segment of the z-axis. Assume that S is a surface of class* C^2 *enclosing* Ξ. *Replace in Theorem 8.1 the LSW functions* $u_{mn}\left(\vec{x};\omega\right)$ *by* $u_{m|m|}\left(\vec{x} - z_n\vec{e}_z\right) = h_{|m|}^{(1)}(kr_n)P_{|m|}^{|m|}(\cos\theta)e^{jm\phi}$, $n = 1, 2, \ldots, \infty$, *and* $m \in \Xi$. *Then the resulting systems of functions are complete in* $L^2(S)$.

These theorems state that Eq. (8.6) may be utilized to describe the radiated acoustic field completely. Once again, we adopt the concept of DSW and utilize instead a finite expansion,

$$\hat{p}\left(\vec{x};\omega\right) = \sum_{n=1}^{n_{\max}} \sum_{m=-m_{\max}}^{m_{\max}} b_{mn} h_{|m|}^{(1)}(kr_n)P_{|m|}^{|m|}(\cos\theta_n)e^{im\phi}, \tag{8.7}$$

where b_{mn} are the expansion coefficients, $n_{\max}$ is the number of auxiliary sources and $m_{\max}$ is the order of expansion, and the total number of expansion terms is $J = n_{\max}(2m_{\max} + 1)$. For simplicity, we consider the case where the auxiliary sources are distributed along a segment of the z-axis with its center at the origin of the coordinate system,

$$z_n = z_0 \cos\beta_n, \tag{8.8}$$

where $n = 1, 2, \ldots, n_{\max}$, z_0 is chosen such that all auxiliary sources are inside Ω_s (some of them can be close to the boundary surface S), and β_n are given by

$$\beta_n = \frac{\pi}{2n_{\max}} + \frac{\pi(n-1)}{n_{\max}}. \tag{8.9}$$

The z_nth auxiliary source is at $\vec{x} - z_n\vec{e}_z$, which is expressible in the cylindrical coordinates as

$$r_n = \sqrt{r^2\sin^2\theta + (r^2\cos^2\theta - z_n)^2}, \tag{8.10}$$

$$\sin\theta_n = \frac{r\sin\theta}{r_n}, \tag{8.11}$$

$$r = \sqrt{x^2 + y^2}. \tag{8.12}$$

8.3 Distributed Point Sources

DPS is known as fundamental solutions to the Helmholtz equation, which are given by

$$\varphi_n\left(\vec{x}\,\middle|\,\vec{x}_n;\omega\right) = G\left(\vec{x}\,\middle|\,\vec{x}_n;\omega\right), \tag{8.13}$$

where $n = 1, 2, \ldots, \infty$, and $\left\{\vec{x}_n\right\}^{\mathrm{T}}_{1\times\infty}$ is a set of discrete point sources distributed on a closed surface S of class C^2. Suppose that $\left\{\varphi_n^-\left(\vec{x}\,\middle|\,\vec{x}_n;\omega\right)\right\}^{\mathrm{T}}_{1\times\infty}$ denote the fundamental solutions with point sources $\left\{\vec{x}_n^-\right\}^{\mathrm{T}}_{1\times\infty}$ distributed on the interior surface S^-, and $\left\{\varphi_n^+\left(\vec{x}\,\middle|\,\vec{x}_n;\omega\right)\right\}^{\mathrm{T}}_{1\times\infty}$ indicate those with the sources $\left\{\vec{x}_n^+\right\}^{\mathrm{T}}_{1\times\infty}$ distributed on the exterior surface S^+. The completeness of DPS as given by Eq. (8.13) is provided by the following theorem, which has been proven by Doicu et al. [125]

Theorem 8.3 *Consider Ω_i a bounded domain of class C^2. Let the set $\left\{\vec{x}_n^-\right\}^{\mathrm{T}}_{1\times\infty}$ be dense on a surface S^- enclosed in Ω_i and let set $\left\{\vec{x}_n^+\right\}^{\mathrm{T}}_{1\times\infty}$ be dense on a surface S^+ enclosing Ω_i. Assume that k is not an eigenvalue of the boundary value problem for the interior region Ω_i. Replace in Theorem 8.1 the radiating spherical wave functions $u_{mn}\left(\vec{x};\omega\right) = h_n^{(1)}(kr)P_n^{|m|}(\cos\theta)e^{jm\phi}$, $n = 0, 1, 2, \ldots, \infty$, and $m = -n$ to n by the functions $\left\{\varphi_n^-\left(\vec{x}\,\middle|\,\vec{x}_n;\omega\right)\right\}^{\mathrm{T}}_{1\times\infty}$ $n = 0, 1, \ldots, \infty$, and the regular spherical wave functions $u_{mn}\left(\vec{x};\omega\right) = j_n^{(1)}(kr)P_n^{|m|}(\cos\theta)e^{jm\phi}$, where $j_n^{(1)}(kr)$ is the spherical Bessel functions of the first kind, $n = 0, 1, 2, \ldots, \infty$, and $m = -n$ to n by $\left\{\varphi_n^+\left(\vec{x}\,\middle|\,\vec{x}_n;\omega\right\}^{\mathrm{T}}_{1\times\infty}$ $n = 0, 1, \ldots, \infty$. Then, the resulting systems of functions are complete in $L^2(S)$.*

We adopt the DPS concept in reconstruction, but instead use a finite expansion,

$$\hat{p}\left(\vec{x};\omega\right) = \sum_{n=1}^{n_{\max}} c_n h_0^{(1)}\left(k\left|\vec{x} - \vec{x}_n\right|\right), \tag{8.14}$$

where c_n are the expansion coefficients, $n_{\max}$ is the number of point sources, and

$$h_0^{(1)}\left(k\left|\vec{x}-\vec{x}_j\right|\right)=\frac{e^{ik\left|\vec{x}-\vec{x}_j\right|}}{\left|\vec{x}-\vec{x}_j\right|}, \tag{8.15}$$

is the fundamental solution (or the free-space Green's function) to the Helmholtz equation. Hence the auxiliary sources in DPS are a set of point sources distributed on a smooth surface S^- inside Ω, but close to the boundary surface S.

Note that DPS formulation (8.14) is the same as those of the so-called equivalent source methods [108–119], and LSW is the same as HELS. Here LSW, DSW, and DPS will be adopted in the HELS expansion to reconstruct the acoustic field generated by a vibrating object in free space. Their results will be validated against the benchmark values and their performances be examined.

8.4 Regularization for LSW, DSW, and DPS Expansions

It is well known that the rate of convergence of any expansion depends on the complexity of the source boundary and frequency [124, 128–130]. Although reconstruction of acoustic quantities may be done by HELS at any frequencies, the accuracy in reconstruction may deteriorate with an increase in the frequency. This is because at high frequencies, SNR is usually very low such that the high-order terms in the HELS expansion may be contaminated by background noise. To avoid distortions in reconstruction due to noise contamination, we must truncate the HELS expansion by eliminating the high-order terms. However, the high-order terms are critical in depicting the details of acoustic quantities at high frequencies and an omission of these terms will make it impossible to obtain the details in reconstruction. It is emphasized that this high-frequency difficulty exists in other methods, for example, BEM, whose performance deteriorates greatly at high frequencies.

Despite the fact that discrete sources methods have been extensively studied in the forward problems such as scattering and prediction of acoustic fields, they have not been tested in backward problems such as reconstruction of acoustic fields, with the exception of HELS. Since the matrix involved in HELS is relatively small, it is possible to utilize a direct regularization method such as TR. For a problem that involves a large matrix, for example, in three-dimensional simulations, an iterative regularization method may be a better alternative.

Success in regularization depends to a large extent on choice of regularization parameter. Based on the type of information available on a targeted solution, the parameter-choice methods (PCM) are classified as a priori, which is independent of the actual data, and a posteriori, which is dependent on the actual data [131]. The former includes heuristic or error-free [132] methods that do not require the knowledge of the noise level in the input data and seek to predict this information from actual data. Note that for an infinite-dimensional compact operator, error-free

PCM may fail to yield a convergent regularization parameter, namely, to provide a regularized solution that will converge to the exact solution as the noise level tends to zero [133].

In practice, there is always noise in the input data and its level is unknown a priori. Hence we must resort to the error-free PCM, even though it may occasionally fail to yield a convergent regularization parameter. Of course, if noise level can be estimated a priori, we can use Morozov's discrepancy principle [106, 134] to determine the correct regularization parameter and get satisfactory reconstruction. Alternatively, we can impose constraints on the norm of a regularized solution as suggested by Isakov and Wu [74] to find a convergent regularization parameter. The trouble is that the right constraint for the norm of the exact solution is hard to find.

Our objective is to examine the effects of different expansions on resultant reconstruction and, more importantly, to identify the expansion that can produce the most accurate and efficient reconstruction.

Specifically, we consider three expansions: LSW, DSW, and DPS in HELS to reconstruct the acoustic pressure radiated from an arbitrary source in free space. In particular, we use TR, MTR, and damped singular value decomposition (DSVD) in regularization scheme with its regularization parameter determined by an error-free PCM such as GCV, L-curve criterion, and quasi-optimality criterion (QOC) [135]. Reconstructed acoustic quantities are validated with respect to the benchmark data measured at the same locations as the reconstruction points.

The L-curve criterion [136] relies on a parametric plot of the norm of a regularized solution versus the residual norm in a log–log scale with respect to the regularization parameter. The corner of an L-curve (which is defined as the point of maximum curvature) separates the horizontal part (where regularization errors dominate) from the vertical part (where perturbation errors dominate), and represents a balance between the regularization and perturbation errors.

8.5 Performances of LSW, DSW, and DPS Expansions

Here we examine the performances of HELS through LSW (8.5), DSW (8.7), and DPS (8.14) expansions to reconstruct the acoustic pressures generated by a JBL® speaker that consists of a woofer, mid-ranger, and tweeter inside a fully anechoic chamber (see Fig. 8.1). In particular, we examine the convergence rates of these expansions and condition numbers of the corresponding transfer matrices. The faster the convergence rates and the smaller the condition numbers are, the more efficient the numerical computations and the more accurate the HELS solutions become.

In experiments the speaker was driven by an HP 8904A Multi-Function Synthesizer DC-600 kHz and a McIntosh MC352 Power Amplifier to produce white noise. The acoustic pressures were measured by an array of 56 PCB T130D21 free-field

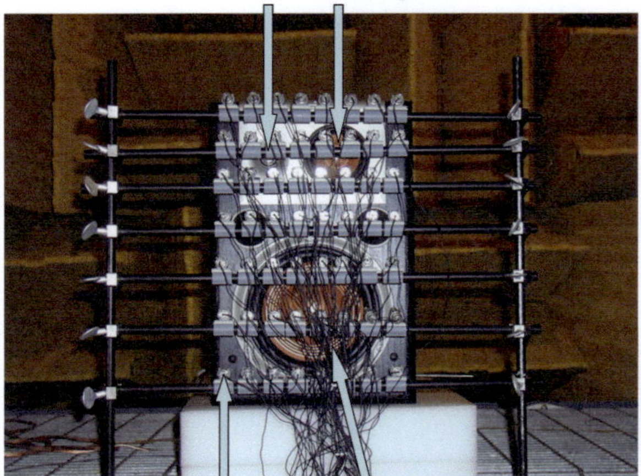

Fig. 8.1 Test setup for reconstructing the acoustic pressure emitted from a JBL® speaker that consists of a woofer, mid-ranger, and tweeter inside a fully anechoic chamber. Input data were collected by an array of microphones

microphones (see Fig. 8.1). The input data were sent to a personal computer through the Larson Davis digital Sensor System Model 100 for analog to digital conversion.

Test procedures were as follows. First, the radiated acoustic pressures were measured on a planar surface Γ at 1 cm clearance in front of the speaker. These data were taken as input to HELS using LSW, DSW, and DPS expansions, respectively, to reconstruct the acoustic pressures on a surface S at 0.5 cm clearance in front of the speaker up to 3,275 Hz (see Fig. 8.2). The reason for selecting this surface S was that there was no way of measuring the acoustic pressures on the speaker membrane directly.

Next, the acoustic pressures on this surface S were measured using the same microphone array, and these benchmark values were compared with the reconstructed acoustic pressures at the same locations. The measurement points on Γ and S were equidistant. The origin of the coordinate system was set at the geometric center of the speaker. In particular, the auxiliary source for LSW was placed at the geometric center of the speaker, those for DSW were distributed along a vertical axis between the front surface and center of the speaker box, and those for DPS were distributed on a plane next to the front surface of the speaker (see Fig. 8.2).

For simplicity without loss of generality, we consider reconstructing the acoustic pressure in front of the speaker. The characteristic dimension was $a = \sqrt{(0.28/2)^2 + (0.42/2)^2} = 0.28$ m. So for the highest frequency of 3,275 Hz, the maximum dimensionless frequency was $ka_{\max} \approx 16.6$.

Since we were only interested in reconstructing the acoustic pressure, it was acceptable to gauge the number of expansion terms with respect to the maximum dimensionless frequency $ka_{\max}$. From [37] we learn that the total number of expansion terms for LSW is $J = (N+1)^2$; here, N is the order of expansion. In general, we may set $N = ka_{\max} \approx 17$. So $J = 324$. Accordingly, we need at least

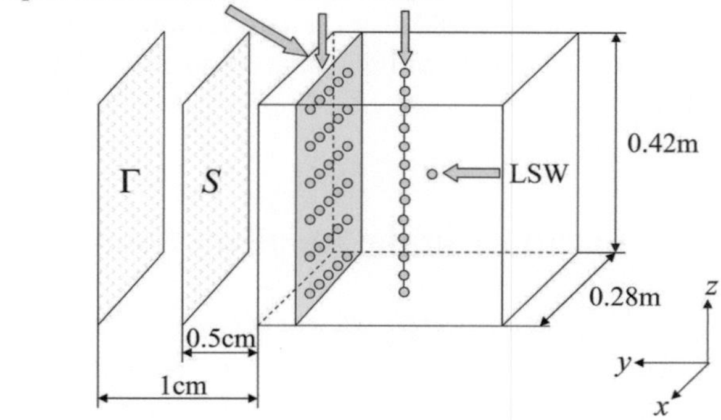

Fig. 8.2 Schematic of the locations of the auxiliary sources for LSW, DSW, and DPS expansions inside the JBL® speaker. The measurement surfaces Γ and *S* were in front of the speaker at, respectively, 1 and 0.5 cm away

324 measurement points on Γ to cover the specified frequency range. In practice, we may have to truncate the expansion to reduce distortion due to a low SNR at high frequencies. In this experiment, we set $N = 9$, so $J = 100$.

The number of expansion terms in DSW is $J = n_{max}(2m_{max} + 1)$, where n_{max} is the number of auxiliary sources and m_{max} is the order of expansion, which is smaller than N in LSW. There are no known theories or methodologies that we can use to estimate the optimum values of n_{max} and m_{max} for arbitrarily shaped sources. In general, the values of n_{max} and m_{max} depend on the complexities of source geometry and the highest frequency of interest. To achieve the best results, it is a good idea to set distances among neighboring auxiliary sources to be less than one wavelength of the highest frequency of interest and distribute the auxiliary sources evenly on a conformal surface inside the source boundary.

For example, the front surface of the speaker is of dimensions $0.28 \times 0.42\ \text{m}^2$, the highest frequency is $f_{max} = 3{,}275$ Hz, and the acoustic wavelength is $\lambda_{min} = c/f_{max} \approx 0.104$ m. Therefore, the estimated number of auxiliary sources for DSW is $n_{max} = (0.28/0.104) \times (0.42/0.104) \approx 11$. Since the speaker in free space is often modeled as a dipole, we set the highest order of expansion for DSW at $m_{max} = 4$. Accordingly, the number of expansion terms is $J = 99$. Therefore we need to take 100 measurement points of the acoustic pressures on Γ to guarantee satisfactory reconstruction of the acoustic pressures in the specified frequency range.

In DPS, the auxiliary sources are distributed uniformly on a surface conformal to a source boundary from the inside. However, the optimal number and locations of the auxiliary sources are unknown a priori, whose determination is a topic of research by itself and will not be considered here. Since the front surface of the speaker is planar, it is sufficient to distribute the auxiliary sources on a plane with $J = n_x \times n_z$, where n_x and n_z are, respectively, the numbers of sources in the x- and z-axis directions. Here we set $n_x = 10$ and $n_z = 10$, so $J = 100$.

In this experiment 112 measurement points of the acoustic pressures were taken on Γ and S, respectively, which were enough for LSW, DSW, and DPS expansions.

It is emphasized that the number of measurement points is not as critical as it seems. The controlling factor is SNR. If SNR is low, there is no way of obtaining good reconstruction because the critical near-field information will be buried in background noise. Accordingly, the number of expansion terms must be significantly reduced to avoid distortion in reconstructed images. Under this condition, the reconstructed results will not be good no matter how many measurement points are taken.

8.6 Locations of the Auxiliary Sources

Selection of the auxiliary source locations can be crucial to the success of reconstruction. Although there is no known theory that can depict exactly the interrelationship between locations of auxiliary sources and rate of convergence of resultant expansion and reconstruction accuracy, the following guideline is clear: the analytic continuation of solution to the Helmholtz equation requires that the surface on which auxiliary sources are distributed must enclose all singularities of an acoustic field.

However the singularities for a given acoustic field are unknown a priori. To gain a good understanding of the singularities locations, we start from an arbitrarily selected auxiliary surface, and measure the acoustic pressures on surfaces Γ and S, respectively. Next, we substitute the data measured on Γ to reconstruct the acoustic pressures on S and calculate the mean relative errors in reconstruction with respect to the benchmark data on S. Finally, we move the auxiliary surface to a different location, and repeat these processes again. Note that there is no need to remeasure the acoustic pressures on S. This iteration is continued until the mean relative errors in reconstruction are minimized, and the corresponding locations of the auxiliary sources are optimized for a given frequency and a set of measurements.

It is emphasized that in practice we only measure the acoustic pressures on the surface Γ. The reason for taking an additional set of measurements on S is to develop a guideline for selecting the optimal location of an auxiliary surface for a specific expansion function. Needless to say, the impact of the auxiliary source locations for different expansions is different. Hence, by taking an independent set of measurements on S, we can validate the reconstructed acoustic pressures, find the optimal location of an auxiliary surface for a given expansion, and study the sensitivity of the auxiliary surface on the reconstruction accuracy using this expansion.

Figure 8.3 depicts the mean relative errors in reconstructing acoustic pressures on S using LSW, DSW, and DPS, respectively, with respect to the auxiliary sources distributed on a plane at $y = -y_0$, where y_0 varies from -0.7 to 0.13 m, for a fixed frequency of 1,690 Hz. Results show that DSW is relatively insensitive to the auxiliary source locations as compared to LSW and DPS are. At the optimum auxiliary surfaces, LSW and DSW can yield nearly the same level of accuracy in reconstruction, whereas DPS produces a slightly lower accuracy in reconstruction.

It is interesting to observe that LSW places its optimum location of the auxiliary surface near the origin of the coordinate system, DSW moves its optimum auxiliary surface slightly away from the origin of the coordinate system toward the front

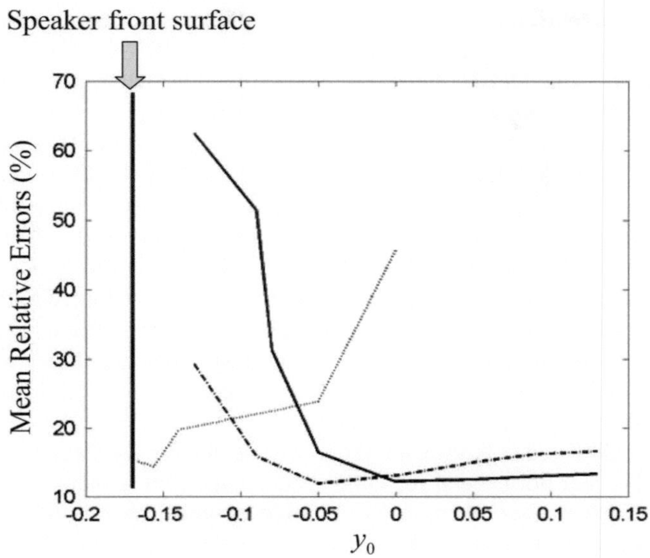

Fig. 8.3 Comparison of the auxiliary surface locations at 1,690 Hz. *Continuous line*: for LSW; *broken line*: DSW; and *dotted line*: DPS

surface of the speaker, and DPS places its optimum auxiliary surface right behind the front surface of the speaker system.

Note that the standoff distance and frequency can also affect the reconstruction accuracy. However, their effects are negligible compared to those of the auxiliary source surface location. So we focus on the determination of optimal locations of auxiliary source surfaces by minimizing the mean relative errors in reconstruction with respect to the benchmark values measured on *S*.

8.7 Condition Number of the Transfer Matrices

Reconstruction of acoustic radiation from any source is an ill-posed problem. As a result, the transfer matrix in HELS may be ill conditioned. Ill conditioning of any matrix is measured by the 2-norm condition number defined as the ratio of the largest to smallest singular values of the matrix. This can be done prior to taking any measurements, if the frequency and measurement and reconstruction locations are specified. In many situations, if the condition number is in the order of $O(10^3)$ or higher, the matrix may be ill conditioned; if the condition number is in the order of $O(10^2)$ or lower, the matrix is more or less well conditioned. Regularization may be omitted if the transfer matrix is well conditioned, but must be implemented if the matrix is ill conditioned.

In our experiments, the condition numbers of the transfer matrices for a selected standoff distance and frequency were found to be: $O(10^1)$ for DSW, $O(10^2) \sim O(10^3)$ for LSW, and $O(10^5)$ for DPS. Therefore, for the same frequency and measurement and reconstruction locations, DSW offers the best-conditioned transfer matrix among all three expansions. This is expected because DSW uses a lower-order spherical Hankel function ($m_{\max} = 4$) than LSW does ($N = 9$). The small singular values are always associated with high-order expansion terms, and the condition number of a transfer matrix containing the high-order terms is much larger than that of a transfer matrix containing the low-order terms. The reason for DPS to produce an ill-conditioned transfer matrix may be attributed to the fact that the sound field produced by the present speaker system cannot be adequately described by a distribution of point sources because there are three speakers that emit sounds simultaneously. Consequently, the resultant transfer matrix in DPS becomes rank deficient and ill conditioned.

From the calculated condition numbers, we see that regularization is needed for DPS and LSW, but not needed for DSW. Since condition numbers are calculated before measurements are taken, no information on noise level in the input data is available. Consequently, we have to resort to an error-free PCM in regularization.

One of our objectives is to examine the performances of HELS using different expansions to reconstruct acoustic radiation from an arbitrary source. To this end, we first calculate the ideal regularization parameter by minimizing reconstruction errors with respect to the benchmark data on S. This process allows for assessing not only the reconstruction accuracy but also the impact of reducing the measurement number on reconstruction using various expansions. This latter is of great importance since in practice fewer measurement points mean bigger savings in time and costs.

8.8 Effect of Measurement Number

The effect of the number of measurement points on reconstruction accuracy is examined. Figures 8.4, 8.5, and 8.6 describe the mean relative errors in reconstructing the acoustic pressure through LSW, DSW, and DPS expansions, respectively, under different numbers of measurement points. Since the frequency range is relatively low, SNR is relatively high. So the more the measurements are taken, the higher the accuracy in regularized reconstruction becomes.

Note that regularization can significantly enhance the reconstruction accuracy, especially at higher frequencies. This is obvious in Figs. 8.4 and 8.6 since the transfer matrices for LSW and DPS are ill conditioned. However, the impact of regularization on reconstruction accuracy for DSW is not as drastic (see Fig. 8.5) because its transfer matrix is more or less well conditioned. Note that we have used ideal regularization parameters for all three expansions to maximize these effects. The calculated regularization parameters for LSW and DPS increased monotonically with frequency from 0.002 to 0.05, whereas that remained negligibly small at

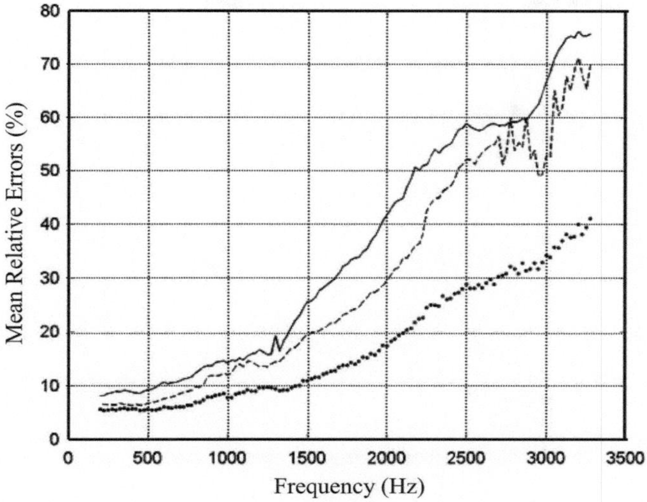

Fig. 8.4 Comparison of the mean relative errors in reconstructing the acoustic pressure by using LSW expansion. *Continuous line*: 56 measurement points with an ideal regularization parameter; *broken line*: 112 measurement points without using regularization; *dotted line*: 112 measurement points with an ideal regularization parameter

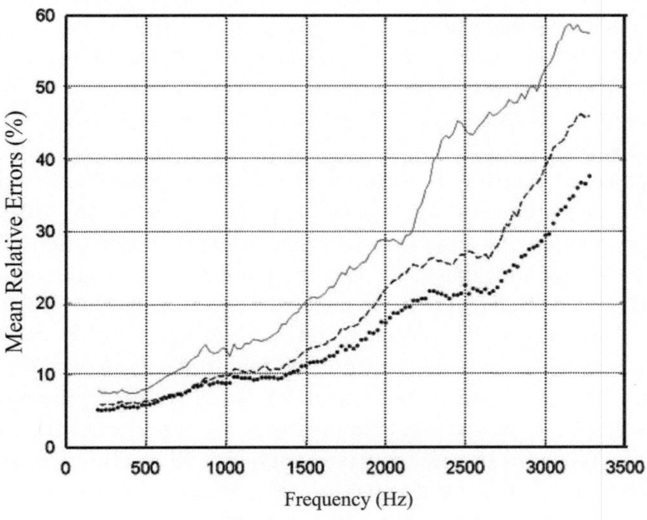

Fig. 8.5 Comparison of the mean relative errors in reconstructing the acoustic pressure by using DSW expansion. *Continuous line*: 56 measurement points with an ideal regularization parameter; *broken line*: 112 measurement points without using regularization; *dotted line*: 112 measurement points with an ideal regularization parameter

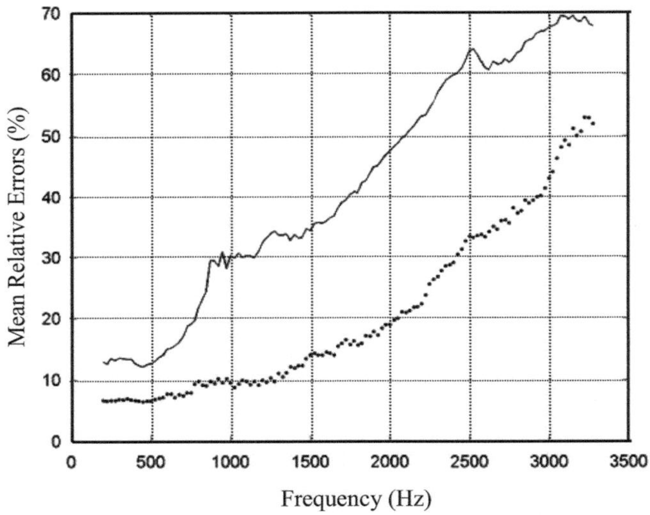

Fig. 8.6 Comparison of the mean relative errors in reconstructing the acoustic pressure by using DPS expansion. *Continuous line*: 56 measurement points with an ideal regularization parameter; *dotted line*: 112 measurement points with an ideal regularization parameter

0.0008 for DSW, which meant that there was almost no need for regularization in DSW within this frequency range.

Results demonstrate that even with an ideal regularization parameter, DPS failed to yield a compatible level of accuracy as compared to those of DSW and LSW under the same set of input data. Moreover, DPS is more sensitive to PCM than DSW and LSW are.

It is emphasized that for ill-conditioned transfer matrices, just increasing the numbers of measurement points and expansion terms in HELS without implementing regularization will only further distort reconstruction. This is seen in Fig. 8.6 using the DPS expansion. As the number of measurement points was doubled, the mean relative errors in reconstruction exceeded 200 % when no regularization was used (the corresponding curve was omitted in Fig. 8.6 for clarity). The reason for that was because the high-order terms in DPS expansion were contaminated by noise embedded in measured data, and these errors were significantly amplified as the acoustic pressures were projected back toward the source surface. When the transfer matrix is not highly ill conditioned, as in the case of LSW expansion, increasing the measurement number can improve reconstruction accuracy to certain frequency without regularization (see Fig. 8.4). If the transfer matrix is more or less well conditioned, as in the case of DSW expansion, increasing the number of measurement points allows for an increase in the number of expansion terms, which enhances the reconstruction accuracy even without regularization (see Fig. 8.5).

8.9 Choice of Regularization

In many engineering applications, the noise level embedded in the input data is unknown a priori. Thus we must rely on an error-free PCM in regularization. In this study, we want to find out if there exists an optimal regularization with an error-free PCM for each of DSW, LSW, and DPS expansions in HELS. To this end, we examine performances of all possible combinations of TR and its modification implemented by utilizing GCV and DSVD, together with various penalty functions with respect to pressure, normal velocity, or both, and error-free PCM such as GCV, L-curve, and QOC to select the best regularization parameter. Results show that for some expansions, it is possible to find the optimal regularization with an error-free PCM that can produce an almost ideal regularization parameter over a wide frequency range; but for other expansion such optimal combinations cannot be found. For brevity, we summarize the most important results here:

1. The optimal regularization for DSW is MTR implemented through DSVD, and the best regularization parameter can be provided by L-curve together with an energy norm as its penalty function.
2. The optimal regularization for LSW is TR implemented by GSVD with its regularization parameter determined by GCV using an energy norm as its penalty function. Depicted in Fig. 8.7 is the comparison of the mean relative errors in reconstruction using LSW and TR with its regularization parameter determined by different error-free PCMs. It is clear that the regularization parameter given by GCV is almost identical to the ideal value over the specified frequency range, that provided by L-curve is close to an ideal one, but those produced by QOC are way off the target. Figure 8.8 shows the regularization parameters given by GCV, L-curve, and QOC for TR in LSW versus the frequency. Results illustrate that GCV yields nearly the ideal regularization parameters, L-curve gives a regularization parameter close to the ideal one, but the regularization parameter provided by QOC is off by at least two orders of magnitude of an ideal value.
3. For DPS, it is not possible to find one regularization scheme that can produce satisfactory reconstruction over a wide frequency range. In fact, we must utilize different combinations of regularization, penalty function, and error-free PCM to select an optimal regularization parameter for different frequency.

Figure 8.9 summarizes the results of this investigation on determining optimal choices of regularization schemes for DSW, LSW, and DPS expansions in HELS. Comparing Fig. 8.9 with Figs. 8.4 and 8.5 demonstrates that using the optimal regularization schemes, for example, TR implemented by DSVD with its regularization parameter specified by L-curve for DSW, and TR with its regularization parameter determined by GCV for LSW, we can obtain the same level of reconstruction accuracy as that produced by an ideal regularization. When we only rely on a single regularization scheme, for example, TR and GCV for DPS, the mean relative reconstruction errors can be very large, especially at higher frequencies. This can be seen by comparing the mean relative errors in Fig. 8.9 with those in Fig. 8.5.

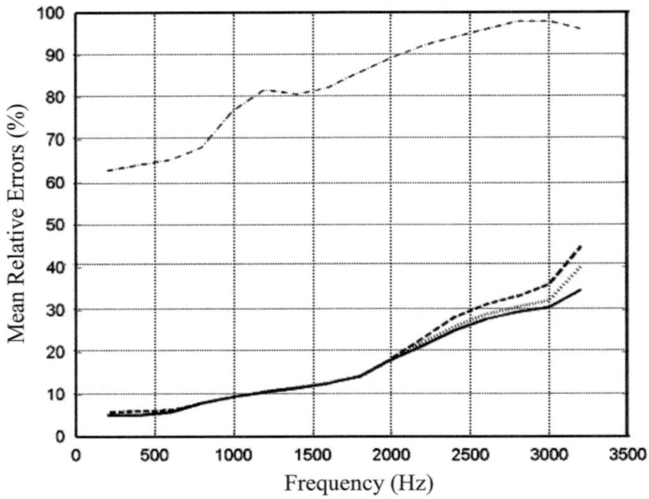

Fig. 8.7 Comparison of the mean relative errors in reconstructing the acoustic pressure by using LSW expansion in HELS with TR and various error-free PCM. *Broken line with dots*: QOC; *broken line*: L-curve; *dotted line*: GCV; *continuous line*: Ideal case

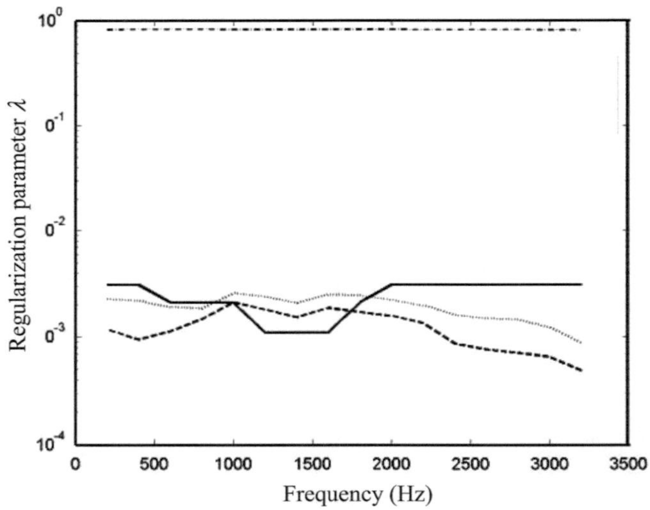

Fig. 8.8 Comparison of regularization parameters calculated by using various error-free PCMs for reconstructing the acoustic pressure using LSW in HELS. *Broken line with dots*: QOC; *broken line*: L-curve; *dotted line*: GCV; *continuous line*: Ideal case

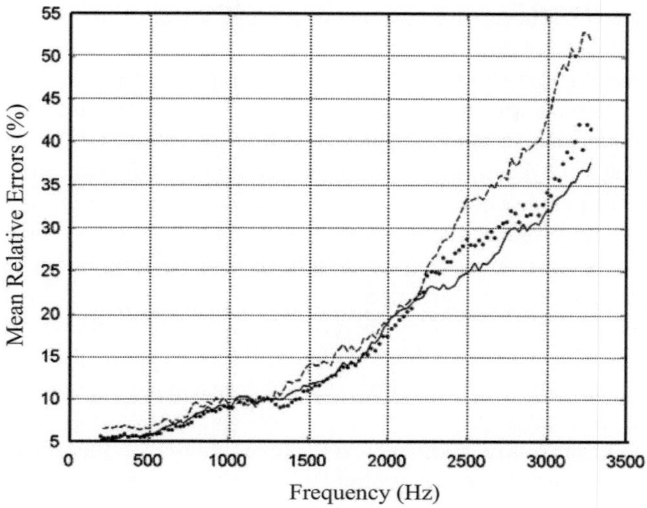

Fig. 8.9 Comparison of the mean relative errors in reconstructing the acoustic pressure using different expansions with regularization strategies. *Continuous line*: DSW using DSVD and L-curve; *dotted line*: LSW using TR and GCV; *broken line*: DPS using TR and GCV

It must be emphasized that there is no single regularization strategy that can yield the best reconstruction for all sources under all circumstances. For example, although TR and DSVD plus L-curve work well for DSW in the present case, it may not work well in a different scenario or in a different frequency range. The best regularization strategy is always case dependent. Also, we must keep in mind that an error-free PCM can fail to yield a convergent regularization parameter at all. On the other hand, it is always advantageous to take double-layer measurements whenever possible. These double-layer measurements can help us to device optimal regularization schemes and produce the best reconstruction of the acoustic quantities.

To summarize, results show that DSW leads to the best-conditioned transfer matrix, is the least sensitive to choices of auxiliary surface locations, and yields most satisfactory reconstruction over a wide frequency range. LSW is the second best choice of expansion for HELS, its transfer matrix can be weakly ill conditioned, and its optimal auxiliary source location can be close to the geometric center of a source. If the optimal location for the auxiliary source is selected, LSW can yield nearly the same level of accuracy in reconstruction as DSW does. Moreover, it is possible to improve the reconstruction accuracy in LSW by increasing the measurement points taken at very close range to the source surface, even without regularization. DPS gives a highly ill-conditioned transfer matrix and is very sensitive to the auxiliary surface location. The reason for that may be due to the fact that the acoustic pressure radiated from an arbitrary source may not be adequately described by a distribution of point sources. When DPS expansion is used in HELS, it is necessary to implement regularization in order to obtain a convergent reconstruction.

The optimal regularization for DSW is MTR implemented through DSVD plus L-curve to determine the regularization parameter with an energy norm as its penalty function. The optimal regularization for LSW is TR implemented by DSVD plus GCV to determine its regularization parameter with an energy norm as its penalty function. There is no single optimal regularization scheme that can produce satisfactory reconstruction over a wide frequency range for DPS. In this case, different regularization schemes must be used at different frequencies in order to produce satisfactory reconstruction.

It is emphasized that it may not be possible to find the optimal regularization schemes for DSW and LSW that will work for all scenarios. In other words, there is no single regularization strategy that can guarantee good reconstruction for all sources under all conditions.

Finally, it is always a good idea to take double-layer measurements. Such measurements can help us to determine the optimal auxiliary surface location for particular expansion functions in HELS and to select the optimal regularization scheme that includes choosing penalty functions and error-free PCMs to yield the best regularization parameter.

Problems

8.1. What is LSW? What does it attempt to do? What are the differences between LSW and the Helmholtz integral formulation?

8.2. What are the differences between LSW and the HELS method? Will LSW suffer from the same difficulties as those of the Rayleigh hypotheses for a corrugated surface?

8.3. What is DSW? What does it attempt to do? What are the differences between DSW and the original HELS method?

8.4. What are the differences between DSW and the HELS method using the DSW expansion?

8.5. Will DSW have the same difficulties as those of the Rayleigh hypotheses for a corrugated surface? Will the HELS method using DSW expansion have the same difficulties as DSW for a corrugated surface?

8.6. What is DPS? What does it attempt to do? What is the difference between DPS and DSW?

8.7. What are the differences between DPS and the HELS method using the DPS expansion?

8.8. What are the optimal regularization schemes for the HELS method using LSW, DSW, and DPS expansions?

8.9. What are the performances of the HELS method using LSW, DSW, and DPS expansions?

8.10. What are the impacts of various parameters such as the locations of the auxiliary sources, number of measurement points, and choices of regularization schemes on the performances of the HELS method using LSW, DSW, and DPS expansions?

Chapter 9
Transient HELS

Most vibrating structures are subject to impulsive or transient force excitations in practice. Oftentimes transient excitations are unknown and therefore the resultant acoustic field cannot be predicted. Even if the excitations are given, prediction of a transient acoustic field produced by an arbitrarily shaped source is very difficult. The scarcity in literature on predicting, not to mention reconstructing a transient acoustic field, is the testimony of how challenging this problem is.

One possibility of determining the transient acoustic field generated by an arbitrary object is to reconstruct the acoustic quantities in the frequency domain first, and take an inverse Fourier transform to retrieve the time-domain signals. Wang is the first to reconstruct a transient acoustic field in this manner [137]. Needless to say, numerical computations involved in this process are very intensive, if possible at all.

Another possibility is to utilize the so-called non-stationary spatial transformation of sound field (NS-STSF) [138]. NS-STSF is based on the time-domain holography (TDH) that processes the acoustic pressures measured by a planar array of microphones with the neighboring microphones separated by one-half the wavelength of a target acoustic wave. Basically, TDH produces "a sequence of snapshots of instantaneous pressure over the array area, the time separation between subsequent snapshots being equal to the sampling interval in A/D conversion. Similarly, the output of TDH is a time sequence of snapshots of a selected acoustic quantity in a calculation plane parallel to the measurement plane" [138]. Therefore, what one sees is a series of the acoustic pressure images in the frequency domain at fixed time instances over the recorded measurement time period. As pointed out in Sect. 5.3, NS-STSF is actually non-stationary acoustical holography because it gauges with respect to the acoustic frequency or the acoustic wavelength, not the spatial frequency or the spatial wavelength.

In this chapter we develop the transient NAH formulations by using the HELS method to visualize acoustic waves traveling in both space and time. Note that Hansen [101] has used a spherical wave expansion to predict time-domain acoustic radiation by scanning the acoustic pressure over a minimal spherical surface

S.F. Wu, *The Helmholtz Equation Least Squares Method*, Modern Acoustics and Signal Processing, DOI 10.1007/978-1-4939-1640-5_9

enclosing target sources. The major difference between Hansen's work and the present one is that the former is based on infinite series of the spherical Hankel functions and spherical harmonics, and expansion coefficients are determined using the orthogonal property of the spherical harmonics; while the latter utilizes a finite expansion and expansion coefficients are determined by matching the expansion solution to the measured data and the errors involved in this process are minimized through regularization. This infinite series is called Rayleigh series and Sect. 4.2 has discussed in detail the differences between the Rayleigh series and the HELS formulations. We have learned that the Rayleigh series is in general invalid for reconstructing the acoustic field on a corrugated or arbitrarily shaped surface based on the acoustic pressure specified on a measurement surface above the source surface.

Theoretically, the transient acoustic field generated by an arbitrary object can be calculated by using the Kirchhoff–Helmholtz integral formulation, provided that the normal component of the surface velocity is specified. For an arbitrarily shaped object, there is no analytic solution to this integral formulation. Hence numerical solutions are sought. A direct approach is to discretize the Kirchhoff–Helmholtz integral formulation in both spatial and temporal domains simultaneously. Such an approach is unrealistic in practice because the corresponding numerical computations are prohibitively expensive and time consuming. One alternative is to find numerical solutions to the radiated acoustic quantities in the frequency domain first, and take an inverse Fourier transform to obtain the time-domain signals [137]. Needless to say, numerical computations involved are intensive, if possible at all. The reality is that in most cases the normal surface velocity is not specified. Thus these numerical solutions strategies, no matter how plausible they are, cannot be utilized.

In Chap. 9 explicit formulations for reconstructing the transient acoustic field generated by an arbitrarily shaped 3D object in free space subject to an arbitrarily time-dependent excitation are derived using the Kirchhoff–Helmholtz integral theory. The reconstructed acoustic quantities are expressed in the frequency domain, and the corresponding time-domain quantities are obtained by taking an inverse Fourier transform, which is facilitated by using the residue theorem. The final formulation for reconstructing a transient acoustic quantity is expressed in a convolution integral of the acoustic pressure signal measured in the time-domain and a unit impulse response function.

It is emphasized that these explicit formulations are applicable to an arbitrary object with a uniformly distributed surface velocity. Input data to these explicit formulations are the acoustic pressure signals measured on a hologram surface in the near field of the target object.

For simplicity yet without loss of generality, background noise and interfering signals are assumed negligible as compared to the measured acoustic pressure signals. Reconstruction of the transient acoustic field is carried out by using BEM- [25–27, 91] and HELS [36, 37, 91, 102]-based NAH.

9.1 Transient Acoustic Radiation

To tackle transient acoustic radiation problems, let us first define the Fourier transform as

$$F\left(\vec{x};\omega\right) = \int_{-\infty}^{\infty} f\left(\vec{x};t\right)\mathrm{e}^{i\omega t}dt \text{ and } f\left(\vec{x};t\right) = \frac{1}{2\pi}\int_{-\infty}^{\infty} F\left(\vec{x};\omega\right)\mathrm{e}^{-i\omega t}d\omega, \tag{9.1}$$

where $f\left(\vec{x};t\right)$ is a continuous and bounded function as $t \to \infty$, namely, $\int_{-\infty}^{\infty} \left|f\left(\vec{x};t\right)\right|dt < \infty$.

Assume that the transient acoustic field is generated by an arbitrary object subject to an arbitrarily time-dependent force excitation. Also, assume that the velocity is uniformly distributed on the surface of the object, which has a closed, smooth, and impermeable surface immersed in an inviscid, isotropic, and unbounded fluid medium. This object is initially stationary and excited by an unknown forcing function at $t = t_s$, causing the amplitude of the velocity to rise from 0 to V_s instantly in a specific direction $\vec{e}_c$, where $\vec{e}_c$ is a unit vector at the center of the object,

$$\vec{v}\left(\vec{x}_s;t\right) = V_s\vec{e}_c H(t - t_s), \tag{9.2}$$

where V_s is a constant and $H(t - t_s)$ represents the Heaviside step function defined as

$$H(t - t_s) = \begin{cases} 0, & t < t_s \\ 1/2, & t = t_s \,. \\ 1, & t > t_s \end{cases} \tag{9.3}$$

The derivative of the Heaviside step function is the Dirac delta function [139],

$$H'(t - t_s) = \delta(t - t_s). \tag{9.4}$$

The acoustic pressure $p\left(\vec{x};t\right)$ generated by this accelerated body in free space satisfies the homogeneous wave equation,

$$\nabla^2 p\left(\vec{x};t\right) - \frac{1}{c^2}\frac{\partial^2 p\left(\vec{x};t\right)}{\partial t^2} = 0, \tag{9.5}$$

subject to the Sommerfeld radiation condition,

$$\lim_{\left|\vec{x}\right|\to\infty}\left|\vec{x}\right|\left(\frac{\partial p}{\partial\left|\vec{x}\right|}+\frac{1}{c}\frac{\partial\hat{p}}{\partial t}\right)=0,\quad\text{as}\quad\left|\vec{x}\right|\to\infty. \tag{9.6}$$

In addition, $p\left(\vec{x};t\right)$ satisfies the causality condition,

$$p\left(\vec{x};t\right)\equiv 0,\quad\text{for}\ \ t<t_s. \tag{9.7}$$

In other words, the field is perfectly silent before the body is suddenly excited at $t=t_s$.

To find an integral representation of the wave equation (9.5), we make use of the temporal free-space Green's function

$$g\left(\vec{x};t\middle|\vec{x}_s;t_s\right)=\frac{\delta(t-t_s-R/c)}{R}, \tag{9.8}$$

where $\delta(t-t_s-R/c)$ is the Dirac delta function, $(t-R/c)$ is known as the retarded time because it takes additional time R/c for the acoustic signal to travel from the source at $\vec{x}_s$ to a receiver at $\vec{x}$, here $R=\left|\vec{x}-\vec{x}_s\right|$ is the distance between the source and receiver in field space.

The temporal free-space Green's function satisfies the homogeneous wave equation,

$$\nabla^2 g\left(\vec{x};t\middle|\vec{x}_s;t_s\right)-\frac{1}{c^2}\frac{\partial^2 g\left(\vec{x};t\middle|\vec{x}_s;t_s\right)}{\partial t^2}=-4\pi\delta\left(\vec{x}-\vec{x}_s\right)\delta(t-t_s), \tag{9.9}$$

subject to the initial condition,

$$g\left(\vec{x};t\middle|\vec{x}_s;t_s\right)=\frac{\partial g\left(\vec{x};t\middle|\vec{x}_s;t_s\right)}{\partial t}\equiv 0,\quad\text{for}\ \ t<t_s, \tag{9.10}$$

and the reciprocal relation,

$$g\left(\vec{x};t\middle|\vec{x}_s;t_s\right)=g\left(\vec{x}_s;-t_s\middle|\vec{x};-t\right). \tag{9.11}$$

Physically, Eq. (9.10) states that if the source is excited at t_s, no sound is detected before time t_s. Equation (9.11) is the reciprocity principle, which states that when the source location and emission time are interchanged with the receiver location and time, the effect remains unchanged.

Multiply Eq. (9.5) by $g\left(\vec{x};t\middle|\vec{x}_s;t_s\right)$ and Eq. (9.9) by $p\left(\vec{x};t\right)$ and utilize the chain rule to replace $\nabla\cdot(A\nabla B)$ by $A\nabla^2B-\nabla A\cdot\nabla B$. Doing so yields

$$\nabla\cdot(g\nabla p)-\nabla p\cdot\nabla g-\frac{1}{c^2}\frac{\partial}{\partial t}\left(g\frac{\partial p}{\partial t}\right)+\frac{1}{c^2}\frac{\partial p}{\partial t}\frac{\partial g}{\partial t}=0, \tag{9.12}$$

$$\begin{aligned}&\nabla\cdot(p\nabla g)-\nabla p\cdot\nabla g-\frac{1}{c^2}\frac{\partial}{\partial t}\left(p\frac{\partial g}{\partial t}\right)+\frac{1}{c^2}\frac{\partial p}{\partial t}\frac{\partial g}{\partial t}\\&=-4\pi p\delta\left(\vec{x}-\vec{x}_s\right)\delta(t-t_s),\end{aligned} \tag{9.13}$$

where the arguments of $p\left(\vec{x};t\right)$ and $g\left(\vec{x};t\middle|\vec{x}_s;t_s\right)$ in Eqs. (9.12) and (9.13) are suppressed for brevity. Subtracting Eq. (9.13) from (9.12), we obtain

$$\nabla\cdot(g\nabla p-p\nabla g)-\frac{1}{c^2}\frac{\partial}{\partial t}\left(g\frac{\partial p}{\partial t}-p\frac{\partial g}{\partial t}\right)=4\pi p\delta\left(\vec{x}-\vec{x}_s\right)\delta(t-t_s). \tag{9.14}$$

Integrating Eq. (9.15) over the entire time history and three-dimensional space leads to

$$\begin{aligned}&\iiint\limits_{\Omega s}\int\limits_{-\infty}^{\infty}\nabla\cdot(g\nabla p-p\nabla g)-\frac{1}{c^2}\frac{\partial}{\partial t_s}\left(g\frac{\partial p}{\partial t_s}-p\frac{\partial g}{\partial t_s}\right)dt_sd\Omega_s\\&=4\pi\iiint\limits_{\Omega s}\int\limits_{-\infty}^{\infty}p\delta\left(\vec{x}-\vec{x}_s\right)\delta(t-t_s)dt_sd\Omega_s.\end{aligned} \tag{9.15}$$

The integrations on the right side of Eq. (9.15) are readily obtained by the sifting property of the Dirac delta function (6.4). Changing the order of volume and temporal integrations of the first term on the left side of Eq. (9.15) and using the Gauss theorem, we can replace the volume integral by a surface integral. As for the second term on the left side of Eq. (9.15), the temporal integration and time derivative cancel each other. Therefore, we obtain

$$\begin{aligned}4\pi p\left(\vec{x};t\right)=&\int\limits_{-\infty}^{\infty}\iint\limits_{S}\left(p_s\frac{\partial g}{\partial \mathrm{n}_s}-g\frac{\partial p_s}{\partial \mathrm{n}_s}\right)dSdt_s\\&-\frac{1}{c^2}\iiint\limits_{\Omega s}\left(g\frac{\partial p_s}{\partial t_s}-p\frac{\partial g}{\partial t_s}\right)\Bigg|_{-\infty}^{\infty}d\Omega_s,\end{aligned} \tag{9.16}$$

where a subscript s in Eq. (9.16) indicates that the quantities are evaluated at a surface point. The second term on the right side of Eq. (9.16) is identically zero because of the property of the Dirac delta function (6.4). Changing the order of temporal and surface integrations in the first term on the left side of Eq. (9.16) once again then leads to

$$4\pi p\left(\vec{x};t\right) = \iint_S \int_{-\infty}^{\infty} \left[p \frac{\partial}{\partial \mathrm{n}_s} \frac{\delta(t - t_s - R/c)}{R} - \frac{\partial p_s}{\partial \mathrm{n}_s} \frac{\delta(t - t_s - R/c)}{R} \right] dt_s dS. \quad (9.17)$$

Using the chain rule and property of the Dirac delta function, we can rewrite the first term on the right side of Eq. (9.17) as

$$\begin{aligned}
&\iint_S \int_{-\infty}^{\infty} p \frac{\partial}{\partial \mathrm{n}_s} \frac{\delta(t - t_s - R/c)}{R} dt_s dS \\
&= -\iint_S \int_{-\infty}^{\infty} \frac{p}{R^2} \frac{\partial R}{\partial \mathrm{n}_s} \delta(t - t_s - R/c) dt_s dS - \iint_S \int_{-\infty}^{\infty} \frac{p}{cR} \frac{\partial R}{\partial \mathrm{n}_s} \delta'(t - t_s - R/c) dt_s dS \\
&= -\iint_S \frac{p}{R^2} \frac{\partial R}{\partial \mathrm{n}_s}\bigg|_{t_s = t - R/c} dS + \iint_S \frac{1}{cR} \frac{\partial R}{\partial \mathrm{n}_s} \int_{-\infty}^{\infty} \frac{\partial}{\partial t_s} [p\delta(t - t_s - R/c)] dt_s dS \\
&\quad - \iint_S \frac{1}{cR} \frac{\partial R}{\partial \mathrm{n}_s} \int_{-\infty}^{\infty} \frac{\partial p}{\partial t_s} \delta(t - t_s - R/c) dt_s dS = -\iint_S \frac{p}{R^2} \frac{\partial R}{\partial \mathrm{n}_s}\bigg|_{t_s = t - R/c} dS \\
&\quad + \iint_S \frac{1}{cR} \frac{\partial R}{\partial \mathrm{n}_s} [p\delta(t - t_s - R/c)]\big|_{-\infty}^{\infty} dS - \iint_S \frac{1}{cR} \frac{\partial R}{\partial \mathrm{n}_s} \frac{\partial p}{\partial t_s}\bigg|_{t_s = t - R/c} dS \\
&= -\iint_S \left[\frac{1}{R} \frac{\partial R}{\partial \mathrm{n}_s} \left(\frac{1}{R} + \frac{1}{c} \frac{\partial}{\partial t_s} \right) p \right]\bigg|_{t_s = t - R/c} dS.
\end{aligned} \quad (9.18)$$

Substituting Eqs. (9.18) into (9.17), we obtain

$$\begin{aligned}
p\left(\vec{x};t\right) = &-\frac{1}{4\pi} \iint_S \left[\frac{1}{R} \frac{\partial R}{\partial \mathrm{n}_s} \left(\frac{1}{R} + \frac{1}{c} \frac{\partial}{\partial t_s} \right) p\left(\vec{x}_s;t_s\right) \right]\bigg|_{t_s = t - R/c} dS \\
&- \frac{1}{4\pi} \iint_S \left[\frac{1}{R} \frac{\partial p\left(\vec{x}_s;t_s\right)}{\partial \mathrm{n}_s} \right]\bigg|_{t_s = t - R/c} dS.
\end{aligned} \quad (9.19)$$

Equation (9.19) is known as the Kirchhoff–Helmholtz integral formulation for predicting the transient acoustic pressure in free space. The surface acoustic pressure $p\left(\vec{x}_s;t_s\right)$ on the right side of Eq. (9.19) is related to its normal derivative

$\partial p\left(\vec{x}_s; t_s\right)/\partial \mathrm{n}\left(\vec{x}_s\right)$ via the surface Kirchhoff–Helmholtz integral equation obtained by taking the limit as the field point approaches the surface $\vec{x} \to \vec{x}_s$. The processes of taking this limit are the same as those described in Sect. 6.3 and the result is

$$p\left(\vec{x}_s; t_s\right) = -\frac{1}{2\pi}\iint_{S'}\left[\frac{1}{R_s}\frac{\partial R_s}{\partial \mathrm{n}_{s'}}\left(\frac{1}{R_s} + \frac{1}{c}\frac{\partial}{\partial t_{s'}}\right)p\left(\vec{x}_{s'}; t_{s'}\right)\right]\Bigg|_{t_{s'}=t_s-R_s/c} dS' - \frac{1}{2\pi}\iint_{S'}\frac{1}{R_s}\frac{\partial p\left(\vec{x}_{s'}; t_{s'}\right)}{\partial \mathrm{n}_{s'}}\Bigg|_{t_{s'}=t_s-R_s/c} dS'. \tag{9.20}$$

The normal derivative of the surface acoustic pressure in the second term on the right side of Eq. (9.20) can be rewritten by using the Euler's equation, the initial condition (9.2) and the derivative of the Heaviside step function (9.4) as

$$\frac{\partial p\left(\vec{x}_{s'}; t_{s'}\right)}{\partial \mathrm{n}_{s'}} = -\rho_0\frac{\partial v_{\mathrm{n}}\left(\vec{x}_{s'}; t_{s'}\right)}{\partial t_{s'}} = -\rho_0 V_s\left(\vec{\mathrm{n}}_{s'} \cdot \vec{e}_c\right)\delta(t_s - t_{s'}). \tag{9.21}$$

Substituting Eq. (9.21) into the second term on the right side of Eq. (9.20) and taking the Fourier transform, we obtain

$$P\left(\vec{x}_s; \omega\right) = -\frac{1}{2\pi}\iint_{S'}\frac{\partial R_s}{\partial \mathrm{n}_{s'}}\left(\frac{1 - ikR_s}{R_s^2}\right)P\left(\vec{x}_{s'}; \omega\right)dS' + \frac{\rho_0 V_s}{2\pi}\iint_{S'}\left(\frac{\vec{\mathrm{n}}_{s'} \cdot \vec{e}_z}{R_s}\right)e^{ikR_s}dS'. \tag{9.22}$$

Equation (9.22) is the surface Helmholtz integral equation for solving the surface acoustic pressure, given the initial condition (9.2). Note that the surface velocity V_s is independent of the spatial variable. This often happens in practice when an object is hit by a force and starts to move impulsively. This sudden motion may result in an impulsive-like sound. The Fourier transform of the resultant acoustic pressure is expressible as

$$P\left(\vec{x}_s; \omega\right) = \xi\left(\vec{x}_s; \omega\right)V_s, \tag{9.23}$$

where $\xi\left(\vec{x}_s; \omega\right)$ may be obtained by substituting Eq. (9.23) into (9.22),

$$\xi\left(\vec{x}_s; \omega\right) = \rho_0\iint_{S'}\left(\frac{\vec{\mathrm{n}}_{s'} \cdot \vec{e}_c}{R_s}\right)\mathrm{e}^{ikR_s}dS'/\varsigma(\omega), \tag{9.24a}$$

where

$$\varsigma(\omega) = 2\pi + \iint_{S'} \left(\frac{\vec{n}_{s'} \cdot \vec{e}_R}{R_s^3} \right) (1 - ikR_s) e^{ikR_s} dS'. \tag{9.24b}$$

Once the surface acoustic pressure is specified, the acoustic pressure anywhere in free space is completely determined by the Fourier transformed version of Eq. (9.19),

$$P\left(\vec{x};\omega\right) = \left[\frac{\eta\left(\vec{x};\omega\right)}{\varsigma(\omega)} \right] V_s, \tag{9.25}$$

where

$$\eta\left(\vec{x};\omega\right) = \frac{\rho_0}{4\pi} \iint\limits_{S} \left[\left(\vec{n}_s \cdot \vec{e}_c\right) R^2 \varsigma(\omega) - \left(\vec{n}_s \cdot \vec{e}_R\right)(1 - ikR) \iint\limits_{S'} \left(\frac{\vec{n}_{s'} \cdot \vec{e}_z}{R_s} \right) e^{ikR_s} dS' \right] \frac{e^{ikR}}{R^3} dS. \tag{9.26}$$

Equation (9.25) offers the closed-form solution for the acoustic pressure in the frequency domain generated by an arbitrary object subject to the initial condition (9.2) in free space. The temporal acoustic pressure can be obtained by taking an inverse Fourier transform of Eq. (9.25),

$$p\left(\vec{x};t\right) = \frac{1}{2\pi} \int_{-\infty}^{\infty} P\left(\vec{x};\omega\right) e^{-i\omega t} d\omega\,. \tag{9.27}$$

For an early portion of the transient event, Eq. (9.27) can be evaluated asymptotically by taking the limit as $\omega \to \infty$ [140],

$$\lim_{t \to 0} p\left(\vec{x};t\right) = -i \lim_{\omega \to \infty} P\left(\vec{x};\omega\right). \tag{9.28a}$$

On the other hand, for a latter portion of the transient event, the inverse Fourier transform (9.27) can be evaluated asymptotically by taking the limit as $\omega \to 0$,

$$\lim_{t \to \infty} p\left(\vec{x};t\right) = -i \lim_{\omega \to 0} P\left(\vec{x};\omega\right). \tag{9.28b}$$

These two extreme cases indicate that the early portion of the transient event is governed by the high-frequency content, whereas the late portion of the transient event is controlled by the low-frequency contents of the spectrum. However, these asymptotic solutions are undesirable as far as the tractability of any transient event is concerned. Alternatively, one can utilize the residue theorem to evaluate the inverse Fourier transform (9.27) as discussed below.

9.2 Residue Theorem

Mathematically, the evaluation of an infinite integral such as the one given by Eq. (9.27) can be facilitated by a contour integral. Namely, one can replace the infinite line integral along the real axis by a finite one from $-\mathbf{R}$ to $+\mathbf{R}$ in the complex frequency domain, and close the integration path by a semicircle in the lower half plane in the clockwise direction. The reason for choosing the lower half plane in the complex frequency domain is to ensure that the integration remains finite. The radius $\mathbf{R}$ is then extended to infinity. The integration along the semicircle is finite because by definition the infinite integral satisfies the boundedness condition [see Eq. (9.1)] [141],

$$p\left(\vec{x};t\right) = \left[\frac{1}{2\pi}\oint_C \frac{\eta\left(\vec{x};\omega\right)}{\varsigma(\omega)} e^{-i\omega t} d\omega\right] V_s. \tag{9.29}$$

Equation (9.29) shows that the temporal acoustic pressure is expressible as V_s multiplied by a contour integral of $\eta\left(\vec{x};\omega\right)/\varsigma(\omega)$ with $\eta\left(\vec{x};\omega\right)$ and $\varsigma(\omega)$ being given by Eqs. (9.26) and (9.24b), respectively. The contour integral in Eq. (9.29) can be evaluated by the residue theory.

Figure 9.1 shows this contour integration path $\boldsymbol{C}$. If there are singularities on the real axis, they must be excluded by drawing a small semicircle of radius $r=\varepsilon$. For example, suppose that the integrand has singularities at $\pm x_0$ on the real axis. Then the contour integration path can be broken up into segments from $-\mathbf{R}$ to $(-x_0-\varepsilon)$, a semicircle from $(-x_0-\varepsilon)$ to $(-x_0+\varepsilon)$, a straight line from $(-x_0+\varepsilon)$ to $(x_0-\varepsilon)$, another semicircle from $(x_0-\varepsilon)$ to $(x_0+\varepsilon)$, another straight line from $(x_0+\varepsilon)$ to $+\mathbf{R}$, and a semicircle from $+\mathbf{R}$ to $-\mathbf{R}$. The integration along the small semicircle is with respect to $\varepsilon d\theta$, where θ varies from π to 0, which vanishes in the limit as $\varepsilon \to 0$. The integration along the large semicircle is with respect to $\mathbf{R}d\theta$, where θ varies from π to 0, which is identically zero because the boundedness condition is satisfied as $\mathbf{R} \to \infty$.

Therefore, the infinite line integral in Eq. (9.26) is equivalent to the contour integral in Eq. (9.29), which reduces to the line integral from $-\mathbf{R}$ to $+\mathbf{R}$ with $\mathbf{R} \to \infty$. Meanwhile, this contour integral is equal to the sum of residues enclosed by the contour $\boldsymbol{C}$. Consequently, we obtain

$$p\left(\vec{x};t\right) = \hbar\left(\vec{x}\middle|\vec{x}_s;t\right) V_s, \tag{9.30}$$

Fig. 9.1 Schematic of a contour integral in the complex frequency domain

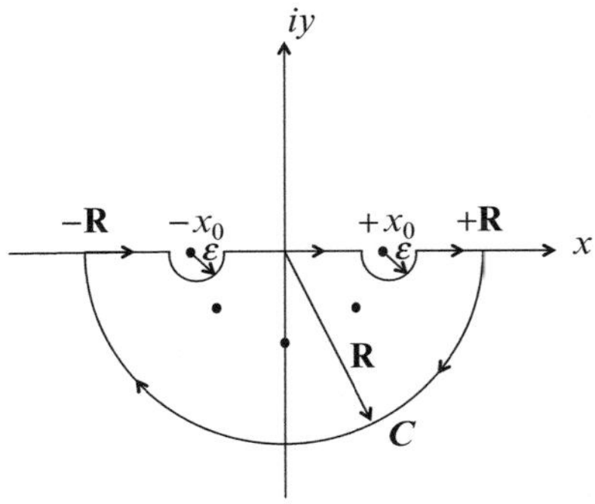

where $\hbar\left(\vec{x}\middle|\vec{x}_s;t\right)$ is the sum of residues enclosed by the contour $\boldsymbol{C}$,

$$\hbar\left(\vec{x}\middle|\vec{x}_s;t\right) = -i\sum_q \frac{\eta\left(\vec{x};\omega_q\right)}{\varsigma'\left(\omega_q\right)}\mathrm{e}^{-i\omega_q(t-t_s-R/c)}H\left(t-t_s-\frac{R}{c}\right), \tag{9.31}$$

where the Heaviside step function appears as the field acoustic pressure is felt only after the source is suddenly excited at $t=t_s$ plus the retarded time r/c, which is needed for the impulsive acoustic signal to travel from the source to any receiver. Also, we have adopted a minus sign because the contour is completed by a semicircle in the lower half plane. The symbol $\varsigma'(\omega_q)$ represents the derivative of $\varsigma(\omega_q)$ with respect to ω, and ω_q is the qth singularity of the ratio $\eta\left(\vec{x};\omega_q\right)/\varsigma(\omega)$, which can be obtained by setting $\varsigma(\omega_q)=0$.

Example 9.1 Consider the case of a sudden-expansion sphere of radius $r=a$ subject to the initial condition (9.2) with $\vec{n}\left(\vec{x}_s\right)\cdot\vec{e}_c = 1$. Suppose that this sudden expansion occurs at $t=t_s$ $(=a/c)$. The surface acoustic pressure in the frequency domain can be obtained by using Eq. (9.23) with $\xi\left(\vec{x}_s;\omega\right)$ given by Eq. (9.24a), which for a spherical surface is given by [142]

$$\xi\left(\vec{x}_s;\omega\right) = \frac{a}{ika-1}.$$

Therefore, Eq. (9.23) gives the surface acoustic pressure in the frequency domain,

$$P\left(\vec{x}_s;\omega\right) = \frac{\rho_0 V_s a \mathrm{e}^{ika}}{1-ika}.$$

Similarly, substituting $\xi\left(\vec{x}_s;\omega\right)$ into Eq. (9.25) yields the field acoustic pressure as

$$P\left(\vec{x};\omega\right) = \frac{\rho_0 V_s}{(1 - ika)}\left(\frac{a^2}{r}\right)e^{ikr}.$$

The temporal acoustic pressure anywhere in free space is given by Eq. (9.27), which can be replaced by the residue theorem through Eq. (9.30), where

$$\eta\left(\vec{x};\omega_q\right) = \rho_0\left(\frac{a^2}{r}\right)e^{i\omega_q(r-a)/c},$$
$$\varsigma(\omega_q) = 1 - ik_q a, \quad \text{and} \quad \varsigma'(\omega_q) = -i\frac{a}{c},$$

where ω_q is the qth root of the characteristic equation, $\varsigma(\omega_q) = 0$. In this case there is only one root, $\omega_1 = -ic/a$. Accordingly, the residue theorem leads to

$$p\left(\vec{x};t\right) = \rho_0 c V_s\left(\frac{a}{r}\right)e^{-(ct-r)/a}H\left(t - \frac{r}{c}\right),$$

which agrees perfectly with the analytic result. This transient sound field is typically seen in an explosion, where the amplitude of the acoustic pressure decays exponentially in all direction.

Example 9.2 Next, consider the case of a sphere of radius $r = a$ that is impulsively accelerated in the z-axis direction such that the normal surface velocity is given by

$$v_n\left(\vec{x}_s;t\right) = V_s\left(\vec{n}_s \cdot \vec{e}_z\right)H(t - t_s),$$

where $t_s = a/c$. Following the same procedures as those in Example 9.1, we obtain

$$\xi\left(\vec{x}_s;\omega\right) = a(ika - 1)/\left[2 - (ka)^2 - i2ka\right].$$

The surface and field acoustic pressures in the frequency domain are given, respectively, by

$$P\left(\vec{x}_s;\omega\right) = \frac{\rho_0 V_s a(1 - ika)\cos\theta}{2 - (ka)^2 - i2ka}e^{ika} \quad \text{and}$$
$$P\left(\vec{x};\omega\right) = \frac{\rho_0 V_s a(1 - ikr)\cos\theta}{2 - (ka)^2 - i2ka}\left(\frac{a}{r}\right)^2 e^{ikr}.$$

Setting the denominator in the above to zero gives two roots in the lower half complex frequency domain, $\omega_1 = (1 - i)c/a$ and $\omega_2 = (-1 - i)c/a$. Accordingly, the temporal acoustic pressure at a field point is found to be

$$p\left(\vec{x};t\right) = \rho_0 c V_s \cos\theta(a/r)\mathrm{e}^{-(ct-r)/a} H(t - r/c)$$
$$\{\cos[(ct-r)/a] - (1 - a/r)\sin[(ct-r)/a]\},$$

which once again agrees perfectly with the analytic result. This transient sound field is typically seen during an impact where the acoustic pressure is highly directional yet decays exponentially.

9.3 Extension to Arbitrary Time-Dependent Excitations

The formulations derived in Sect. 9.2 for predicting the transient acoustic pressure field can be extended to arbitrary time-dependent excitations acting on rigid bodies in free space. To the end, we consider a rigid body subject to a temporal rectangle function, which consists of two unit step functions in the opposite signs.

$$\Delta\vec{v}\left(\vec{x}_s;t\right) = V_s \vec{e}_c[H(t - t_s) - H(t - t_s - \Delta t)], \tag{9.32}$$

where Δt is the gap between two unit step functions.

Following the same procedures as those described in Sect. 9.2, we derive the resultant surface acoustic pressure in the frequency domain as

$$\Delta P\left(\vec{x};\omega\right) = \left[\frac{(1 - \mathrm{e}^{i\omega\Delta t})\eta\left(\vec{x};\omega\right)}{\varsigma(\omega)}\right] V_s. \tag{9.33}$$

The corresponding temporal acoustic pressure anywhere in free space may be obtained by taking the inverse Fourier transform of Eq. (9.33), which can be evaluated via the residue theorem and be expressible as

$$\Delta p\left(\vec{x};t\right) = \left[\hbar\left(\vec{x}\middle|\vec{x}_s;t\right) - \hbar\left(\vec{x}\middle|\vec{x}_s;t - \Delta t\right)\right] V_s, \tag{9.34}$$

where $\hbar\left(\vec{x}\middle|\vec{x}_s;t\right)$ is the same as that given by Eq. (9.31) and $\hbar\left(\vec{x}\middle|\vec{x}_s;t - \Delta t\right)$ is defined as

$$\hbar\left(\vec{x}\middle|\vec{x}_s;t-\Delta t\right) = -i\sum_q \frac{\eta\left(\vec{x};\omega_q\right)}{\varsigma'\left(\omega_q\right)} e^{-i\omega_q(t-\Delta t-t_s-R/c)} H\left(t-\Delta t-t_s-\frac{R}{c}\right). \tag{9.35}$$

Consequently, the transient acoustic pressure radiated from an object subject to a velocity rectangle impulse of constant amplitude is the superposition of two step response functions of the same amplitudes but opposite signs with a separation of Δt in time. Meanwhile, any continuous and arbitrarily time-dependent excitation may be approximated as a sum of rectangle impulses of constant amplitudes with a small duration Δt. Therefore, for an object subject to a continuous and arbitrarily time-dependent velocity excitation, we can write the field acoustic pressure as a sum of individual acoustic pressure pulses,

$$p\left(\vec{x};t\right) = \sum_\ell \left[\hbar\left(\vec{x}\middle|\vec{x}_s;t_\ell\right) - \hbar\left(\vec{x}\middle|\vec{x}_s;t_\ell-\Delta t\right)\right]V_s. \tag{9.36}$$

Equation (9.36) is now ready to be extended to a general, continuous, and time-dependent excitation. For this purpose, we rewrite Eq. (9.34) in the following manner,

$$\Delta p\left(\vec{x};t\right) = \left\{\left[\frac{\hbar\left(\vec{x}\middle|\vec{x}_s;t\right) - \hbar\left(\vec{x}\middle|\vec{x}_s;t-\Delta t\right)}{\Delta t}\right]\Delta t\right\}V_s. \tag{9.37}$$

Equation (9.37) represents an acoustic pressure pulse at a field point $\vec{x}$ and time t due to a velocity rectangle pulse at time t_s. As $\Delta t \to 0$, the square bracket term of Eq. (9.37) becomes an impulse response function. The transient field acoustic pressure at $\vec{x}$ due to all the velocity impulses prior to time t can be expressed as the Duhamel integral [143],

$$p\left(\vec{x};t\right) = \int_0^t h\left(\vec{x}\middle|\vec{x}_s;t-\tau\right)V_s d\tau, \tag{9.38}$$

where $h\left(\vec{x}\middle|\vec{x}_s;t-\tau\right)$ is known as the impulse response function since it is the response to a velocity impulse at time τ, and can be obtained by using the residue theorem as

$$h\left(\vec{x}\middle|\vec{x}_s;t-\tau\right) = -i\sum_q \frac{\eta\left(\vec{x};\omega_q\right)}{\varsigma'\left(\omega_q\right)} e^{-i\omega_q\tau}. \tag{9.39}$$

Equation (9.38) states that the transient acoustic pressure a field point $\vec{x}$ and time t can be expressed as the convolution integral of the impulse response function and the time history of the surface velocity of an object. Sometimes this convolution integral is abbreviated as

$$p\left(\vec{x};t\right)=h\left(\vec{x};t\right)\times v_{\mathrm{n}}\left(\vec{x}_s;t\right), \tag{9.40}$$

where $v_{\mathrm{n}}\left(\vec{x}_s;t\right)=V_s\left(\vec{\mathrm{n}}_s\cdot\vec{e}_c\right)H(t-t_s)$ represents the normal surface velocity of the source and the symbol * indicates the convolution integral given in Eq. (9.38).

9.4 Transient NAH Formulations

The transient formulations developed in Sects. 9.1–9.3 have laid a solid foundation for performing transient NAH. Two types of implementation schemes, namely, the Helmholtz integral formulation and HELS method-based NAH are considered in this section.

9.4.1 Reconstruction Through BEM-Based NAH

Suppose that the input data consist of the acoustic pressure signals measured at $\vec{x}_m^{\Gamma}$ on the hologram surface, $m=1, 2,\ldots,M$, which is positioned around the source surface in the near field. Taking the Fourier transform of the measured acoustic pressure signals and using Eq. (9.25) lead to the following general, discretized BEM-based formulations:

$$\left\{P\left(\vec{x}_m^{\Gamma};\omega\right)\right\}_{M\times 1}=\left\{T_{pv}\left(\vec{x}_m^{\Gamma}\middle|\vec{x}_s;\omega\right)\right\}_{M\times 1}V_s(\omega), \tag{9.41}$$

where $\left\{P\left(\vec{x}_m^{\Gamma};\omega\right)\right\}_{M\times 1}$ is the acoustic pressure measured on the hologram surface in the frequency domain and $\left\{T_{pv}\left(\vec{x}_m^{\Gamma}\middle|\vec{x}_s;\omega\right)\right\}_{M\times 1}$ is the transfer function correlating the measured acoustic pressure at $\vec{x}_m^{\Gamma}$ to the velocity magnitude on the source surface $\vec{x}_s$, whose elements are defined as

$$T_{pv,m}\left(\vec{x}_m^{\Gamma}\middle|\vec{x}_s;\omega\right)=\frac{\eta\left(\vec{x}_m^{\Gamma};\omega\right)}{\varsigma(\omega)}. \tag{9.42}$$

The symbol $V_s(\omega)$ on the right side of Eq. (9.41) indicates the magnitude of the surface velocity, which is frequency dependent but spatially invariant on the source surface, and $R_m = \left| \vec{x}_m^{\Gamma} - \vec{x}_s \right|$. The value of $V_s(\omega)$ may be obtained by taking a pseudo inversion of Eq. (9.41),

$$V_s(\omega) = \left\{ T_{pv}\left(\vec{x}_s \middle| \vec{x}_m^{\Gamma}; \omega \right) \right\}^{\dagger}_{1\times M} \left\{ P\left(\vec{x}_m^{\Gamma}; \omega \right) \right\}_{M\times 1}, \tag{9.43}$$

where

$$\begin{aligned} \left\{ T_{pv}\left(\vec{x}_s \middle| \vec{x}_m^{\Gamma}; \omega \right) \right\}^{\dagger}_{1\times M} = & \left[\left\{ T_{pv}\left(\vec{x}_s \middle| \vec{x}_m^{\Gamma}; \omega \right) \right\}^{H}_{1\times M} \left\{ T_{pv}\left(\vec{x}_m^{\Gamma} \middle| \vec{x}_s; \omega \right) \right\}_{M\times 1} \right]^{-1} \\ & \left\{ T_{pv}\left(\vec{x}_s \middle| \vec{x}_m^{\Gamma}; \omega \right) \right\}^{H}_{1\times M}. \end{aligned} \tag{9.44}$$

In practice Eq. (9.43) must be regularized because the errors involved in the input data may make the pseudo-inversion matrix singular and cause solutions to diverge without a bound. There are many choices for conduct regularization, ranging from the simplest TSVD, L-Curve, to MTR [46, 49, 50], which have been discussed extensively in the past and are omitted here for brevity.

Once the surface velocity is reconstructed, the surface acoustic pressure can be obtained by substituting Eq. (9.43) into Eq. (9.23), and the result is

$$P\left(\vec{x}_s; \omega \right) = \xi\left(\vec{x}_s; \omega \right) \left\{ T_{pv}\left(\vec{x}_s \middle| \vec{x}_m^{\Gamma}; \omega \right) \right\}^{\dagger}_{1\times M} \left\{ P\left(\vec{x}_m^{\Gamma}; \omega \right) \right\}_{M\times 1}. \tag{9.45}$$

Meanwhile, the reconstructed acoustic pressure at any field point $\vec{x}$ can be determined by substituting Eq. (9.43) into Eq. (9.25), which is expressible as

$$P\left(\vec{x}; \omega \right) = \frac{\eta\left(\vec{x}; \omega \right)}{\varsigma(\omega)} \left\{ T_{pv}\left(\vec{x}_s \middle| \vec{x}_m^{\Gamma}; \omega \right) \right\}^{\dagger}_{1\times M} \left\{ P\left(\vec{x}_m^{\Gamma}; \omega \right) \right\}_{M\times 1}. \tag{9.46}$$

The normal component of the particle velocity at $\vec{x}$ can be obtained by taking the normal derivative of Eq. (9.46),

$$V_{\mathrm{n}}\left(\vec{x}; \omega \right) = -i \frac{1}{\rho_0 \omega \varsigma(\omega)} \frac{\partial \eta\left(\vec{x}; \omega \right)}{\partial \mathrm{n}} \left\{ T_{pv}\left(\vec{x}_s \middle| \vec{x}_m^{\Gamma}; \omega \right) \right\}^{\dagger}_{1\times M} \left\{ P\left(\vec{x}_m^{\Gamma}; \omega \right) \right\}_{M\times 1}. \tag{9.47}$$

9.4.2 Reconstruction Through HELS-Based NAH

Alternatively, the expansion theory can be used to reconstruct the acoustic field. One such approach is the so-called HELS method that employs the spherical Hankel functions and spherical harmonics as the basis functions to describe the acoustic quantities [37, 38].

Suppose that the acoustic pressure is specified on a hologram surface Γ in the same way as that depicted in the preceding section. The acoustic pressure and normal component of the particle velocity anywhere in the field, including the source surface, can be reconstructed by using the HELS formulations and the results are

$$P\left(\vec{x};\omega\right) = \left\{G_{pp}\left(\vec{x}\middle|\vec{x}_m^{\Gamma};\omega\right)\right\}_{1\times M}\left\{P\left(\vec{x}_m^{\Gamma};\omega\right)\right\}_{M\times 1}, \tag{9.48}$$

$$V_{\mathrm{n}}\left(\vec{x};\omega\right) = \left\{G_{vp}\left(\vec{x}\middle|\vec{x}_m^{\Gamma};\omega\right)\right\}_{1\times M}\left\{P\left(\vec{x}_m^{\Gamma};\omega\right)\right\}_{M\times 1}, \tag{9.49}$$

where $\left\{G_{pp}\left(\vec{x}\middle|\vec{x}_m^{\Gamma};\omega\right)\right\}_{1\times M}^{\mathrm{T}}$ and $\left\{G_{vp}\left(\vec{x}\middle|\vec{x}_m^{\Gamma};\omega\right)\right\}_{1\times M}^{\mathrm{T}}$ are the transfer functions that correlate $P\left(\vec{x};\omega\right)$ and $V_{\mathrm{n}}\left(\vec{x};\omega\right)$ anywhere in the field to $P\left(\vec{x}_m^{\Gamma};\omega\right)$ on the hologram surface Γ, respectively,

$$\left\{G_{pp}\left(\vec{x}\middle|\vec{x}_m^{\Gamma};\omega\right)\right\}_{1\times M}^{\mathrm{T}} = \left\{\Psi\left(\vec{x};\omega\right)\right\}_{1\times J}^{\mathrm{T}}\left(\left[\Psi\left(\vec{x}_m^{\Gamma};\omega\right)\right]_{J\times M}^{H}\left[\Psi\left(\vec{x}_m^{\Gamma};\omega\right)\right]_{M\times J}\right)^{-1}\left[\Psi\left(\vec{x}_m^{\Gamma};\omega\right)\right]_{J\times M}^{H}, \tag{9.50}$$

$$\left\{G_{vp}\left(\vec{x}\middle|\vec{x}_m^{\Gamma};\omega\right)\right\}_{1\times M}^{\mathrm{T}} = \frac{1}{i\omega\rho_0}\left\{\frac{\partial\Psi\left(\vec{x};\omega\right)}{\partial \mathrm{n}}\right\}_{1\times J}^{\mathrm{T}}\left(\left[\Psi\left(\vec{x}_m^{\Gamma};\omega\right)\right]_{J\times M}^{H}\left[\Psi\left(\vec{x}_m^{\Gamma};\omega\right)\right]_{M\times J}\right)^{-1}\left[\Psi\left(\vec{x}_m^{\Gamma};\omega\right)\right]_{J\times M}^{H}, \tag{9.51}$$

where the elements of the matrix $\left[\Psi\left(\vec{x}_m^{\Gamma};\omega\right)\right]_{J\times M}$ consist of the particular solution to the Helmholtz equation, which are expressible in the spherical coordinates as

$$\Psi_j(r,\theta,\phi;\omega) \equiv \Psi_{nl}(r,\theta,\phi;\omega) = h_n^{(1)}(kr)Y_n^l(\theta,\phi), \tag{9.52}$$

where $h_n^{(1)}(kr)$ and $Y_n^l(\theta,\phi)$ are the spherical Hankel functions of the first kind and the spherical harmonics, respectively, and the indices j, n, and l in Eq. (9.52) are related through $j = n^2 + n + l + 1$, where the order of expansion in the radial function n starts from 0 to N and l ranges from $-n$ to $+n$.

9.4.3 Transient NAH Formulations

Once the acoustic quantities in the frequency domain are determined by utilizing either the BEM- or HELS-based NAH formulations, the corresponding time-domain signals are obtained by taking an inverse Fourier transform of either Eqs. (9.46) and (9.47) or Eqs. (9.48) and (9.49). These equations may be evaluated by using the residue theorem and expressed as a convolution integral (9.40), except that input data consist of the measured acoustic pressure signal $p\left(\vec{x}_m^{\Gamma}; t\right)$ rather than velocity signal on the source surface,

$$p\left(\vec{x}; t\right) = g_{pp}\left(\vec{x}\middle|\vec{x}_m^{\Gamma}; t\right) \times p\left(\vec{x}_m^{\Gamma}; t\right), \tag{9.53}$$

$$v_{\mathrm{n}}\left(\vec{x}; t\right) = g_{vp}\left(\vec{x}\middle|\vec{x}_m^{\Gamma}; t\right) \times p\left(\vec{x}_m^{\Gamma}; t\right), \tag{9.54}$$

where the temporal kernels $g_{pp}\left(\vec{x}\middle|\vec{x}_m^{\Gamma}; t\right)$ and $g_{vp}\left(\vec{x}\middle|\vec{x}_m^{\Gamma}; t\right)$ are expressible, respectively, as

$$g_{pp}\left(\vec{x}\middle|\vec{x}_m^{\Gamma}; t-\tau\right) = -i\sum_q \frac{\eta_{pp}\left(\vec{x}; \omega_q^{pp}\right)}{\varsigma'_{pp}\left(\omega_q^{pp}\right)} \mathrm{e}^{-i\omega_q^{pp}\tau}, \tag{9.55}$$

$$g_{vp}\left(\vec{x}\middle|\vec{x}_m^{\Gamma}; t-\tau\right) = -i\sum_q \frac{\eta_{vp}\left(\vec{x}; \omega_q^{vp}\right)}{\varsigma'_{vp}\left(\omega_q^{vp}\right)} \mathrm{e}^{-i\omega_q^{vp}\tau}, \tag{9.56}$$

where ω_q^{pp} and ω_q^{vp} are, respectively, the roots of the characteristic equations of

$$\varsigma_{pp}\left(\omega_q^{pp}\right) = 0, \tag{9.57}$$

$$\varsigma_{vp}\left(\omega_q^{vp}\right) = 0. \tag{9.58}$$

It is emphasized that there are no closed-form solutions for $g_{pp}\left(\vec{x}\middle|\vec{x}_m^{\Gamma}; t\right)$ and $g_{vp}\left(\vec{x}\middle|\vec{x}_m^{\Gamma}; t\right)$ in general because the source surfaces, measurements, and reconstruction locations are arbitrary. Mathematically, $g_{pp}\left(\vec{x}\middle|\vec{x}_m^{\Gamma}; t\right)$ implies the impulse response function correlating the reconstructed acoustic pressure $p\left(\vec{x}; t\right)$ at $\vec{x}$ to the measured acoustic pressure signal $p\left(\vec{x}_m^{\Gamma}; t\right)$ at $\vec{x}_m^{\Gamma}$. Similarly, $g_{vp}\left(\vec{x} \middle| \vec{x}_m^{\Gamma}; t\right)$ is the impulse response function that correlates the reconstructed normal component of the particle velocity $v_n\left(\vec{x}; t\right)$ at $\vec{x}$ to the measured acoustic pressure signal $p\left(\vec{x}_m^{\Gamma}; t\right)$ at $\vec{x}_m^{\Gamma}$. Note that because the residue theorem is used in Eqs. (9.53) and (9.54) to reconstruct the transient acoustic field, rather than a direct inverse Fourier transform, the conventional discretization and the minimal sampling rate requirement in the time domain are avoided.

9.4.4 Applications of the Transient NAH Formulations

In this section both the integral theory and HELS-based NAH formulations are utilized to reconstruct the transient acoustic pressure fields, and results are compared with the analytic ones.

Example 9.3 (A Sudden-Expansion Sphere) Consider a sudden-expansion sphere of radius $r = a$ subject to the initial condition (9.2) with $\vec{n}\left(\vec{x}_s\right) \cdot \vec{e}_c = 1$. Suppose that sudden expansion occurs at $t = a/c$. The analytic acoustic pressure signal on a hologram surface is taken as the input. For simplicity, we assume that the time history of the acoustic pressure signal measured at any field point is (see Example 9.1)

$$p\left(\vec{x}_m^{\Gamma}; t\right) = \rho_0 c V_s \left(\frac{a}{r_m^{\Gamma}}\right) e^{-\left(ct - r_m^{\Gamma}\right)/a} H\left(t - \frac{r_m^{\Gamma}}{c}\right). \tag{9.59}$$

The reconstructed acoustic pressure signal at any field point $\vec{x}$ can be determined by using Eq. (9.53). Since the normal surface velocity is constant, it suffices to take one measurement on a hologram surface, i.e., $M = 1$. First, we use the BEM-based NAH formulation to reconstruct the acoustic pressure field. Accordingly, the pseudo inversion defined in Eq. (9.43) reduces to

$$T_{pp}^{\dagger}\left(\vec{x}_m^{\Gamma} \middle| \vec{x}_s; \omega_q^{pp}\right) = \frac{\varsigma(\omega)}{\eta\left(\vec{x}_m^{\Gamma}; \omega\right)}. \tag{9.60}$$

Substituting Eq. (9.60) into Eq. (9.46) yields the reconstructed acoustic pressure at any field point $\vec{x}$ in the frequency domain,

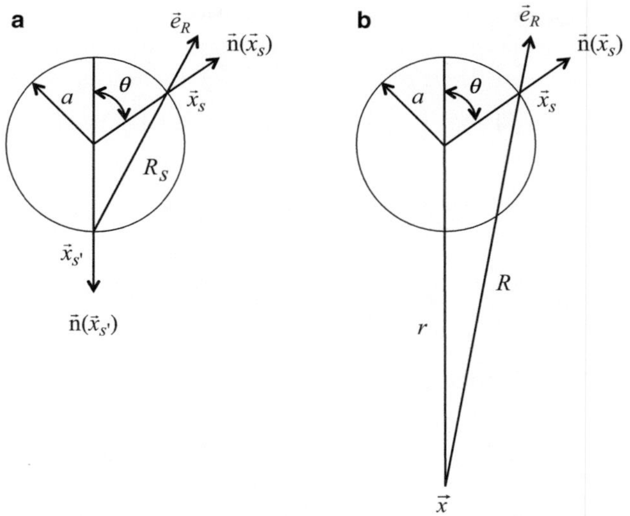

Fig. 9.2 Schematic of points on the surface of the sphere of radius $r = a$ and in the field, and corresponding distances R_s and R, respectively. (**a**): Both points on the source surface; (**b**): One point on the source surface and another in the field

$$P\left(\vec{x};\omega\right) = \frac{\eta\left(\vec{x};\omega\right)}{\eta\left(\vec{x}_m^{\Gamma};\omega\right)} P\left(\vec{x}_m^{\Gamma};\omega\right), \tag{9.61}$$

The temporal acoustic pressure at any field point $p\left(\vec{x};\omega\right)$ may be obtained by Eq. (9.27), which can be evaluated using the residue Eq. (9.53), where the temporal kernel $g_{pp}\left(\vec{x}\middle|\vec{x}_m^{\Gamma};t-\tau\right)$ is

$$g_{pp}\left(\vec{x}\middle|\vec{x}_m^{\Gamma};t-\tau\right) = -i\sum_q \frac{\eta\left(\vec{x};\omega_q^{pp}\right)}{\eta\left(\vec{x}_m^{\Gamma};\omega_q^{pp}\right)}, \tag{9.62}$$

where ω_q^{pp} is the qth root of the characteristic equation $\varsigma_{pp}(\omega_q^{pp}) = 0$. In this case, there is only one root, $\omega_1^{pp} = -ic/a$, so $q = 1$.

Figure 9.2 displays the schematic of relative positions of the locations of surface and field points with respect to a sudden-expansion sphere. The quantity $\xi\left(\vec{x}_s;\omega_1^{pp}\right)$ involved in $\eta\left(\vec{x};\omega_1^{pp}\right)$ and $\eta\left(\vec{x}_m^{\Gamma};\omega_1^{pp}\right)$ is given by Eq. (9.26). The distance between two points on the source surface is $R_s = 2a\cos(\theta/2)$, $\partial R_s/\partial n_s = \cos(\theta/2)$ and $dS' = a^2\sin\theta d\theta d\phi$, with θ varying from 0 to π and ϕ from 0 to 2π. Since integrands are independent of the azimuthal angle ϕ, integration over ϕ can be done separately, yielding 2π. The distance between a surface and field point is $R = \sqrt{r^2 + a^2 + 2ar\cos\theta}$. For simplicity, the radial distance r is assumed much larger than radius a, so $R \approx r$, $e^{ikR} \approx e^{ikR + ika\cos\theta}$ and $\partial R/\partial n_s = \cos\theta$. Detailed integrations for $\xi(a; \omega_1^{pp})$ and $\eta\left(\vec{x}_m^{\Gamma}; \omega_1^{pp}\right)$ are shown in reference [142] and omitted here for brevity.

Substituting $\xi\left(\vec{x}_s; \omega_1^{pp}\right)$ and $\eta\left(\vec{x}_m^{\Gamma}; \omega_1^{pp}\right)$ into Eq. (9.62) yields

$$g_{pp}\left(\vec{x}\middle|\vec{x}_m^{\Gamma}; t-\tau\right) = \frac{(a/r)e^{ik_1^{pp}r}}{\left(a/r_m^{\Gamma}\right)e^{ik_1^{pp}r_m^{\Gamma}}} = \left(\frac{r_m^{\Gamma}}{r}\right)e^{\left(r-r_m^{\Gamma}\right)/a}. \tag{9.63}$$

Substituting Eqs. (9.59) and (9.63) into Eq. (9.53) then leads to

$$p\left(\vec{x}; t\right) = \rho_0 c V_s \left(\frac{a}{r}\right) e^{-(ct-r)/a} H\left(t - \frac{r}{c}\right), \tag{9.64}$$

which matches the analytic solution for a sudden-expansion sphere [142].

Next the HELS-based NAH formulation is used to reconstruct the transient acoustic field. The basis function in the HELS expansion is given by Eq. (9.52). For a sudden-expansion sphere, it suffices to use a one-term expansion. Accordingly, we have $\Psi_1(r, \theta, \phi; \omega) = e^{ikr}/r$. The temporal kernels $g_{pp}\left(\vec{x}\middle|\vec{x}_m^{\Gamma}; t-\tau\right)$ as defined by Eq. (9.62) reduces to

$$\begin{aligned} g_{pp}\left(\vec{x}\middle|\vec{x}_m^{\Gamma}; t-\tau\right) &= -i\frac{\eta_{pp}\left(\vec{x}; \omega_1^{pp}\right)}{\varsigma'_{pp}\left(\omega_1^{pp}\right)} e^{-i\omega_1^{pp}\tau} = \frac{\left(ar_m^{\Gamma}/cr\right)e^{-i[c/(-ia)]\left(r-r_m^{\Gamma}\right)/c}}{a/c} \\ &= \left(\frac{r_m^{\Gamma}}{r}\right)e^{\left(r-r_m^{\Gamma}\right)/a}. \end{aligned} \tag{9.65}$$

Substituting (9.59) and (9.65) into the convolution integral (9.53) yields

$$p\left(\vec{x}; t\right) = g_{pp}\left(\vec{x}\middle|\vec{x}_m^{\Gamma}; t\right) \times p\left(\vec{x}_m^{\Gamma}; t\right) = \rho_0 c V_s e^{-(ct-r)/a}\left(\frac{a}{r}\right) H\left(t - \frac{r}{c}\right), \tag{9.66}$$

which agrees with the analytic solution [142].

Figure 9.3 demonstrates three-dimensional images of acoustic pressure fields at arbitrarily selected time instances $t = 3.24$ (ms), 4.41 (ms), 5.88 (ms), and 7.35 (ms).

Similarly, the normal component of the particle velocity at any field point is reconstructed by using Eq. (9.54) by using $p\left(\vec{x}_m^{\Gamma}; t\right)$ and $g_{vp}\left(\vec{x}\middle|\vec{x}_m^{\Gamma}; t\right)$ given by

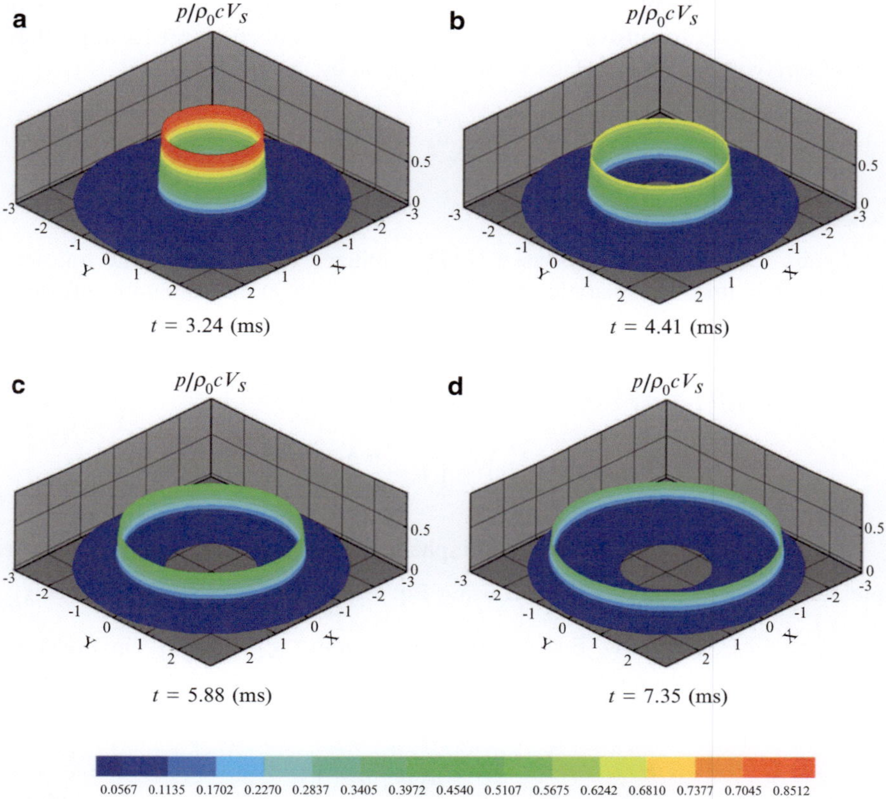

Fig. 9.3 Reconstructed temporal acoustic pressure fields resulting from a sudden-expansion sphere of radius a at different time instances. (**a**): $t = 3.24$ ms; (**b**): $t = 4.41$ ms; (**c**): $t = 5.88$ ms; and (**d**): $t = 7.35$ ms

$$
\begin{aligned}
g_{vp}\left(\vec{x}\middle|\vec{x}_m^{\Gamma}; t-\tau\right) &= -i\frac{\eta_{vp}\left(\vec{x};\omega_1^{vp}\right)}{\varsigma'_{vp}\left(\omega_1^{vp}\right)}\mathrm{e}^{-i\omega_1^{vp}\tau} \\
&= \frac{\left(ar_m^{\Gamma}/\rho_0c^2r\right)\mathrm{e}^{-i[c/(-ia)]\left(r-r_m^{\Gamma}\right)/c}}{a/c} = \frac{1}{\rho_0 c}\left(\frac{r_m^{\Gamma}}{r}\right)\mathrm{e}^{\left(r-r_m^{\Gamma}\right)/a}.
\end{aligned}
\tag{9.67}
$$

Substituting Eqs. (9.59) and (9.67) into Eq. (9.54) leads to

$$
v_{\mathrm{n}}\left(\vec{x};t\right) = g_{vp}\left(\vec{x}\middle|\vec{x}_m^{\Gamma};t\right)\times p\left(\vec{x}_m^{\Gamma};t\right) = V_s\mathrm{e}^{-(ct-r)/a}\left(\frac{a}{r}\right)H\left(t-\frac{r}{c}\right), \tag{9.68}
$$

which reduces to the initial condition (9.2) when r is set on the source surface and $a/c = t_s$.

Example 9.4 (An Impulsively Accelerated Sphere) Consider the case of a sphere of radius $r = a$ that is impulsively accelerated in the z-axis direction (see Example 9.2). Accordingly, the normal surface velocity is given by

$$v_{\mathrm{n}}\left(\vec{x}_s; t\right) = V_s\left(\vec{\mathrm{n}}_s \cdot \vec{e}_z\right)H(t - t_s), \tag{9.69}$$

where $\vec{\mathrm{n}}\left(\vec{x}_s\right) \cdot \vec{e}_z = \cos\theta$.

Again, the integral theory-based NAH is utilized to reconstruct the transient acoustic field first. The analytic acoustic pressure signal at the hologram surface is taken as the input,

$$\begin{aligned} p\left(\vec{x}_m^{\Gamma}; t\right) = \rho_0 c V_s \cos\theta_m^{\Gamma}\left(\frac{a}{r_m^{\Gamma}}\right)\mathrm{e}^{-\left(ct - r_m^{\Gamma}\right)/a} H\left(t - \frac{r_m^{\Gamma}}{c}\right) \\ \left[\cos\left(\frac{ct - r_m^{\Gamma}}{a}\right) - \left(1 - \frac{a}{r_m^{\Gamma}}\right)\sin\left(\frac{ct - r_m^{\Gamma}}{a}\right)\right]. \end{aligned} \tag{9.70}$$

As in the previous sudden-expansion sphere, the reconstructed temporal acoustic pressure $p\left(\vec{x}; \omega\right)$ can be obtained using Eq. (9.53) with its temporal kernel $g_{pp}\left(\vec{x}\middle|\vec{x}_m^{\Gamma}; t - \tau\right)$ given by Eq. (9.62). The quantity $\xi\left(\vec{x}_s; \omega_q^{pp}\right)$ involved in $\eta_{pp}\left(\vec{x}; \omega_q^{pp}\right)$ and $\eta_{pp}\left(\vec{x}_m^{\Gamma}; \omega_q^{pp}\right)$ can be shown as

$$\xi\left(\vec{x}_s; \omega_q^{pp}\right) = \frac{a\left(ik_q^{pp}a - 1\right)}{2 - \left(k_q^{pp}a\right)^2 - i2k_q^{pp}a}. \tag{9.71}$$

In this case $q = 2$, $\omega_1^{pp} = (1 - i)c/a$ and $\omega_1^{pp} = (-1 - i)c/a$. Thus the temporal kernel becomes

$$g_{pp}\left(\vec{x}\middle|\vec{x}_m^{\Gamma}; t - \tau\right) = -i\sum_{q=1}^{2}\left(\frac{r_m^{\Gamma}}{r}\right)^2 \frac{\left(1 - ik_q^{pp}r\right)}{\left(1 - ik_q^{pp}r_m^{\Gamma}\right)}\mathrm{e}^{ik_q^{pp}\left(r - r_m^{\Gamma}\right)}. \tag{9.72}$$

Detailed derivations of Eq. (9.72) are omitted here for brevity. Substituting Eqs. (9.70) and (9.72) into Eq. (9.53) and summing the residues yield

$$p\left(\vec{x}; t\right) = \rho_0 c V_s \cos\theta\left(\frac{a}{r}\right)\mathrm{e}^{-(ct - r)/a} H\left(t - \frac{r}{c}\right)\left[\cos\left(\frac{ct - r}{a}\right) - \left(1 - \frac{a}{r}\right)\sin\left(\frac{ct - r}{a}\right)\right], \tag{9.73}$$

which is the analytic solution to the acoustic pressure due to an impulsively accelerated sphere [142].

Alternatively, the HELS-based NAH can be used to reconstruct the transient acoustic field. Suppose that a two-term HELS expansion is used,

$$\Psi_1(r,\theta,\phi;\omega) = \frac{e^{ikr}}{r} \quad \text{and} \quad \Psi_2(r,\theta,\phi;\omega) = \frac{(kr+i)\cos\theta}{(kr)^2} e^{ikr}. \tag{9.74}$$

Accordingly, the temporal kernel involved in Eq. (9.55) can be written as

$$g_{pp}\left(\vec{x}\middle|\vec{x}_m^{\Gamma};t-\tau\right) = -i\sum_{q=1}^{2} \frac{\eta_{pp}\left(\vec{x};\omega_q^{pp}\right)}{\varsigma'_{pp}(\omega_q^{pp})} e^{-i\omega_q^{pp}\tau}, \tag{9.75}$$

where $q = 1$ and 2, $\omega_1^{pp} = (1-i)c/a$, $\omega_1^{pp} = (-1-i)c/a$, $\eta_{pp}\left(\vec{x};\omega_q^{pp}\right)$, and $\varsigma'_{pp}\left(\omega_q^{pp}\right)$ are given, respectively, by

$$\eta_{pp}\left(\vec{x};\omega_1^{pp}\right) = \frac{(\omega_1^{pp}r/c+i)a^3\cos\theta}{r^2} e^{i\omega_1^{pp}r/c}, \tag{9.76a}$$

$$\eta_{pp}\left(\vec{x};\omega_2^{pp}\right) = \frac{(\omega_2^{pp}r/c+i)a^3\cos\theta}{r^2} e^{i\omega_2^{pp}r/c}, \tag{9.76b}$$

$$\varsigma'_{pp}(\omega_1^{pp}) = -2(a/c)\left(a\omega_1^{pp}+i\right), \tag{9.76c}$$

$$\varsigma'_{pp}(\omega_2^{pp}) = -2(a/c)\left(a\omega_2^{pp}+i\right). \tag{9.76d}$$

Substituting Eqs. (9.73) and (9.75) into Eq. (9.53) and summing the residues yield

$$\begin{aligned} p\left(\vec{x};t\right) &= g_{pp}\left(\vec{x}\middle|\vec{x}_m^{\Gamma};t\right) \times p\left(\vec{x}_m^{\Gamma};t\right) \\ &= \rho_0 c V_s \cos\theta\left(\frac{a}{r}\right) e^{-(ct-r)/a} H\left(t-\frac{r}{c}\right)\left[\cos\left(\frac{ct-r}{a}\right) - \left(1-\frac{a}{r}\right)\sin\left(\frac{ct-r}{a}\right)\right], \end{aligned} \tag{9.77}$$

which matches the analytic solution for the temporal acoustic pressure emitted by an impulsively accelerated sphere of radius a in free space [142].

Figure 9.4 shows three-dimensional images of the acoustic pressure fields at arbitrarily selected time instances [149].

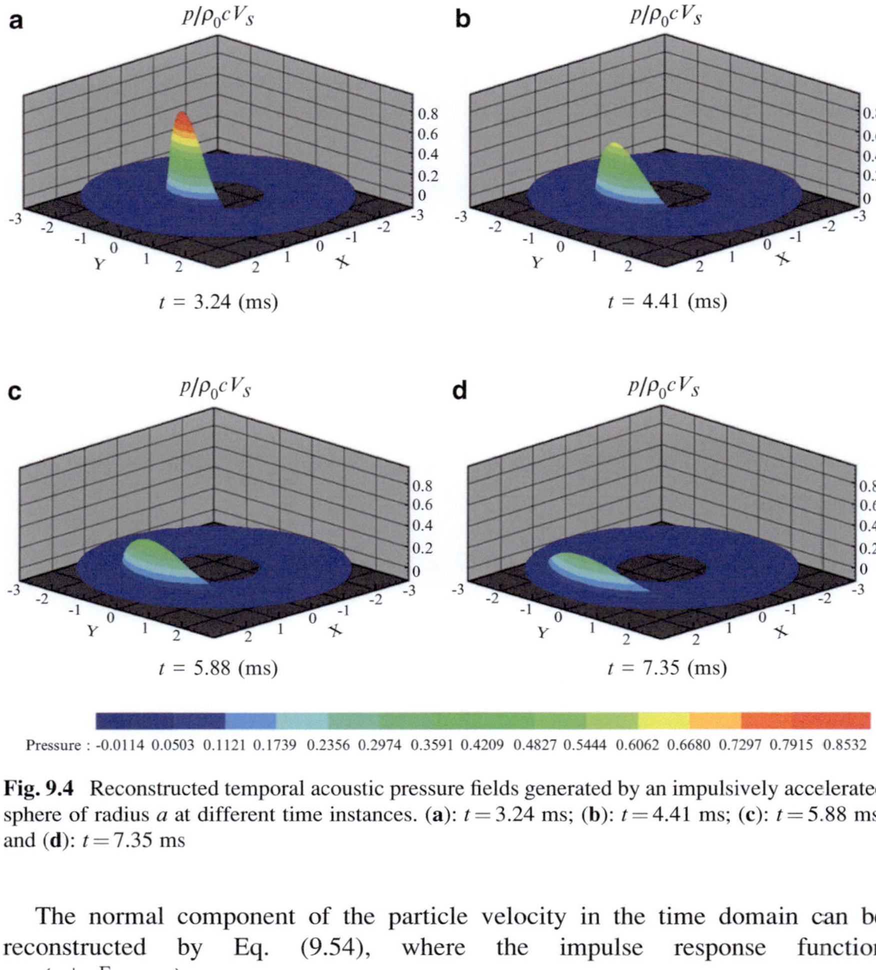

Fig. 9.4 Reconstructed temporal acoustic pressure fields generated by an impulsively accelerated sphere of radius a at different time instances. (**a**): $t=3.24$ ms; (**b**): $t=4.41$ ms; (**c**): $t=5.88$ ms; and (**d**): $t=7.35$ ms

The normal component of the particle velocity in the time domain can be reconstructed by Eq. (9.54), where the impulse response function $g_{vp}\left(\vec{x}\middle|\vec{x}_m^{\Gamma};t-\tau\right)$ is given by

$$g_{vp}\left(\vec{x}\middle|\vec{x}_m^{\Gamma};t-\tau\right)=-i\sum_{q=1}^{2}\frac{\eta_{vp}\left(\vec{x};\omega_q^{vp}\right)}{\varsigma_{vp}'(\omega_q^{vp})}\mathrm{e}^{-i\omega_q^{vp}\tau}, \tag{9.78}$$

where

$$\eta_{vp}\left(\vec{x};\omega_1^{vp}\right)=\frac{(\omega_1^{vp}r/c+i)a^3\cos\theta}{\rho_0 cr^2}\mathrm{e}^{i\omega_1^{vp}r/c}, \tag{9.79a}$$

$$\eta_{vp}\left(\vec{x};\omega_2^{vp}\right)=\frac{(\omega_2^{vp}r/c+i)a^3\cos\theta}{\rho_0 cr^2}e^{i\omega_2^{vp}r/c},\tag{9.79b}$$

$$\varsigma_{vp}^{'}\left(\omega_1^{vp}\right)=-2(a/c)\left(a\omega_1^{vp}+i\right),\tag{9.79c}$$

$$\varsigma_{pv}^{'}\left(\omega_2^{vp}\right)=-2(a/c)\left(a\omega_2^{vp}+i\right).\tag{9.79d}$$

where $\omega_1^{pp}=(1-i)c/a$ and $\omega_1^{pp}=(-1-i)c/a$.

Substituting Eqs. (9.73) and (9.78) into Eq. (9.54) yields the normal component of the particle velocity anywhere in the field,

$$\begin{aligned}v_n\left(\vec{x};t\right)&=g_{vp}\left(\vec{x}\middle|\vec{x}_m^{\Gamma};t\right)\times p\left(\vec{x}_m^{\Gamma};t\right)\\&=V_s\cos\theta\left(\frac{a}{r}\right)e^{-(ct-r)/a}H\left(t-\frac{r}{c}\right)\left[\cos\left(\frac{ct-r}{a}\right)-\left(1-\frac{a}{r}\right)\sin\left(\frac{ct-r}{a}\right)\right],\end{aligned}\tag{9.80}$$

which reduces the initial condition (9.2) when the distance is set to $r=a$ and $t=a/c$.

Example 9.5 (An Impulsively Accelerated Baffled Piston on a Sphere) Consider acoustic radiation from a piston mounted on a sphere of radius a. In general, the acoustic pressure generated by a spherical source in the frequency domain can be described by an infinite series of the spherical Hankel functions of the first kind and the spherical harmonics [99],

$$P(r,\theta,\phi;\omega)=\sum_{n=0}^{\infty}\sum_{l=-n}^{n}A_{nl}h_n^{(1)}(kr)Y_n^l(\theta,\phi),\tag{9.81}$$

where the expansion coefficients A_{nl} can be obtained by the orthonormal property of the spherical harmonics. Suppose that the normal surface velocity is given in the boundary condition. Then the coefficients A_{nl} can be obtained by the Euler's equation (9.21) and the orthonormal property of the spherical harmonics,

$$A_{nl}=i\frac{\rho_0 c}{h_n^{(1)'}(ka)}\int_0^{2\pi}\int_0^{\pi}V_s(a,\theta,\phi;\omega)Y_n^{l*}(\theta,\phi)\sin\theta d\theta d\phi,\tag{9.82}$$

where $V_s(a,\theta,\phi;\omega)$ is specified on the surface of the sphere; $h_n^{(1)'}(ka)=(c/\omega)[dh_n^{(1)}(kr)/dr]|_{r=a}$ is the normal derivative evaluated on the surface of the sphere.

Once the expansion coefficients A_{nl} are specified, the acoustic pressure at any field point in the frequency domain can be determined by Eq. (9.81). The temporal acoustic pressure can be obtained by taking the inverse Fourier transform and evaluated by using Eq. (9.53).

For simplicity, the piston is assumed axisymmetric with respect to the polar axis at $\theta = 0$, and is impulsively accelerated at $t = t_s$. Moreover, the normal surface velocity is non-zero over a vertex angle, $\pm\theta_0$, and zero elsewhere,

$$v_{\rm n}(a, \theta; t) = V_s[H(\theta + \theta_0) - H(\theta - \theta_0)]H(t - t_s), \tag{9.83}$$

where $t_s = a/c$.

Accordingly, Eq. (9.81) is reduced to [99]

$$P(r, \theta; \omega) = \sum_{n=0}^{\infty} A_n h_n^{(1)}(kr) Q_n^{(1)}(\cos\theta), \tag{9.84}$$

where $Q_n^{(1)}(\cos\theta)$ are the Legendre functions of the first kind.

Note that there is no closed-form solution for the radiated acoustic pressure signal $P(r,\theta;\omega)$ in this case. Hence numerical solutions are sought. As an example, a circular piston with a vertex angle of $\pm\theta_0 = \angle 15^\circ$ is considered in this section. Theoretically, the normal surface velocity given by Eq. (9.83) requires an infinite series to depict the sharp edges at $\theta_0 = \pm\angle 15^\circ$. For the purpose of demonstrating the application of the transient NAH formulations, a finite expansion is utilized to approximate the velocity profile as specified in Eq. (9.83),

$$v_{\rm n}(a, \theta; t) = V_s \sum_{n=1}^{N} B_n Q_n^{(1)}(\cos\theta) H(t - a/c), \tag{9.85}$$

where N is finite. The larger the value of N is, the better the approximation to the velocity profile is, but the more intensive numerical computations are. For simplicity yet without loss of generality, $N = 11$ is selected in this numerical example. The expansion coefficients B_n can be determined by using the orthonormal property of the Legendre functions [99],

$$B_n = \left(\frac{2n+1}{2}\right) \int_{-\theta_0}^{\theta_0} Q_n^{(1)*}(\cos\theta) \sin\theta d\theta. \tag{9.86}$$

Accordingly, the expansion coefficients A_n for the acoustic pressure, Eq. (9.84), can be obtained by using the orthonormal property of the Legendre functions and boundary condition

$$A_n = i \frac{(2n+1)\rho_0 c V_s}{2h_n^{(1)'}(ka)} \sum_{n'=1}^{N} \left[\int_{-\theta_0}^{\theta_0} Q_{n'}^{(1)*}(\cos\theta) \sin\theta d\theta \right]. \tag{9.87}$$

The temporal acoustic pressure at any field point can be determined by taking the inverse Fourier transform of Eq. (9.84) and facilitated by Eq. (9.53). The resultant

Table 9.1 Singularities of the impulse response functions for reconstructing the transient acoustic field generated by a partially impulsively accelerating piston mounted on a sphere of radius a with a vertex angle of $\pm\theta° = \angle 15°$

No.	Singularities ω_q^{vp}
1	$-5.53363E+02-i3.28058E+03$
2	$-1.41101E+03-i2.46182E+03$
3	$-1.87779E+03-i1.80707E+03$
4	$-2.16515E+03-i1.19157E+03$
5	$-2.32467E+03-i5.92509E+02$
6	$-2.37603E+03+i0.00000E+00$
7	$-2.32467E+03+i5.29509E+02$
8	$-2.16515E+03+i1.19157E+03$
9	$-1.87779E+03+i1.80707E+03$
10	$-1.41101E+03+i2.46182E+03$
11	$-5.53363E+02+i3.28058E+03$

Table 9.2 Comparison of the expansion coefficients C_j reconstructed by the HELS method and the analytic ones for an impulsively accelerating piston mounted on a sphere of radius a with a vertex angle of $\pm\theta° = \angle 15°$

C_j	Reconstructed values	Benchmark values
C_1	$+6.69999E-02$	$+6.70000E-02$
C_2	$+1.87500E-01$	$+1.87500E-01$
C_3	$+2.70633E-01$	$+2.70633E-01$
C_4	$+3.00783E-01$	$+3.00781E-01$
C_5	$+2.74016E-01$	$+2.74016E-01$
C_6	$+1.98730E-01$	$+1.98730E-01$
C_7	$+9.34531E-02$	$+9.34529E-02$
C_8	$-1.76239E-02$	$-1.76239E-02$
C_9	$-1.10301E-01$	$-1.10301E-01$
C_{10}	$-1.65869E-01$	$-1.65869E-01$
C_{11}	$-1.75139E-01$	$-1.75140E-01$

acoustic pressure signals on the hologram surface can be taken as input to Eqs. (9.53) and (9.54) to reconstruct the acoustic pressure and particle velocity.

In this example the reconstructed acoustic pressure and particle velocity are obtained using the HELS-based NAH. Numerical computations involved in the BEM-based NAH are excessively intensive as compared to those of the HELS-based NAH and are omitted here for brevity.

Specifically, Eq. (9.54) is used to reconstruct the normal surface velocity with its temporal kernel $g_{vp}\left(\vec{x}\middle|\vec{x}_m^{\Gamma};t\right)$ determined by Eq. (9.56). Table 9.1 lists the singularities of $g_{vp}\left(\vec{x}\middle|\vec{x}_m^{\Gamma};t\right)$ that are obtained by using Eq. (9.58), namely, ω_q^{vp}, $q=1$ to 11, in this case.

Substituting ω_q^{vp} into Eq. (9.54) and evaluating the residues give the reconstructed normal surface velocity. Table 9.2 shows the comparison of the reconstructed expansion coefficients with benchmark values. Results indicate that the accuracy in the reconstructed expansion coefficients is guaranteed up to the 5th decimal point.

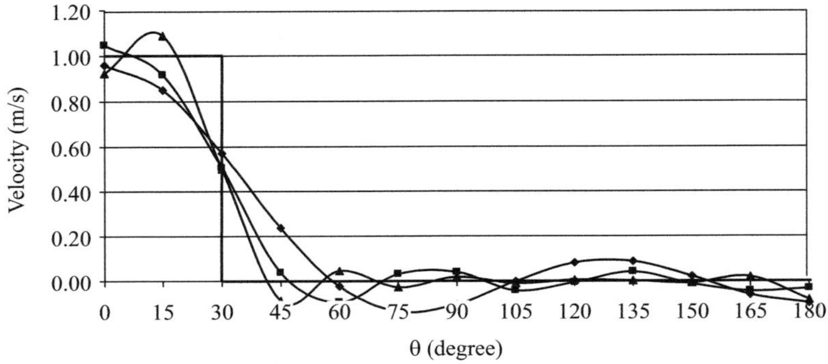

Fig. 9.5 Comparison of the reconstructed normal surface velocity distributions along the generator of the surface of a partially impulsively accelerated sphere of radius a. *dashed line*: Exact velocity profile; *filled diamond*: Reconstructed with $J=4$; *filled square*: Reconstructed with $J=8$; *filled triangle*: Reconstructed with $J=11$

Figure 9.5 displays the comparison of the normal surface velocity reconstructed by using Eq. (9.54) under various numbers of the expansion terms with Eq. (9.83). Results indicate that the reconstructed normal surface velocity converges to the correct velocity profile as the number of expansion terms increases from $J=4$, 8, and 11.

Figure 9.6 demonstrates the acoustic pressure fields reconstructed by using Eq. (9.53) at four different time instances: $t=3.24$ ms; $t=4.41$ ms; $t=5.88$ ms; and $t=7.35$ ms [145].

Example 9.6 (An Impulsively Accelerated Baffled Circular Disk) Finally, consider reconstruction of transient acoustic radiation from a non-spherical object. Specifically, Eq. (9.53) is utilized to reconstruct the acoustic pressure generated by an impulsive accelerated circular disk of radius a mounted on an infinite baffled. The normal surface velocity of this baffled disk is given by

$$v_{\text{n}}\left(\vec{x}_s;t\right) = V_s H(a-r)H(t). \tag{9.88}$$

The procedures for reconstruction are exactly the same as those described in Example 9.5 and not repeated here. The reconstructed temporal acoustic pressures are obtained by Eq. (9.53). Figure 9.7 shows the locations at which the input acoustic pressure signals are collected. Assume that the baffled circular disk is axisymmetric with respect to the z-axis. The temporal acoustic pressure signals on the hologram surface are given by

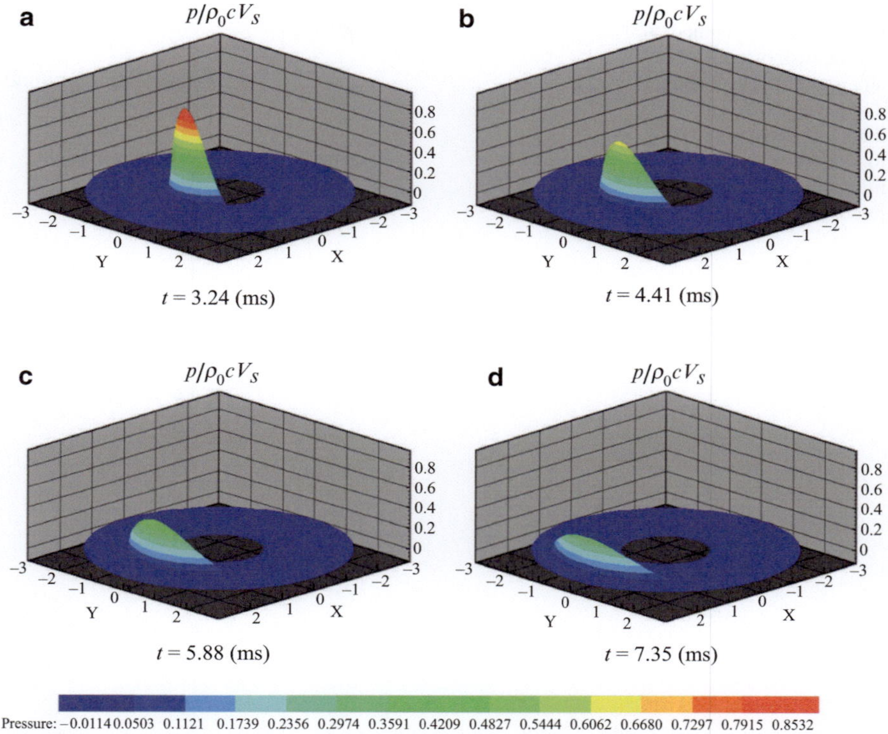

Fig. 9.6 Reconstructed acoustic pressure fields generated by a partially impulsively accelerated sphere of radius a at different time instances. (**a**): $t = 3.24$ ms; (**b**): $t = 4.41$ ms; (**c**): $t = 5.88$ ms; and (**d**): $t = 7.35$ ms

$$p\left(\vec{x}_m; t\right) = \begin{cases} 0, & ct < z_m \\ 0, & w > a, z_m \le ct < \sqrt{(a-w)^2 + z_m^2} \\ \rho_0 c V_s, & w < a, z_m \le ct < \sqrt{(a-w)^2 + z_m^2} \\ \left(\dfrac{\rho_0 c V_s}{\pi}\right) \cos^{-1}\left[\dfrac{\left(c^2t^2 - z_m^2 + w^2 - a^2\right)}{2w\sqrt{c^2t^2 - z_m^2}}\right]; & \sqrt{(a-w)^2 + z_m^2} \le ct < \sqrt{(a+w)^2 + z_m^2} \\ 0 & ct \ge \sqrt{(a+w)^2 + z_m^2} \end{cases}, \tag{9.89}$$

Note that in this case the standoff distance z_m is of no concern because the input data given by Eq. (9.89) are analytic. The array of microphones extends to twice the diameter of the baffled disk with respect to its geometric center. The total number of microphones is 40 and microphone spacing is $\Delta = a/10$. Figure 9.8 illustrates the

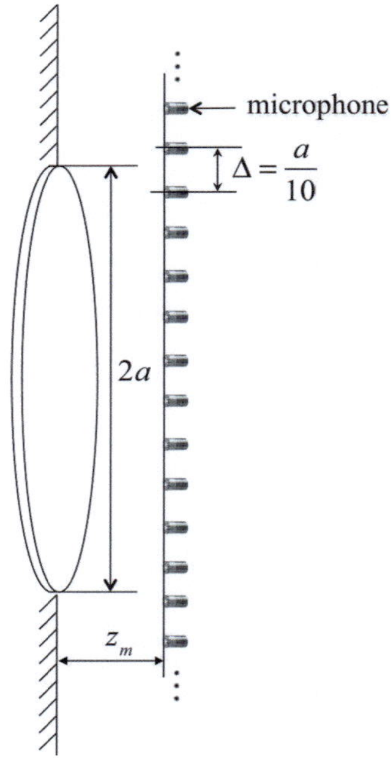

Fig. 9.7 Schematic of an impulsively accelerated baffled disk of radius a and an array of microphones that covers twice the diameter of the disk. The total number of microphones is 40 and the microphone spacing is $a/10$

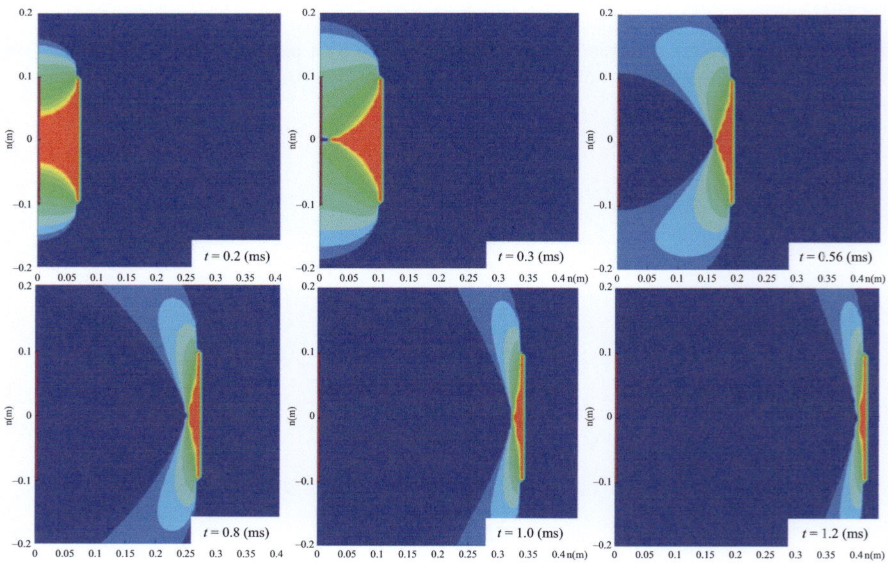

Fig. 9.8 Reconstructed acoustic pressure fields generated by an impulsively accelerated baffled disk of radius a at different time instances. Clockwise: $t = 0.2$ ms, $t = 0.3$ ms, $t = 0.56$ ms, $t = 0.8$ ms, $t = 1.0$ ms, and $t = 1.2$ ms

transient acoustic pressure fields captured at six time instances $t = 0.2$ ms, 0.3 ms, 0.56 ms, 0.8 ms, 1.0 ms, and 1.2 ms, respectively.

Problems

9.1. Show that the Kirchhoff–Helmholtz integral formulation for predicting transient acoustic radiation is given by Eq. (9.19).

9.2. Show that the Kirchhoff–Helmholtz integral equation for determining the transient surface acoustic pressure is given by Eq. (9.20).

9.3. Show that the Fourier transformed version of the Kirchhoff–Helmholtz integral formulation for predicting the acoustic pressure produced by an impulsively accelerating body is given by Eq. (9.25).

9.4. Show that the transient acoustic pressure radiated from an impulsively accelerating object can be written as Eq. (9.30) through the residue theorem.

9.5. Consider an explosion that occurs in free space at $t = t_0$. Assume that the particle velocity rises from near zero to a very high constant value (like a step function) omnidirectionally. Determine the resultant transient acoustic pressure anywhere in the field.

9.6. Consider the case in which an object is impacted by an external force and accelerates in a particular direction in free space. Assume that this impact occurs at $t = t_0$ and the velocity of the entire body rises from zero to a constant value (like a step function) in this direction. Determine the resultant transient acoustic pressure anywhere in the field.

9.7. Show that the transient acoustic pressure field generated by an arbitrarily shaped rigid body subject to an arbitrarily time-dependent excitation is expressible as a convolution integral given by Eq. (9.45).

9.8. Similarly, show that the normal component of the particle velocity in free space generated by an arbitrarily shaped rigid body subject to an arbitrarily time-dependent excitation can be written as a convolution integral given by Eq. (9.46).

9.9. Continue Problem 9.5. Assume that the time histories of the acoustic pressure signals that are measured at two arbitrary points in free space are $p\left(\vec{x}_m; t\right) = (Q/r_m)\mathrm{e}^{-(ct-r_m)/a}$, where Q is a constant and a is the characteristic dimension of the initial explosion region, $m = 1$ and 2. Determine the transient acoustic pressure field anywhere.

9.10. Continue Problem 9.6. Assume that the time histories of the acoustic pressure signals that are measured at two arbitrary points in free space are given

$$p\left(\vec{x}^{\Gamma}_m; t\right) = \left(\frac{Q}{r_m}\right)\mathrm{e}^{-(ct-r_m)/a}\cos\theta_m H\left(t - \frac{r_m}{c}\right)$$
$$\left[\cos\left(\frac{ct - r_m}{a}\right) - \left(1 - \frac{a}{r_m}\right)\sin\left(\frac{ct - r_m}{a}\right)\right],$$

where Q is a constant, θ indicates the polar angle, a is the characteristic dimension of the rigid body, and $m = 1$ and 2. Determine the transient acoustic pressure field anywhere.

Chapter 10
Panel Acoustic Contribution Analysis Using HELS

In this chapter we show how to use the HELS method to assess the relative contributions of individual panels of a complex vibrating structure toward SPL at any field point, for example, in diagnosing vehicle interior noise or reducing noise emission from any vibrating machinery. Being able to identify the major contributors of acoustic emission is the first step toward an effective noise reduction of a vibrating structure in engineering applications.

Theoretically, panel acoustic contribution analysis can be accomplished by calculating the acoustic power flows from individual panels of a vibrating structure and rank their contributions toward the SPL value at a field point of interest. It is important to understand that in practice, the excitations and boundary conditions of a vibrating structure are unknown a priori. Therefore, there is no way to predict acoustic radiation, not to mention calculating the relative contributions from individual panel surfaces of a vibrating structure.

This difficulty can be circumvented by utilizing NAH technology to reconstruct the vibro-acoustic quantities on the surface of a vibrating structure, based on the acoustic pressure measured around the target structure. Once the surface acoustic quantities are specified, the acoustic power flows from individual panel surfaces can be calculated, and their relative contributions be assessed and ranked. Note that one can select any approach to implement NAH. Here we choose the HELS method to accomplish this goal.

S.F. Wu, *The Helmholtz Equation Least Squares Method*, Modern Acoustics and Signal Processing, DOI 10.1007/978-1-4939-1640-5_10

10.1 The HELS-Based Panel Acoustic Contribution Analysis

The most important aspect of panel contributions analysis is the establishment of the direct correlations between acoustic power flows from individual panels of a vibrating structure to the SPL values at a field point [146]. To derive these correlations, we start from the definition of the SPL value at any field point.

$$L_p\left(\vec{x};\omega\right) = 10\log_{10}\left[\frac{\hat{p}^{\,2}_{\mathrm{rms}}\left(\vec{x};\omega\right)}{\hat{p}^{\,2}_{\mathrm{ref}}}\right], \tag{10.1}$$

where $\hat{p}^{\,2}_{\mathrm{rms}}\left(\vec{x};\omega\right)$ implies the mean-squared acoustic pressure at any field point $\vec{x}$ of interest, and $\hat{p}_{\mathrm{ref}} = 20$ (μPa) is the reference acoustic pressure for air used in practice. By definition $\hat{p}^{\,2}_{\mathrm{rms}}\left(\vec{x};\omega\right)$ is linked to the complex amplitude of the acoustic pressure $\hat{p}\left(\vec{x};\omega\right)$ at $\vec{x}$ through

$$\hat{p}^{\,2}_{\mathrm{rms}}\left(\vec{x};\omega\right) = \frac{\left|\hat{p}\left(\vec{x};\omega\right)\right|^2}{2} = \frac{1}{2}\mathrm{Re}\left|\hat{p}\left(\vec{x};\omega\right)\hat{p}^*\left(\vec{x};\omega\right)\right|. \tag{10.2}$$

The question is: "How is $\hat{p}\left(\vec{x};\omega\right)$ at any field point $\vec{x}$ related to the vibro-acoustic responses of a vibrating structure?" Answer to this question is the key to the analysis of the structure-borne sound radiation. Currently, this problem is being tackled by a brute force way in the industry to locate the noise sources, and establish correlations between structural vibrations and sound radiation. Some typical approaches include the uses of scanning an intensity probe over target vibrating structures, transfer path analysis (TPA) [147], and acoustic reciprocity principle (ARP) [148]. These approaches are measurement based. In particular, TPA and ARP require taking two sets of measurements:

1. Measure the transfer functions between the excitations at ad hoc positions on test structures and the SPL values at designated field points for TPA; or the transfer functions between the volume velocity of a known point source placed at the designated field point and the normal surface velocities and acoustic pressures on the panels of a test structure under a laboratory condition for ARP.
2. Measure the vibration responses of the structures under the actual operating conditions for TPA, or the normal surface velocities and acoustic pressures on the panels of the structures under the actual operating conditions.

The relative panel acoustic contributions toward the SPL values at designated field points are obtained by multiplying the transfer functions specified during the first round of measurements by the vibro-acoustic responses on the panel surfaces of the structure obtained during the second round of measurements. Needless to say, these types of approaches are not only time consuming but also ambiguous. This is because the TPA formulations are ad hoc in nature, and test conditions in two rounds of measurements for ARP are not exactly the same. Furthermore, the panel acoustic contributions analyses are restricted to the preselected field points at which the transfer functions are measured. If panel contributions toward the SPL values at the field points that are not included in the first round of measurements are needed, the same measurement procedures must be repeated. So the existing approaches are labor intensive, and their results might be called into question.

Our goal is to derive formulations that enable one to correlate the acoustic power flow from any vibrating panel to any field point of interest in the most cost-effective manner. In particular, these correlations must be valid for SPL values at any number of field points without the need to retake input data, thus significantly reducing the measurement complexities and time.

To this end, we reconstruct $\hat{p}\left(\vec{x};\omega\right)$ by using the HELS formulations. Section 3.5 exhibits that the complex amplitude of the acoustic pressure at a field point $\vec{x}$ can be reconstructed based on the measurements of the acoustic pressures in the near field of a vibrating structure,

$$\hat{p}\left(\vec{x}^{\,\mathrm{rec},\alpha};\omega\right)=\left[V_p^{\beta}\right]_{1\times1}\left[F_p^{\beta}\right]_{1\times1}\left\{\Sigma_p^{\beta}\right\}_{1\times M}^{-\mathrm{T}}\left[U_p^{\beta}\right]_{M\times M}^{H}\left\{\hat{p}\left(\vec{x}_m^{\,\mathrm{meas}};\omega\right)\right\}_{M\times1}, \tag{10.3}$$

where $[V_p^{\beta}]_{1\times1}$ and $[U_p^{\beta}]_{M\times M}^{H}$ are the right and left unitary orthonormal matrices, respectively, for the transfer function [see Eq. (3.16)] correlating the measured acoustic pressure $\left\{\hat{p}\left(\vec{x}_m^{\,\mathrm{meas}};\omega\right)\right\}_{M\times1}$ to reconstructed field acoustic pressure $\hat{p}\left(\vec{x};\omega\right)$, $[F_p^{\beta}]_{1\times1}$ represents the low-pass filter defined in Eq. (3.68), and $\{\Sigma_p^{\beta}\}_{1\times M}^{-\mathrm{T}}$ contains the inverted singular values of this transfer matrix.

In order to correlated the field acoustic pressures $\hat{p}\left(\vec{x};\omega\right)$ to the vibro-acoustic quantities on the surface of a vibrating structure, we first reconstruct the acoustic pressure on the surface of a target structure by using Eq. (3.66),

$$\left\{\hat{p}\left(\vec{x}_s^{\,\mathrm{rec},\alpha};\omega\right)\right\}_{S\times1}=\left[V_p^{\lambda}\right]_{S\times S}\left[F_p^{\lambda}\right]_{S\times S}\left[\Sigma_p^{\lambda}\right]_{S\times M}^{-1}\left[U_p^{\lambda}\right]_{M\times M}^{H}\left\{p\left(\vec{x}_m^{\,\mathrm{meas}};\omega\right)\right\}_{M\times1}, \tag{10.4}$$

where $\vec{x}_s^{\,\mathrm{rec},\alpha}$ represents a surface point; $[V_p^{\lambda}]_{1\times1}$ and $[U_p^{\lambda}]_{M\times M}^{H}$ stand for the right and left unitary orthonormal matrices, respectively, for the transfer function given by Eq. (3.16) that correlates the measured acoustic pressure $\left\{\hat{p}\left(\vec{x}_m^{\,\mathrm{meas}};\omega\right)\right\}_{M\times1}$ to reconstructed acoustic pressure $\left\{\hat{p}\left(\vec{x}_s^{\,\mathrm{rec},\alpha};\omega\right)\right\}_{S\times1}$ on the surface of a vibrating structure; $[F_p^\lambda]_{1\times 1}$ indicates the low-pass filter defined in Eq. (3.68); and $\{\Sigma_p^\lambda\}_{1\times M}^{-\mathrm{T}}$ contains the inverted singular values of this transfer matrix.

Next, we invert Eq. (10.4) as follows:

$$\left\{\hat{p}\left(\vec{x}_m^{\,\mathrm{meas}};\omega\right)\right\}_{M\times 1}=\left[U_p^\lambda\right]_{M\times M}\left[\Sigma_p^\lambda\right]_{M\times S}\left[F_p^\lambda\right]_{S\times S}^{-1}\left[V_p^\lambda\right]_{S\times S}^{H}\left\{\hat{p}\left(\vec{x}_s^{\,\mathrm{rec},\alpha};\omega\right)\right\}_{S\times 1}. \tag{10.5}$$

Substituting Eq. (10.5) into (10.3) yields

$$\hat{p}\left(\vec{x}^{\,\mathrm{rec},\alpha};\omega\right)=\left\{T_{\mathrm{pp}}\left(\vec{x}^{\,\mathrm{rec},\alpha}\middle|\vec{x}_s^{\,\mathrm{rec},\alpha};\omega\right)\right\}_{1\times S}^{\mathrm{T}}\left\{\hat{p}\left(\vec{x}_s^{\,\mathrm{rec},\alpha};\omega\right)\right\}_{S\times 1}, \tag{10.6}$$

where $\left\{T_{\mathrm{pp}}\left(\vec{x}^{\,\mathrm{rec},\alpha}\middle|\vec{x}_s^{\,\mathrm{rec},\alpha};\omega\right)\right\}_{1\times S}^{\mathrm{T}}$ implies the transfer function that correlates the reconstructed field acoustic pressure $\hat{p}\left(\vec{x}^{\,\mathrm{rec},\alpha};\omega\right)$ at any field point $\vec{x}^{\,\mathrm{rec},\alpha}$ to $\hat{p}\left(\vec{x}_s^{\,\mathrm{rec},\alpha};\omega\right)$ at a surface point $\vec{x}_s^{\,\mathrm{rec},\alpha}$,

$$\begin{aligned}&\left\{T_{\mathrm{pp}}\left(\vec{x}^{\,\mathrm{rec},\alpha}\middle|\vec{x}_s^{\,\mathrm{rec},\alpha};\omega\right)\right\}_{1\times S}^{\mathrm{T}}\\&=\left[V_p^\beta\right]_{1\times 1}\left[F_p^\beta\right]_{1\times 1}\left\{\Sigma_p^\beta\right\}_{1\times M}^{-\mathrm{T}}\left[U_p^\beta\right]_{M\times M}^{H}\left[U_p^\lambda\right]_{M\times M}\left[\Sigma_p^\lambda\right]_{M\times S}\left[F_p^\lambda\right]_{S\times S}^{-1}\left[V_p^\lambda\right]_{S\times S}^{H}.\end{aligned} \tag{10.7}$$

In a similar manner, we can reconstruct the normal surface velocity by using Eq. (3.67),

$$\left\{\hat{v}_{\mathrm{n}}\left(\vec{x}_s^{\,\mathrm{rec},\alpha};\omega\right)\right\}_{S\times 1}=\left[V_v^\lambda\right]_{S\times S}\left[F_v^\lambda\right]_{S\times S}\left[\Sigma_v^\lambda\right]_{S\times M}^{-1}\left[U_v^\lambda\right]_{M\times M}^{H}\left\{p\left(\vec{x}_m^{\,\mathrm{meas}};\omega\right)\right\}_{M\times 1}, \tag{10.8}$$

where $[V_v^\lambda]_{1\times 1}$ and $[U_v^\lambda]_{M\times M}^H$ are the right and left unitary orthonormal matrices, respectively, for the transfer function given by Eq. (3.18) that correlates $\left\{\hat{p}\left(\vec{x}_m^{\,\mathrm{meas}};\omega\right)\right\}_{M\times 1}$ to reconstructed normal surface velocity $\left\{\hat{v}_{\mathrm{n}}\left(\vec{x}_s^{\,\mathrm{rec},\alpha};\omega\right)\right\}_{S\times 1}$ on the surface of a vibrating structure, $[F_v^\lambda]_{1\times 1}$ indicates the low-pass filter defined in Eq. (3.69), and $\{\Sigma_v^\lambda\}_{1\times M}^{-\mathrm{T}}$ contains the inverted singular values of this transfer matrix. Once again, we invert Eq. (10.8) as follows:

$$\left\{\hat{p}\left(\vec{x}_m^{\,\mathrm{meas}};\omega\right)\right\}_{M\times 1}=\left[U_v^\lambda\right]_{M\times M}\left[\Sigma_v^\lambda\right]_{M\times S}\left[F_v^\lambda\right]_{S\times S}^{-1}\left[V_v^\lambda\right]_{S\times S}^{H}\left\{\hat{v}_{\mathrm{n}}\left(\vec{x}_s^{\,\mathrm{rec},\alpha};\omega\right)\right\}_{S\times 1}. \tag{10.9}$$

Substituting Eq. (10.9) into (10.3) leads to the relationship between the field acoustic $\hat{p}\left(\overrightarrow{x}^{\mathrm{rec},\alpha};\omega\right)$ and the reconstructed normal surface velocity $\hat{v}_{\mathrm{n}}\left(\overrightarrow{x}_{s}^{\mathrm{rec},\alpha};\omega\right)$,

$$\hat{p}\left(\overrightarrow{x}^{\mathrm{rec},\alpha};\omega\right)=\left\{T_{\mathrm{pv}}\left(\overrightarrow{x}^{\mathrm{rec},\alpha}\middle|\overrightarrow{x}_{s}^{\mathrm{rec},\alpha};\omega\right)\right\}_{1\times S}^{\mathrm{T}}\left\{\hat{v}_{\mathrm{n}}\left(\overrightarrow{x}_{s}^{\mathrm{rec},\alpha};\omega\right)\right\}_{S\times 1},\qquad(10.10)$$

where $\left\{T_{\mathrm{pv}}\left(\overrightarrow{x}^{\mathrm{rec},\alpha}\middle|\overrightarrow{x}_{s}^{\mathrm{rec},\alpha};\omega\right)\right\}_{1\times S}^{\mathrm{T}}$ is the transfer function given by

$$\begin{aligned}&\left\{T_{\mathrm{pv}}\left(\overrightarrow{x}^{\mathrm{rec},\alpha}\middle|\overrightarrow{x}_{s}^{\mathrm{rec},\alpha};\omega\right)\right\}_{1\times S}^{\mathrm{T}}\\&=\left[V_{p}^{\beta}\right]_{1\times 1}\left[F_{p}^{\beta}\right]_{1\times 1}\left\{\Sigma_{p}^{\beta}\right\}_{1\times M}^{-\mathrm{T}}\left[U_{p}^{\beta}\right]_{M\times M}^{H}\left[U_{v}^{\lambda}\right]_{M\times M}\left[\Sigma_{v}^{\lambda}\right]_{M\times S}\left[F_{v}^{\lambda}\right]_{S\times S}^{-1}\left[V_{v}^{\lambda}\right]_{S\times S}^{H}.\end{aligned}\qquad(10.11)$$

Equations (10.6) and (10.10) demonstrate that the field acoustic pressure $\hat{p}\left(\overrightarrow{x}^{\mathrm{rec},\alpha};\omega\right)$ can be correlated to the reconstructed surface acoustic pressure $\hat{p}\left(\overrightarrow{x}_{s}^{\mathrm{rec},\alpha};\omega\right)$ and the reconstructed normal surface velocity $\hat{v}_{\mathrm{n}}\left(\overrightarrow{x}_{s}^{\mathrm{rec},\alpha};\omega\right)$.

Taking the complex conjugate of the reconstructed field acoustic pressure $\hat{p}\left(\overrightarrow{x}^{\mathrm{rec},\alpha};\omega\right)$ in Eq. (10.2), we obtain

$$\hat{p}^{*}\left(\overrightarrow{x}^{\mathrm{rec},\alpha};\omega\right)=\left\{\hat{v}_{\mathrm{n}}\left(\overrightarrow{x}_{s}^{\mathrm{rec},\alpha};\omega\right)\right\}_{1\times S}^{H}\left[T_{\mathrm{pv}}\left(\overrightarrow{x}^{\mathrm{rec},\alpha}\middle|\overrightarrow{x}_{s}^{\mathrm{rec},\alpha};\omega\right)\right]_{S\times 1}^{H}.\qquad(10.12)$$

Substituting Eqs. (10.6) and (10.12) into Eq. (10.3) yields the mean-squared acoustic pressure at any field point $\overrightarrow{x}$,

$$\begin{aligned}&\hat{p}_{\mathrm{rms}}^{2}\left(\overrightarrow{x};\omega\right)\\&=\mathrm{Re}\left\langle\left[T_{\mathrm{pp}}\left(\overrightarrow{x}^{\mathrm{rec},\alpha}\middle|\overrightarrow{x}_{s}^{\mathrm{rec},\alpha};\omega\right)\right]_{1\times S}\left\{\hat{I}_{\mathrm{av,n}}\left(\overrightarrow{x}_{s}^{\mathrm{rec},\alpha};\omega\right)\right\}_{S\times S}\left[T_{\mathrm{pv}}\left(\overrightarrow{x}^{\mathrm{rec},\alpha}\middle|\overrightarrow{x}_{s}^{\mathrm{rec},\alpha};\omega\right)\right]_{S\times 1}^{H}\right\rangle,\end{aligned}\qquad(10.13)$$

where $\hat{I}_{\mathrm{av,n}}\left(\overrightarrow{x}_{s}^{\mathrm{rec},\alpha};\omega\right)$ is the normal component of the time-averaged acoustic intensity given by

$$\left\{\hat{I}_{\mathrm{av,n}}\left(\overrightarrow{x}_{s}^{\mathrm{rec}};\omega\right)\right\}_{S\times S}=\frac{1}{2}\mathrm{Re}\left\langle\left\{\hat{p}\left(\overrightarrow{x}_{s}^{\mathrm{rec},\alpha};\omega\right)\right\}_{S\times 1}\left\{\hat{v}_{\mathrm{n}}\left(\overrightarrow{x}_{s,n}^{\mathrm{rec}};\omega\right)\right\}_{1\times S}^{H}\right\rangle.\qquad(10.14)$$

Substituting Eq. (10.13) into (10.1) then leads to

$$L_p\left(\overset{\rightarrow\text{rec},\alpha}{x};\omega\right) = 10\log_{10}\left[\frac{\text{Re}\left\langle\left[T_{\text{pp}}\left(\overset{\rightarrow\text{rec},\alpha}{x}\middle|\overset{\rightarrow\text{rec},\alpha}{x_s};\omega\right)\right]_{1\times S}\left\{\hat{I}_{\text{av,n}}\left(\overset{\rightarrow\text{rec},\alpha}{x_s};\omega\right)\right\}_{S\times S}\left[T_{\text{pv}}\left(\overset{\rightarrow\text{rec},\alpha}{x}\middle|\overset{\rightarrow\text{rec},\alpha}{x_s};\omega\right)\right]^{H}_{S\times 1}\right\rangle}{\hat{p}^2_{\text{ref}}}\right]. \tag{10.15}$$

Equation (10.15) indicates that the SPL value at any field point can be correlated to the normal component of the time-averaged acoustic intensity on the surfaces of a vibrating structure. Suppose that the structure consists of individual panels. By summing the normal components of the time-averaged acoustic intensities on these panels, one can calculate the acoustic power flows from individual panels and rank their contributions to the SPL value at any field point of interest. This is the essence of the HELS-based panel acoustic contribution analysis.

It is emphasized that no assumption is made in deriving Eq. (10.1) to Eq. (10.15), and the panel contributions analyses are valid for any number of field points based on a single set of input data. In other words, analyses can be repeated to reveal the critical panels responsible for acoustic radiation anywhere without the need to retake the data, thus significantly reducing the overall cost and effort in noise diagnosis.

10.2 Procedures for Conducting HELS-Based Panel Acoustic Contributions Analyses

The procedures for carrying out the HELS-based panel acoustic contributions analyses for a vibrating structure in either interior or exterior region are as follows:

1. Follow all the guidelines listed in Sect. 3.5 to prepare for the measurement setup to measure the near-field acoustic pressures generated by a vibrating structure.
2. Sometimes it might be desirable to take the measurements of the acoustic pressures at a few field points of interest. The spectra of the measured acoustic pressures can be used as benchmarks to validate the reconstruction accuracy. However, this is not a required step. It is completely up to the user who is conducting the panel acoustic contributions analyses. In any event, measure the field acoustic pressure $\left\{\hat{p}\left(\overset{\rightarrow\text{meas}}{x_m};\omega\right)\right\}_{M\times 1}$.
3. For a large structure, multiple patches of measurements are required. So it is important to set at least one fixed reference point as the patch measurements are moved from one location to another. Measure the acoustic pressure

$\hat{p}\left(\vec{x}_{\text{ref}}^{\text{meas}};\omega\right)$ at the reference point.

4. Measure the transfer function between $\left\{\hat{p}\left(\vec{x}_{\text{m}}^{\text{meas}};\omega\right)\right\}_{M\times 1}$ at each measurement point and $\hat{p}\left(\vec{x}_{\text{ref}}^{\text{meas}};\omega\right)$ at the reference point as the measurement patches are moved from one place to another.
5. Multiply the transfer functions by the reference acoustic pressure $\hat{p}\left(\vec{x}_{\text{ref}}^{\text{meas}};\omega\right)$, which is equivalent to taking measurements at all points $\left\{\hat{p}\left(\vec{x}_{m}^{\text{meas}};\omega\right)\right\}_{M\times 1}$ simultaneously.
6. Use Eqs. (10.4) and (10.8) to reconstruct the surface acoustic pressure $\hat{p}\left(\vec{x}_{s}^{\text{rec},\alpha};\omega\right)$ and normal surface velocity $\hat{v}_{\text{n}}\left(\vec{x}_{s}^{\text{rec},\alpha};\omega\right)$.
7. Use Eq. (10.14) to calculate the reconstructed normal component of the time-averaged acoustic intensity $\hat{I}_{\text{av,n}}^{s}\left(\vec{x}_{s}^{\text{rec},\alpha};\omega\right)$ at all surface points $\vec{x}_{s}^{\text{rec},\alpha}$ $s=1$, 2, . . ., S.
8. Divide the structure into any number of panel surfaces, $\ell = 1, 2, \ldots, \mathbf{N}$, and associate the normal surface acoustic intensities $\hat{I}_{\text{av,n}}^{\ell}\left(\vec{x}_{s}^{\text{rec},\alpha};\omega\right)$ with **N** panel surfaces.
9. Use Eqs. (10.7) and (10.9) to establish transfer functions $\left\{T_{\text{pp}}\left(\vec{x}^{\text{rec},\alpha}\middle|\vec{x}_{s}^{\text{rec},\alpha};\omega\right)\right\}_{1\times S}^{\text{T}}$ and $\left\{T_{\text{pv}}\left(\vec{x}^{\text{rec},\alpha}\middle|\vec{x}_{s}^{\text{rec},\alpha};\omega\right)\right\}_{1\times S}^{\text{T}}$.
10. Use Eq. (10.15) to calculate the SPL value at any field point $L_p\left(\vec{x}^{\text{rec},\alpha};\omega\right)S$.
11. Calculate the acoustic power flow $\mathbf{P}_{\text{av},\ell}(\omega)$ from each panel by multiplying the normal surface acoustic intensity $\hat{I}_{\text{av,n}}^{\ell}\left(\vec{x}_{s}^{\text{rec},\alpha};\omega\right)$ by the area of the panel ΔS_{ℓ}, $\ell = 1, 2, \ldots, \mathbf{N}$. The sum of the acoustic power $\mathbf{P}_{\text{av},\ell}(\omega)$ from the individual panel surfaces should be equal to the total acoustic power $\mathbf{P}_{\text{av}}(\omega)$ generated by the entire surface of the vibrating structure.

$$\mathbf{P}_{\text{av}}(\omega) = \sum_{\ell=1}^{\mathbf{N}} \mathbf{P}_{\text{av},\ell}(\omega) = \sum_{\ell=1}^{\mathbf{N}} \left[\hat{I}_{\text{av,n}}^{\ell}(\omega)\Delta S_{\ell}\right], \tag{10.16}$$

where N is the total number of panels of a structure, and $\mathbf{P}_{\text{av},\ell}(\omega)$ is the acoustic power flow from the ℓth panel surface of area ΔS_{ν}.

12. Calculate contributions from the individual panel surfaces toward the SPL value at any field point,

$$L_p^\ell\left(\vec{x}^{\text{rec},\alpha};\omega\right) = 10\log_{10}\left[\frac{T_{\text{pp}}^\ell\left(\vec{x}^{\text{rec},\alpha}\middle|\vec{x}_s^{\text{rec},\alpha};\omega\right)\hat{I}_{\text{av,n}}^\ell\left(\vec{x}_s^{\text{rec},\alpha};\omega\right)T_{\text{pv}}^\ell\left(\vec{x}^{\text{rec},\alpha}\middle|\vec{x}_s^{\text{rec},\alpha};\omega\right)}{\hat{p}_{\text{ref}}^2}\right], \tag{10.17}$$

where $L_p^\ell\left(\vec{x}^{\text{rec},\alpha};\omega\right)$ represents the contribution from the ℓth panel surface of area ΔS_ν.

13. Calculate the acoustic power ranking for the ℓth panel surface of the vibrating structure,

$$\mathfrak{R} = \frac{L_p^\ell\left(\vec{x}^{\text{rec},\alpha};\omega\right)\Delta S_\ell}{L_p\left(\vec{x}^{\text{rec},\alpha};\omega\right)S} \times 100\%, \tag{10.18}$$

where $L_p^\ell\left(\vec{x}^{\text{rec},\alpha};\omega\right)\Delta S_\ell$ indicates the acoustic power produced by the ℓth panel surface and delivered to the field point $\vec{x}$ via the transfer functions $\left\{T_{\text{pp}}\left(\vec{x}^{\text{rec},\alpha}\middle|\vec{x}_s^{\text{rec},\alpha};\omega\right)\right\}_{1\times S}^{\text{T}}$ and $\left\{T_{\text{pv}}\left(\vec{x}^{\text{rec},\alpha}\middle|\vec{x}_s^{\text{rec},\alpha};\omega\right)\right\}_{1\times S}^{\text{T}}$, and $L_p\left(\vec{x}^{\text{rec},\alpha};\omega\right)S$ implies the total acoustic power generated by the entire vibrating structure and delivered to the same field point $\vec{x}$ through transfer functions $\left\{T_{\text{pp}}\left(\vec{x}^{\text{rec},\alpha}\middle|\vec{x}_s^{\text{rec},\alpha};\omega\right)\right\}_{1\times S}^{\text{T}}$ and $\left\{T_{\text{pv}}\left(\vec{x}^{\text{rec},\alpha}\middle|\vec{x}_s^{\text{rec},\alpha};\omega\right)\right\}_{1\times S}^{\text{T}}$.

Equation (10.18) clearly demonstrates that the acoustic power received at any field point $\vec{x}$ consists of two components: (1) the acoustic power generated by individual panel surfaces and (2) delivery of the acoustic power from an individual panel surface to a specific field point through the transfer functions $\left\{T_{\text{pp}}\left(\vec{x}^{\text{rec},\alpha}\middle|\vec{x}_s^{\text{rec},\alpha};\omega\right)\right\}_{1\times S}^{\text{T}}$ and $\left\{T_{\text{pv}}\left(\vec{x}^{\text{rec},\alpha}\middle|\vec{x}_s^{\text{rec},\alpha};\omega\right)\right\}_{1\times S}^{\text{T}}$. The former describes the effectiveness of acoustic power generation by an individual panel, and the latter depicts that of delivering this acoustic power to any field point. Both of them are critical from the viewpoint of structure-borne sound radiation. Hence, attention must be paid to both of them in order to reduce the structure-borne sound radiation in the most cost-effective manner.

10.3 Stories of Panel Acoustic Contributions Analyses

The following are true stories about engineers battling with vehicle interior noise reduction problems in the automobile industry.

10.3.1 Story 1: Sound Transmission Paths

Diagnosis and analysis of sound transmission paths into the vehicle passenger compartment have always been one of the major noise, vibration, and harshness challenges facing the automobile OEM (original equipment manufacturer) and the parts suppliers. Oftentimes, challenges come at the end of a production cycle and before an official launch of a vehicle or at recall of a vehicle due to customers' complaints of an unacceptable interior noise level. Consequently, engineers and managers are under tremendous pressure to meet certain deadline to resolve noise issues. In these cases, measurements of SPL values and spectra at the driver's and passengers' ears positions are taken, followed by a trial-and-error approach depending on one's noise abatement experiences. Hopefully, this quick fix would be enough to meet the NVH criteria and the problem is solved. If not, more elaborate noise diagnosis would be carried out and more effective noise mitigations be tested. This process continues until NVH requirements are met, and problems are solved.

This scenario occurs quite often in practice with minor variations here and there. The point is that oftentimes engineers rely solely on the measurements of SPL values and spectra in locating sound source and sound transmission paths into the vehicle passenger compartment. This may be adequate in dealing with airborne sound, but inappropriate in dealing with structure-borne sound and its transmission path.

Here is the story.

In 2004, a major automobile OEM in Japan was wrestling with locating sound transmission paths into the passenger compartment so as to reduce vehicle interior noise. Specifically, the test driver of a new vehicle reported high rumbling sounds, and engineers took measurements at the driver's ears positions and confirmed this rumbling noise problem. Then engineers measured the SPL values and spectra near the interior surface of the roof and were able to correlate the sounds measured there to those measured at the drivers' ears positions. This was a very common practice in noise diagnosis in industry. Therefore, they concluded that rumbling sounds were transmitted through the roof into the passenger compartment, but they were not sure which roof panel was the culprit. To locate this panel, engineers at this OEM used heavy sandbags to suppress vibrations of the roof panel one at a time, which once again was a common practice in industry. Surprisingly, it made no difference whatsoever where heavy sandbag was placed. Rumbling sounds remained the same as before, even after the entire roof surface was covered with heavy sandbags.

At the time the present author happened to be in Tokyo for a lecture tour on using NAH as an effective noise diagnostic tool. During Q&A session, an engineer used this case as an example and asked how to identify noise transmission paths into the passenger compartment.

This is a general question. The general answer is that, in dealing with structure-borne sound problem, one should focus on the acoustic power radiation from a vibrating surface, which is the product of the normal component of the time-averaged acoustic intensity and surface area, rather than the acoustic pressure, which is a scalar quantity that indicates the level of SPL value, but does not reveal the source of the acoustic energy or the acoustic power flow. Since the acoustic power is the area integration of the normal component of the time-averaged acoustic intensity, which is one-half of the real part of the product of the normal surface velocity and acoustic pressure, all we need to do is to determine the surface acoustic pressure and normal surface velocity. Once this is done, we can calculate the net outflow of the acoustic energy.

However, it is difficult at least at the present to measure the normal component of the time-averaged acoustic intensity on the surface of a vibrating structure directly. The intensity probe that is commonly used in practice can only measure the normal component of the acoustic intensity at certain distances away from a vibrating surface. This may create problems, especially in a relatively reverberant environment in which the interferences of sound reflections and reverberation tend to make the readings of the acoustic intensities unstable and unreliable. One solution is to utilize the NAH technology to reconstruct the normal component of the acoustic intensity on the surface of a vibrating structure.

For example, the panel acoustic contribution analysis discussed in Sects. 10.1 and 10.2 can be directly applied to a vehicle. In fact, such a panel contribution analysis has been done on a full-size vehicle in collaborations with engineers at a major automobile OEM in the USA. In that case, the acoustic pressure $\left\{\hat{p}\left(\vec{x}_m^{\,\text{meas}};\omega\right)\right\}_{M\times 1}$ inside the vehicle passenger compartment was measured at very close distances with respect to the interior surfaces by using a 4×8 conformal padel array of microphones. A total of 25 patches of measurements were taken, resulting in 800 measurement points covering the entire interior surface of the vehicle. The vehicle was running at two constant speeds, 60 and 100 kph, with its rear wheels mounted on a chassis dynamometer with coarse rolls to simulate the rough road surface condition.

The measured acoustic pressures $\left\{\hat{p}\left(\vec{x}_m^{\,\text{meas}};\omega\right)\right\}_{M\times 1}$, $m = 1, 2, \ldots, 800$, were taken as input to Eqs. (10.4) and (10.8) to reconstruct the surface acoustic pressure $\hat{p}\left(\vec{x}_s^{\,\text{rec},\alpha};\omega\right)$ and normal surface velocity $\hat{v}_{\text{n}}\left(\vec{x}_s^{\,\text{rec},\alpha};\omega\right)$. Once this is done, Eq. (10.14) is employed to calculate the normal component of the time-averaged surface acoustic intensity.

To illustrate this type of approach to identify transmission paths into the vehicle passenger compartment, the present author showed the reconstructed surface

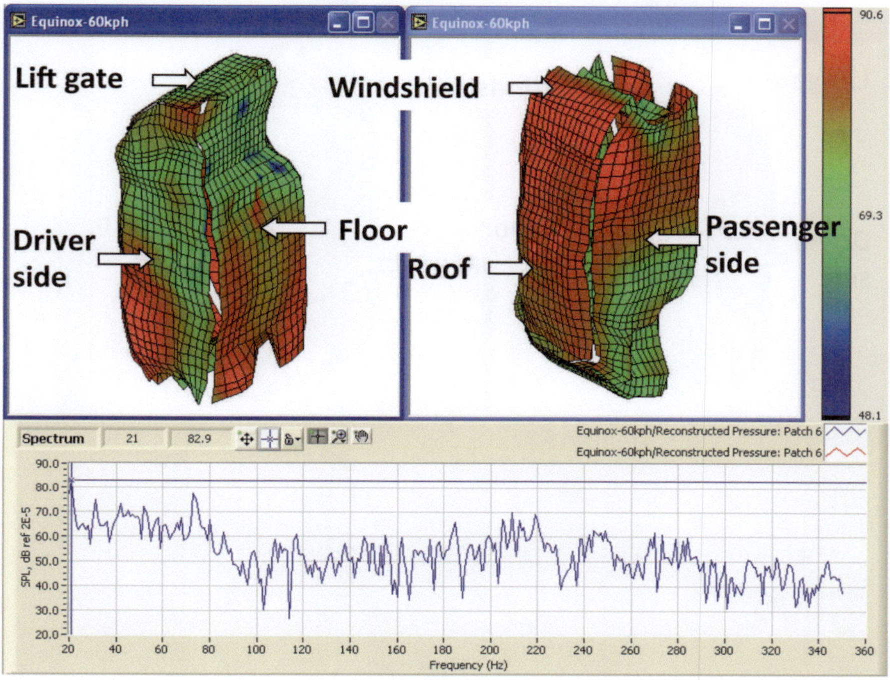

Fig. 10.1 The reconstructed acoustic pressure distribution at 21 Hz over the interior surface of a full-size vehicle with rear wheels mounted on a chassis dynamometer with coarse rolls that were running at 60 kph. The color scale of the acoustic pressure was from *red* (90 dB) to *blue* (40 dB)

acoustic pressures (see Fig. 10.1), the normal component of the surface velocities (see Fig. 10.2) and the normal component of the time-averaged acoustic intensities (see Fig. 10.3) on the interior surfaces of the vehicle at 21 Hz, where there was a dominant peak in the acoustic pressure spectrum. Results showed that the reconstructed acoustic pressure on the roof surface was indeed quite high [the color scale for the acoustic pressure was from red (90 dB) to blue (40 dB)]. However, this did not mean that there was an acoustic source on the roof surface because the normal surface velocity on the roof surface was identically zero [the color scale for the normal surface velocity was from red (0.2 m/s) to blue (0 m/s)]. In other words, the roof surface was motionless. The reconstructed normal component of the acoustic intensity on the root surface was zero as well, which indicated that there was no flow of the acoustic energy from the roof surface [the color scale for the normal component of the time-averaged acoustic intensity ranged from red ($+5 \times 10^{-3}$ W/m^2) to blue (-5×10^{-3} W/m^2)].

For proprietary reasons, the image of the test vehicle has been deleted in this example.

These results demonstrated that the roof surface just reflected the incident sound waves. It neither produced nor transmitted the rumbling sound. However, the

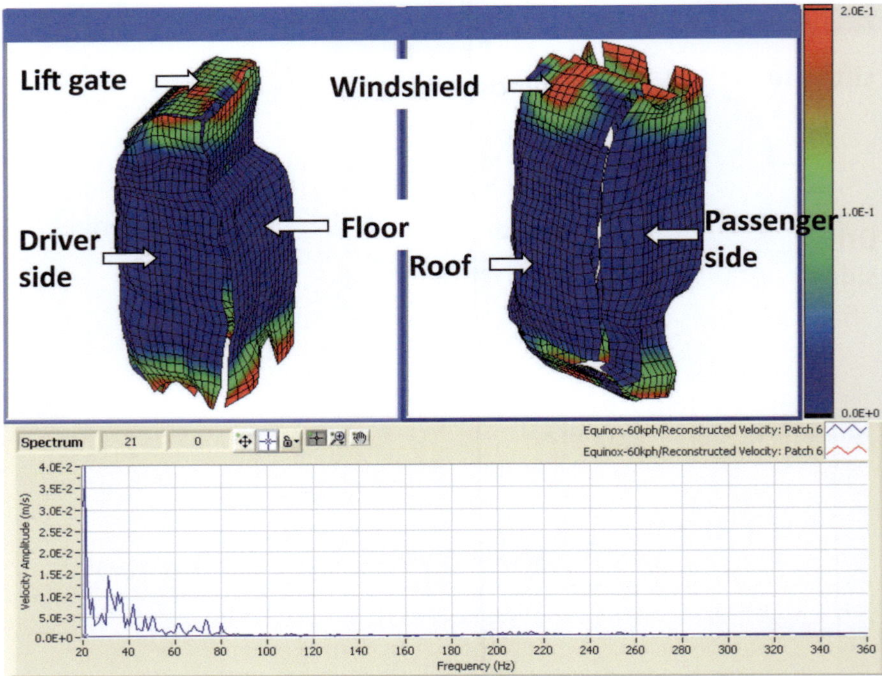

Fig. 10.2 The reconstructed normal component of the velocity distribution at 21 Hz on the entire interior surface of a full-size vehicle with its rear wheels mounted on a chassis dynamometer with coarse rolls that were running at 60 kph. The color scale of the normal surface velocity was from *red* (0.2 m/s) to *blue* (0 m/s)

driver's ears happened to be close to the roof surface, on which the amplitude of the acoustic pressure was very high. So it gave the (wrong) impression that the sound was coming through from the roof surface.

This was why suppressing the vibration of the roof panels with heavy sandbags came to no avail as demonstrated in the vehicle interior noise analysis conducted at a major automobile OEM in Japan.

The more likely transmission paths for this rumbling sound in this case might be through the front and end portion of the vehicle as indicated by the red areas where the normal component of the time-averaged acoustic intensity was the highest (see Fig. 10.3).

Figures 10.4, 10.5, and 10.6 display the reconstructed acoustic pressure, normal component of the velocity, and normal component of the time-averaged acoustic intensity distributions on the entire interior surface of the vehicle at 23 Hz. Once again, the amplitudes of the acoustic pressures were very high on the roof surface. However, the normal surface velocity and normal component of the time-averaged acoustic intensity were zero on the roof surface. This indicated that the roof merely reflected sound at this low frequency. The rumbling sound was transmitted through the panels on which the normal component of the time-averaged acoustic intensity was positive, for example, at the front and tail parts of the vehicle.

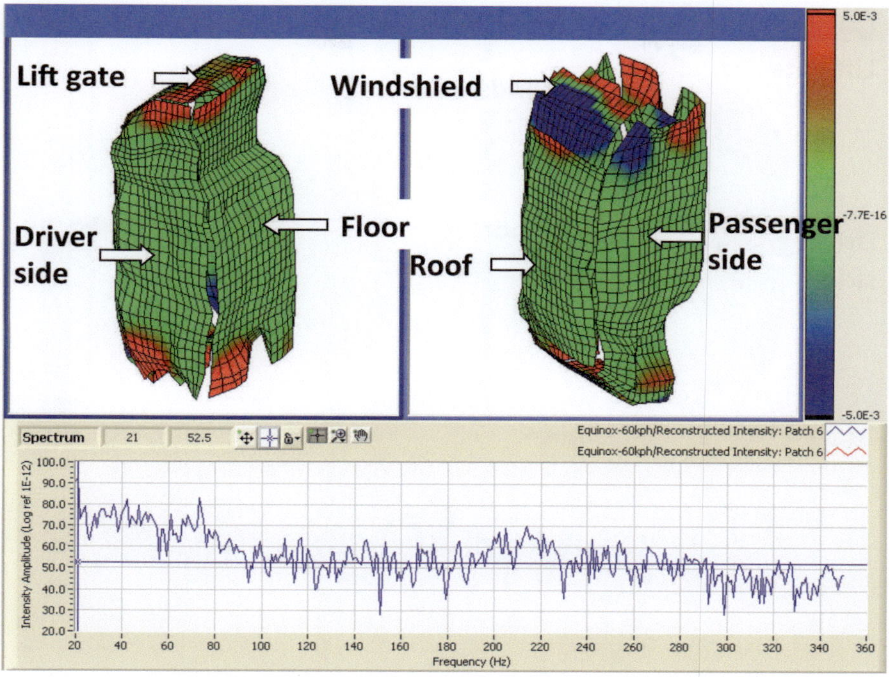

Fig. 10.3 The reconstructed normal component of the time-averaged acoustic intensity distribution at 21 Hz on the entire interior surface of a full-size vehicle with its rear wheels mounted on a chassis dynamometer with coarse rolls that were running at 60 kph. The color scale of the normal surface velocity was from *red* (+5 × 10^{-3} W/m^2) to *blue* (−5 × 10^{-3} W/m^2)

To check the accuracy in the reconstructed acoustic quantities, the acoustic pressure spectra at the drivers' and passengers' ears positions were reconstructed, and the results were compared to those of the benchmark values (see Figs. 10.7 and 10.8). For clarity in comparisons, the spectra zoomed in from 100 to 350 Hz range were displayed. Comparisons were made at four locations: (1) front driver (designated as FL), (2) front passenger (designated as FR), (3) rear left passenger (designated as RL), and (4) rear right passenger (designated as RR).

The agreements between the reconstructed and measured acoustic pressure spectra at these four critical locations were remarkably good, which gave the confidence for conducting the panels acoustic contributions analyses shown below.

10.3.2 Story 2: Panel Acoustic Contribution Analyses

The first story was related to the second one because engineers at this automobile OEM in the beginning also thought the rumbling sound was transmitted through the roof until they saw the analysis results presented in Figs. 10.1, 10.2, 10.3, 10.4,

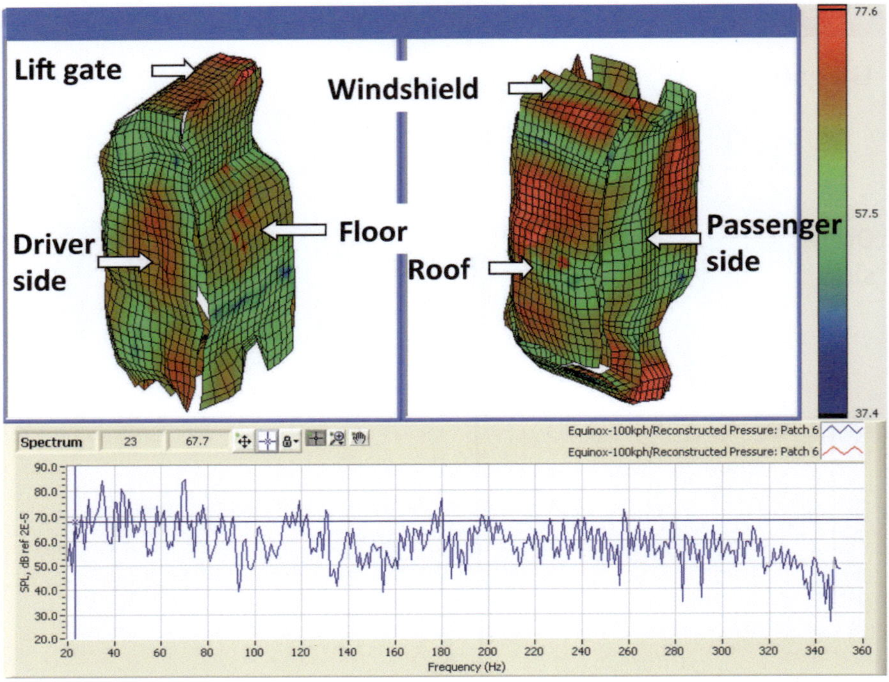

Fig. 10.4 The reconstructed acoustic pressure distribution at 23 Hz over the interior surface of a full-size vehicle with rear wheels mounted on a chassis dynamometer with coarse rolls that were running at 100 kph. The color scale of the acoustic pressure was from *red* (90 dB) to *blue* (40 dB)

10.5, 10.6, 10.7, and 10.8. Once they realized that noise transmission paths could be identified by using the HELS-based NAH, they wanted to examine the relative acoustic contributions from ten specific vehicle panels toward the SPL values measured at driver's, front passenger's, and rear passenger's ears positions. These panels included three on the front floor, six inside the trunk floor, and one for the jack storage area (see Fig. 10.9). Engineers wanted to identify the relative acoustic contributions from these panels toward the SPL values at the driver, front passenger, and rear passenger positions.

As mentioned in Sect. 10.2 (Step 8), users can designate any number of panels for the panel acoustic contributions analyses. There is no need to consider all panels of a target vibrating structure. However, measurements of the acoustic pressures must cover the entire source surface.

Following the steps outlined in Sect. 10.2, engineers reconstructed the surface acoustic pressures and normal surface velocities and calculated acoustic power flows from the designated ten panels. Their relative acoustic contributions at the designated field points were then assessed. It is emphasized that these panel

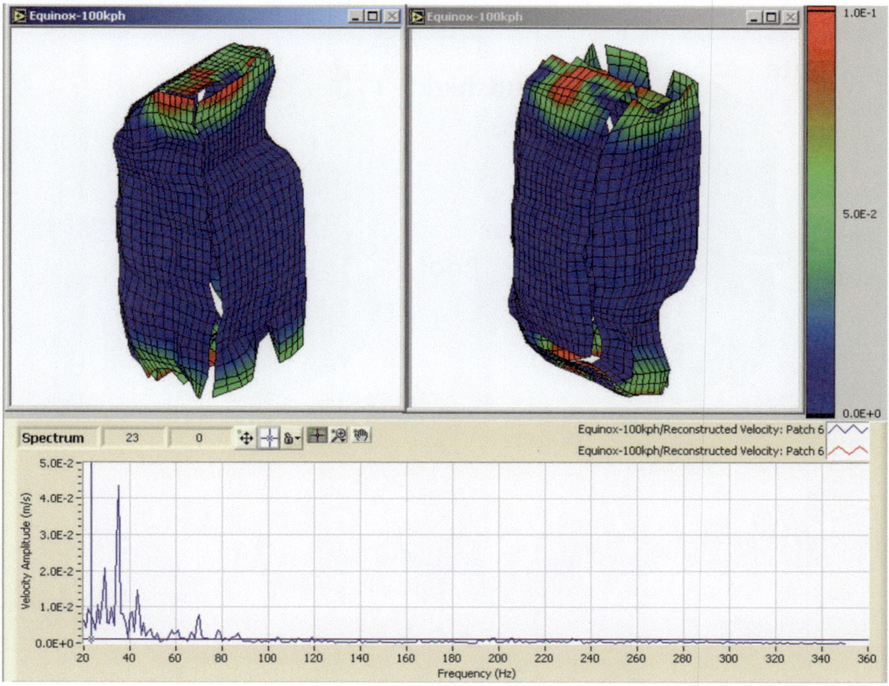

Fig. 10.5 The reconstructed normal component of the velocity distribution at 23 Hz on the entire interior surface of a full-size vehicle with its rear wheels mounted on a chassis dynamometer with coarse rolls that were running at 100 kph. The color scale of the normal surface velocity was from *red* (0.2 m/s) to *blue* (0 m/s)

acoustic contributions analyses can be continued at any number of field points without the need to retake any measurement again. All the input data are collected in the beginning. The panel acoustic contributions analyses can be performed at any frequency so long as input data were collected according to the guidelines specified in Sects. 5.2 and 5.3.

For brevity, representative panel acoustic contributions analyses are shown to illustrate the applications of the formulations given in Sect. 10.1. It is emphasized that these formulations are of generality because they are derived from the definition of SPL, and no assumptions are made.

Figure 10.10 showed that the jack storage area was the major contributor toward the SPL values at the designated field points at 104 Hz. Here, the ranking was normalized with respect to the highest contributor and multiplied by 10, and total values were summed together. In this manner, jack storage area was the major contributor, followed by Lift Gate Driver Side 2 and Lift Gate Middle 1 panels that were tied for the second major contributor toward SPL values at designated field points at 104 Hz. The users can design their own way to display order ranking for

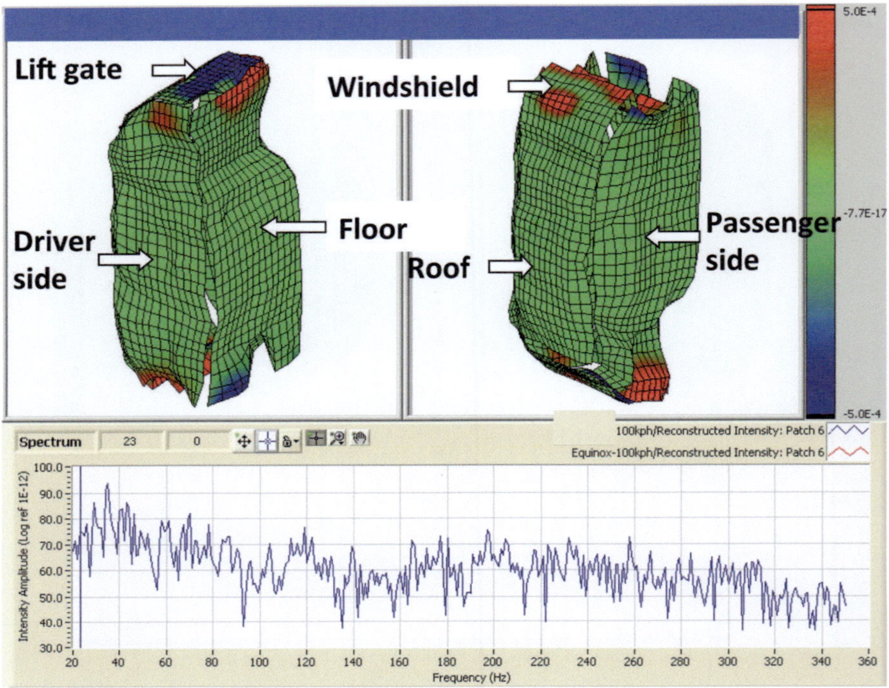

Fig. 10.6 The reconstructed normal component of the time-averaged acoustic intensity distribution at 23 Hz on the entire interior surface of a full-size vehicle with its rear wheels mounted on a chassis dynamometer with coarse rolls that were running at 100 kph. The color scale of the normal surface velocity was from *red* ($+5 \times 10^{-3}$ W/m^2) to *blue* (-5×10^{-3} W/m^2)

the major acoustic contributor. Results show that each panel can have different contributions toward the SPL values at different field points. This allows the user to focus on any specific panel to come up with desired noise reduction measures at the desired locations.

Figures 10.11, 10.12, 10.13, and 10.14 demonstrate similar panel acoustic contributions analyses results at 116 Hz, 235 Hz, 241 Hz, and 255 Hz, respectively. These frequencies represent peak amplitudes in the acoustic pressure spectra measured at the designated field points (see Figs. 10.7 and 10.8). These frequencies are chosen on purpose to illustrate the fact that different panels can become the major contributors at different frequencies. In this example, jack storage area was found to be the major contributor to the SPL values at the designated field points for the majority of frequencies. Once engineers at this automobile OEM applied noise reduction measures to jack storage area, the overall vehicle interior noise was significantly reduced.

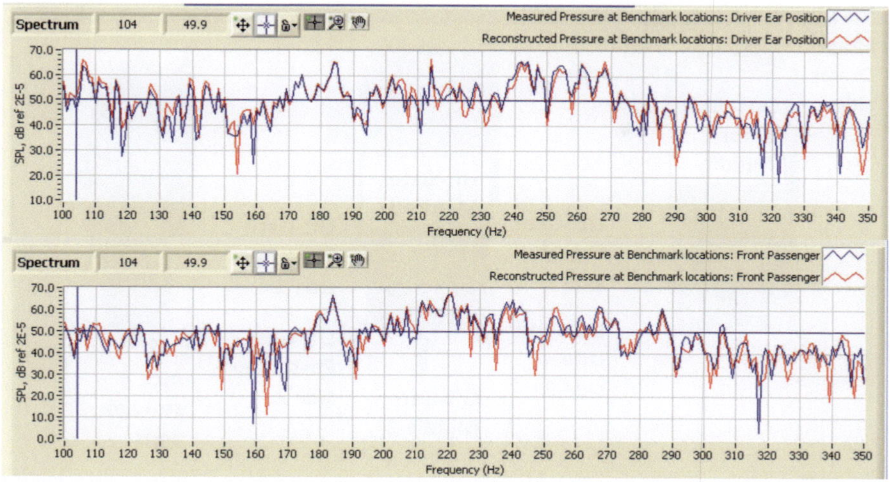

Fig. 10.7 Comparisons of reconstructed acoustic pressure spectra with benchmark data measured at two locations in the 100–350 Hz range: (1) driver designated as FL (*above*) and (2) front passenger designated as FR (*below*)

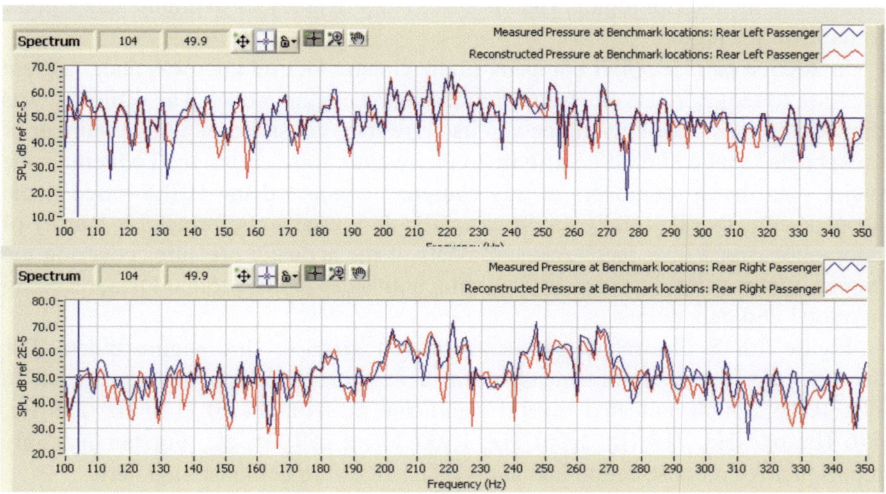

Fig. 10.8 Comparisons of reconstructed acoustic pressure spectra with benchmark data measured at two locations in 100–350 Hz: (1) rear left passenger designated as RL (*above*) and (2) rear right passenger designated as RR (*below*)

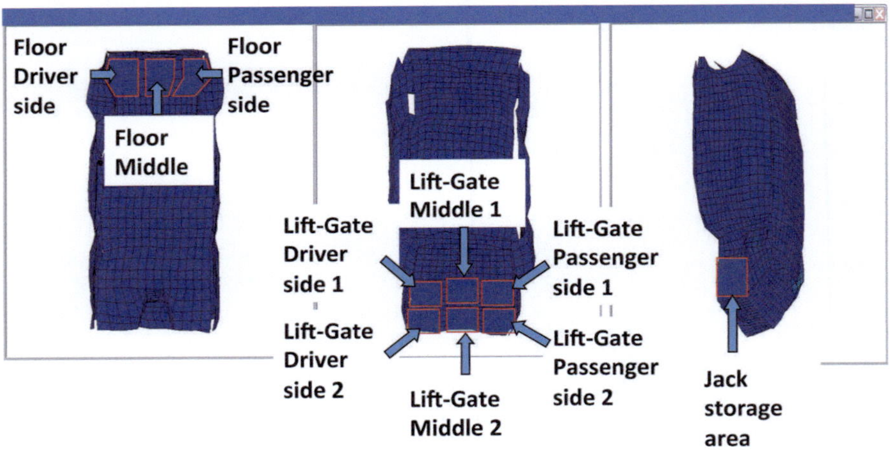

Fig. 10.9 Locations of the designated panels for which the relative acoustic contributions toward the SPL values at the driver, front passenger, and rear passenger positions

10.3.3 Story 3: Engine Block Noise Analyses

The first two stories have shown the applications of panel acoustic contributions analyses in the interior region. Here is a true story of analyses of engine block noise in the exterior region.

In 2005, a parts supplier company in the Greater Detroit area was under a great pressure to reduce noise emission from a specific engine model. The difficulty was that this full-size gasoline engine and its drive train consisted of a large number of components that could all contribute to the overall noise emission. To identify the major contributor, engineers at this company conducted the HELS-based panel acoustic contributions analyses. Specifically, they used a conformal 7×8 array of microphones and took patch measurements of the acoustic pressures covering the engine intake manifold, front driving train, both sides, and bottom oil pan, resulting in a total 280 data points.

Figure 10.15 illustrated the reconstruction grids of this engine block. The acoustic pressure spectrum measured at a designated field point was shown in Fig. 10.16, which indicated several significant peaks at 64 Hz, 129 Hz, 258 Hz, 516 Hz, 645 Hz, 710 Hz, 1,161 Hz, 1,937 Hz, 2,259 Hz, etc. For brevity, only representative reconstruction and analysis results were demonstrated here.

As pointed out in Story 1, in analyzing the structure-borne sound radiation, we must focus on the normal component of the time-averaged acoustic intensity distribution on a source surface, rather than the surface acoustic pressure distribution. This is because the normal surface acoustic intensity depicts the acoustic energy flow. If the normal surface acoustic intensity is positive, then there is a net acoustic energy outflow, and this surface may be the major contributor for acoustic

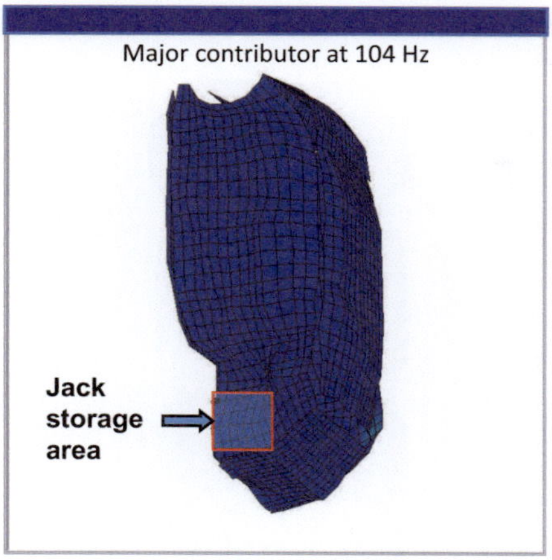

	FR	RR	FL	RL	Total
Floor-Driver side	8.8	8.5	8.9	8.9	35.2
Floor-Middle	8.0	7.7	8.3	8.2	32.4
Floor-Passenger Side	7.0	6.5	7.4	7.3	28.4
Jack Storage Area	10.0	10.0	10.0	10.0	40.0
Lift Gate-Driver Side 1	8.7	8.5	8.9	8.8	34.8
Lift Gate-Driver Side 2	8.9	8.6	9.0	9.0	35.6
Lift Gate-Middle 1	8.8	8.7	9.0	9.0	35.6
Lift Gate-Middle 2	8.4	8.1	8.6	8.5	33.6
Lift Gate-Passenger Side 1	7.9	7.5	8.2	8.1	31.6
Lift Gate-Passenger Side 2	6.3	5.6	6.8	6.6	25.2

Fig. 10.10 The jack storage area was the major acoustic contributor toward the SPL values at 104 Hz at the front passenger's (designated as FR), rear right passenger's (designated as RR), driver's (designated as FL), and rear left passenger's (designated as RL) ears positions (*above*). The relative panel acoustic contributions are normalized with respect to the highest value and multiplied by 10, and the total values are summed together (*below*)

radiation. If the normal surface acoustic intensity is negative, there is a net acoustic energy inflow, and the corresponding surface is known as an acoustic sink, meaning it actually absorbs sound. When the normal surface acoustic intensities show both positive and negative values, the outflow acoustic energy tends to cancel inflow acoustic energy. As a result, the acoustic contribution from this surface can be minimal.

For example, at 129 Hz, the normal surface acoustic intensity on the top was negative and those on the front and side surfaces were both positive and negative, whereas that on the bottom was positive (Fig. 10.17). Therefore, the bottom surface

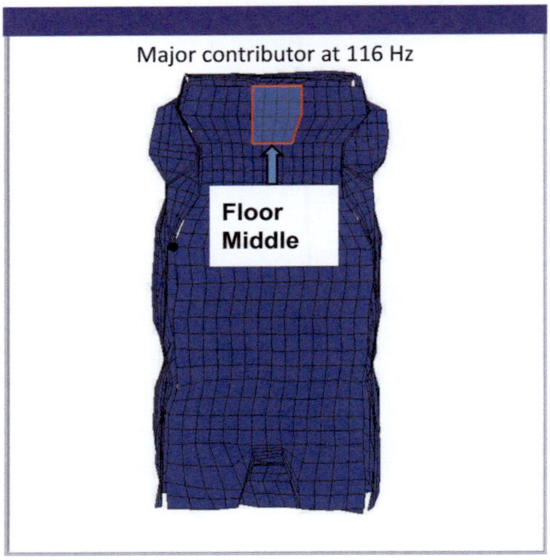

	FR	RR	FL	RL	Total
Floor-Driver side	7.0	6.5	7.4	7.3	28.2
Floor-Middle	10.0	10.0	10.0	10.0	40.0
Floor-Passenger Side	8.7	8.5	8.9	8.8	34.9
Jack Storage Area	8.9	8.6	9.0	9.0	35.5
Lift Gate-Driver Side 1	8.8	8.7	9.0	9.0	35.5
Lift Gate-Driver Side 2	8.4	8.1	8.6	8.5	33.6
Lift Gate-Middle 1	7.9	7.5	8.2	8.1	31.7
Lift Gate-Middle 2	6.3	5.6	6.8	6.6	25.3
Lift Gate-Passenger Side 1	0.0	0.0	0.0	0.0	0
Lift Gate-Passenger Side 2	0.0	0.0	0.0	0.0	0

Fig. 10.11 The Floor Middle panel was the major acoustic contributor toward the SPL values at 116 Hz at the front passenger's (designated as FR), rear right passenger's (designated as RR), driver's (designated as FL), and rear left passenger's (designated as RL) ears positions (*above*). Relative panel acoustic contributions are normalized with respect to the highest value and multiplied by 10, and the total values are summed together (*below*)

where oil pan was located was the major contributor, and the top, front, and side surfaces virtually did not contribute any acoustic radiation. Similar results were obtained for other frequencies. Figure 10.18 demonstrated that at 645 Hz the engine noise was primarily emitted from the bottom surface because the normal surface acoustic intensity was consistently positive. Meanwhile, the normal surface acoustic intensities on the top, front, and side surfaces were either negative or both positive and negative.

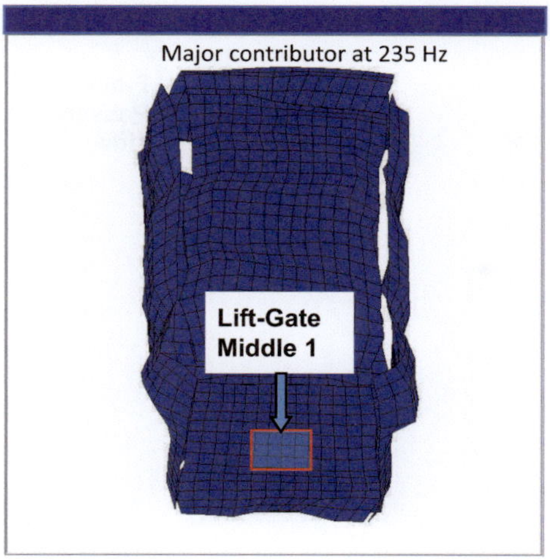

	FR	RR	FL	RL	Total
Floor-Driver side	6.8	6.1	7.1	7.2	27.2
Floor-Middle	9.3	9.1	9.4	9.4	37.2
Floor-Passenger Side	9.7	9.7	9.8	9.8	39.0
Jack Storage Area	7.7	7.2	7.9	8.0	30.8
Lift Gate-Driver Side 1	9.7	9.6	9.7	9.7	38.7
Lift Gate-Driver Side 2	9.8	9.7	9.8	9.8	39.1
Lift Gate-Middle 1	10.0	10.0	10.0	10.0	40.0
Lift Gate-Middle 2	9.9	9.8	9.9	9.9	39.5
Lift Gate-Passenger Side 1	9.6	9.69	9.7	9.7	38.6
Lift Gate-Passenger Side 2	8.5	8.2	8.6	8.7	34.0

Fig. 10.12 The Lift Gate Middle 1 panel was the major acoustic contributor toward the SPL values at 235 Hz at the front passenger's (designated as FR), rear right passenger's (designated as RR), driver's (designated as FL), and rear left passenger's (designated as RL) ears positions (*above*). The relative panel acoustic contributions are normalized with respect to the highest value and multiplied by 10, and the total values are summed together (*below*)

Table 10.1 shows the total normal surface acoustic intensity values and the corresponding contributions in percentages. Results indicate that the bottom surface is the major contributor for sound radiation, whereas the top and side surfaces are actually acoustic sinks, and the front surface contributes about 1/6 of those from the bottom surface. With this insight, engineers replaced the stamped oil pan by a composite one, and the resultant engine noise was significantly reduced.

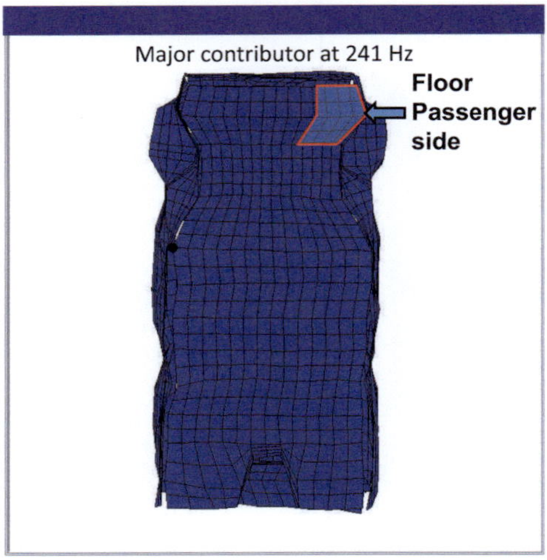

	FR	RR	FL	RL	Total
Floor-Driver side	6.8	5.3	7.9	7.8	27.8
Floor-Middle	9.3	9.0	9.6	9.6	37.5
Floor-Passenger Side	10.0	10.0	10.0	10.0	40.0
Jack Storage Area	9.6	9.4	9.7	9.7	38.4
Lift Gate-Driver Side 1	9.8	9.7	9.8	9.8	39.1
Lift Gate-Driver Side 2	9.1	8.7	9.4	9.4	36.6
Lift Gate-Middle 1	8.1	7.2	8.8	8.7	32.8
Lift Gate-Middle 2	5.3	3.1	6.9	6.8	22.1
Lift Gate-Passenger Side 1	8.2	7.4	8.8	8.8	33.2
Lift Gate-Passenger Side 2	9.8	9.7	9.9	9.9	39.3

Fig. 10.13 The Floor Passenger Side was the major acoustic contributor toward the SPL values at 241 Hz at the front passenger's (designated as FR), rear right passenger's (designated as RR), driver's (designated as FL), and rear left passenger's (designated as RL) ears positions (*above*). The relative panel acoustic contributions are normalized with respect to the highest value and multiplied by 10, and the total values are summed together (*below*)

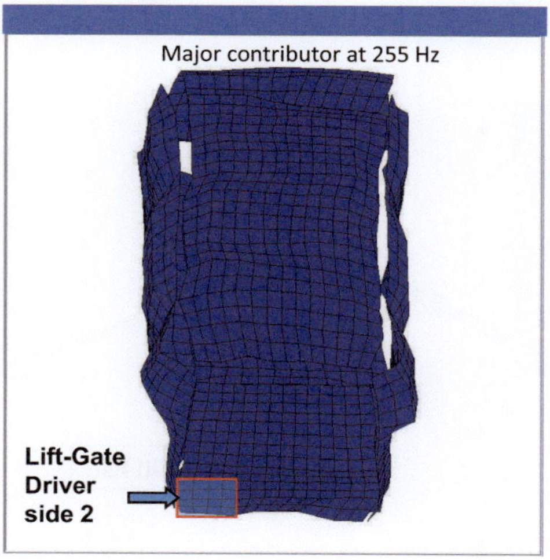

	FR	RR	FL	RL	Total
Floor-Driver side	5.6	8.9	9	8.8	32.3
Floor-Middle	7.4	9.3	9.4	9.3	35.4
Floor-Passenger Side	6.9	9.2	9.3	9.1	33.8
Jack Storage Area	9.7	9.9	9.9	9.9	39.4
Lift Gate-Driver Side 1	9.4	9.8	9.9	9.8	38.9
Lift Gate-Driver Side 2	10.0	10.0	10.0	10.0	40.0
Lift Gate-Middle 1	5.9	5.9	6.4	5.5	23.7
Lift Gate-Middle 2	5.8	8.9	9.1	8.8	32.6
Lift Gate-Passenger Side 1	7.9	9.5	9.5	9.4	36.3
Lift Gate-Passenger Side 2	5.1	8.7	8.9	8.6	31.3

Fig. 10.14 The Lift Gate Driver Side 2 was the major acoustic contributor toward the SPL values at 241 Hz at the front passenger's (designated as FR), rear right passenger's (designated as RR), driver's (designated as FL), and rear left passenger's (designated as RL) ears positions (*above*). The relative panel acoustic contributions are normalized with respect to the highest value and multiplied by 10, and the total values are summed together (*below*)

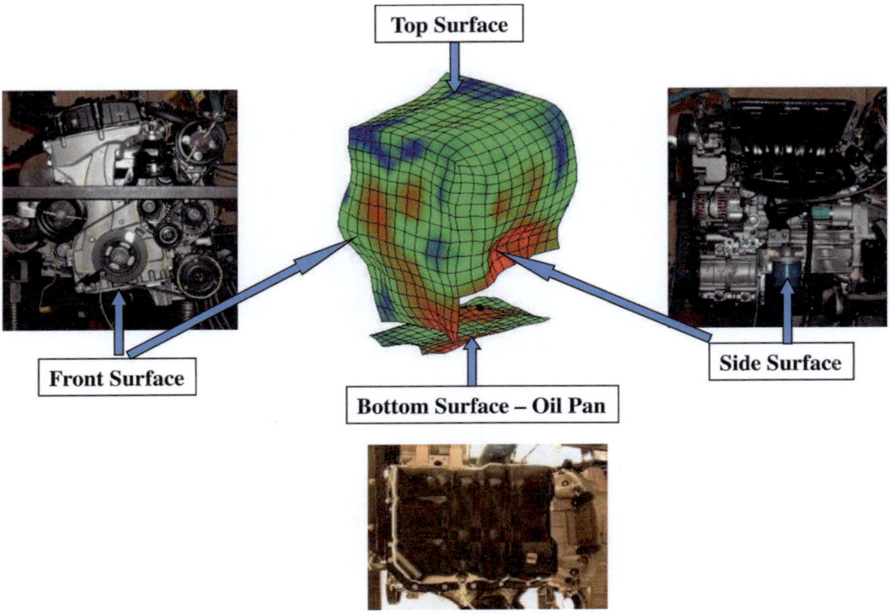

Fig. 10.15 Overall view of the reconstruction grids and different sides of the gasoline engine block

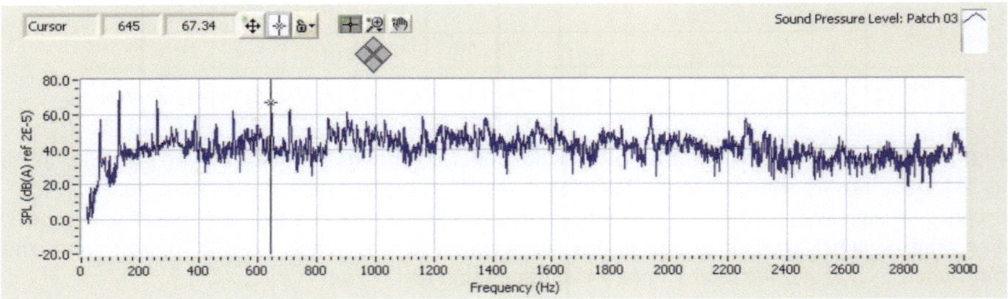

Fig. 10.16 A-weighted acoustic pressure spectrum measured at a designated field point for this gasoline engine. Several significant narrow-band peaks were identified at 64 Hz, 129 Hz, 258 Hz, 516 Hz, 645 Hz, 710 Hz, 1,161 Hz, 1,937 Hz, 2,259 Hz, etc. were identified

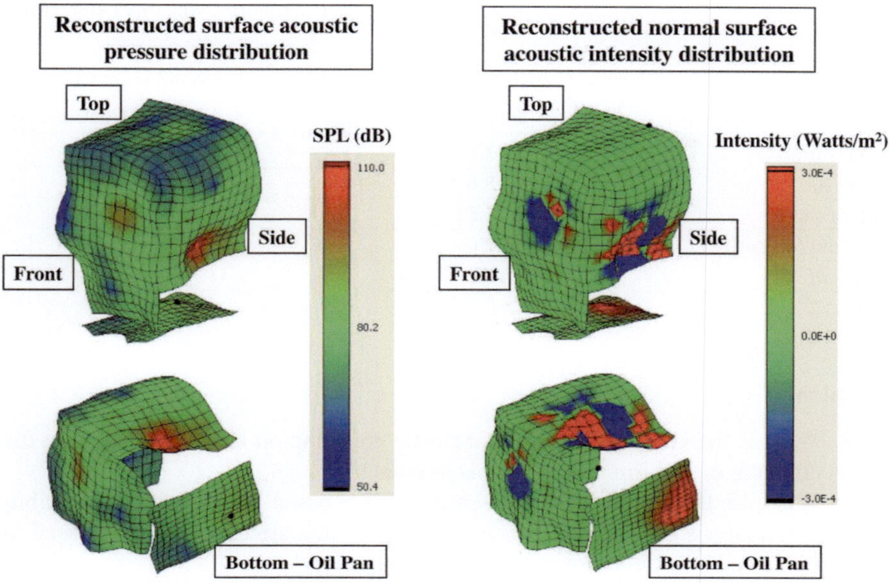

Fig. 10.17 Reconstructed surface acoustic pressure distributions on the engine block (*left*) and the normal surface acoustic intensity distributions on the engine block (*right*) at 129 Hz. The bottom surface was the major contributor of sound radiation because the normal surface acoustic intensity was consistently positive, whereas those on the top, front, and side surfaces were either negative or both positive and negative

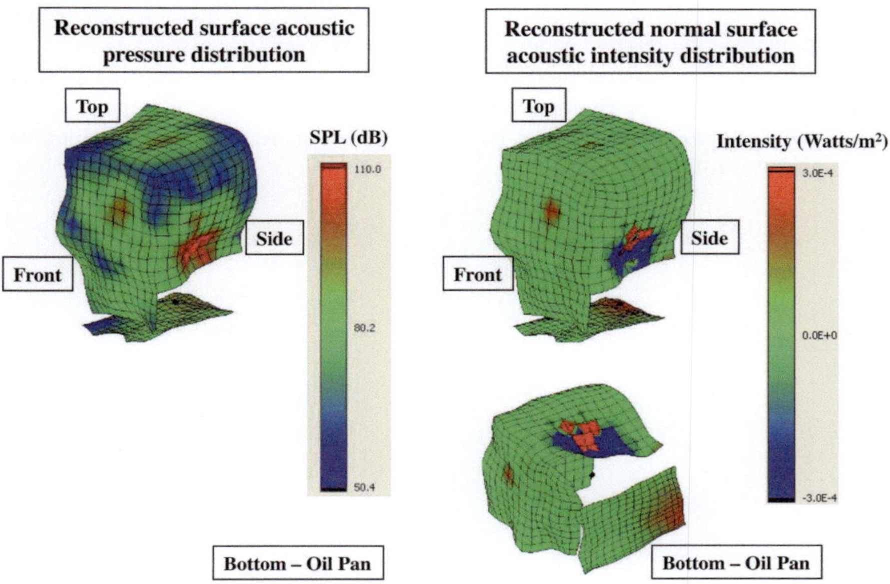

Fig. 10.18 Reconstructed surface acoustic pressure distributions on the engine block (*left*) and the normal surface acoustic intensity distributions on the engine block (*right*) at 645 Hz. The bottom surface was the major contributor of sound radiation because the normal surface acoustic intensity was consistently positive, whereas those on the top, front, and side surfaces were either negative or both positive and negative

Table 10.1 Relative panel acoustic contributions of the engine block to overall noise emission

Surface of the engine	Sound power (μ Watts)	% Contribution
Top surface	−1.8	−10.60
Side surface	−18.6	−109
Front surface	5.5	32.20
Bottom (oil pan)	32.0	187
Overall	*17.1*	*100*

Problems

10.1. What is the HELS-based panel acoustic contribution analysis? What are the advantages and limitations of this approach?
10.2. What are the differences between the HELS-based panel acoustic contribution analysis and the existing methodologies that are currently being used in industry to tackle various noise diagnosis issues?
10.3. Why do we need examine the acoustic power flows from individual panels to assess their relative contributions?
10.4. How are the acoustic power flows from individual panels of a vibrating structure related to the SPL values at any field points?
10.5. Why is it not appropriate to use the acoustic pressure measurements alone to analyze the relative panel acoustic contributions?
10.6. When will it be appropriate to use the acoustic pressure measurements alone to analyze the relative acoustic contributions? Why?
10.7. What are the general procedures involved in conducting the HELS-based panel acoustic contributions analyses?
10.8. What should we learn from the stories of the panel acoustic contributions analyses told in Chap. 10?

References

1. *The Nobel Prize in Physics 1971*. The Nobel Foundation (nobelprize.org)
2. Y.N. Denisyuk, On the reflection of optical properties of an object in a wave field of light scattered by it. Dokl. Akad. Nauk SSSR **144**, 1275–1278 (1962)
3. E.N. Leith, J. Upatnieks, Reconstructed wavefronts and communication theory. J. Opt. Soc. Am. **52**, 1123–1130 (1962)
4. F. Thurstone, Ultrasound holography and visual reconstruction. Proc. IEEE Symp. Biomed. Eng. **1**, 12 (1966)
5. B.P. Hildebrand, B.B. Brenden, *An Introduction to Acoustical Holography* (Plenum, New York, 1974)
6. J.W. Goodman, *Introduction to Fourier Optics* (McGraw-Hill, New York, 1968)
7. E.G. Williams, J.D. Maynard, Holographic imaging without the wavelength resolution limit". Phys. Rev. Lett. **45**, 554–557 (1980)
8. E.G. Williams, *Fourier Acoustics: Sound Radiation and Nearfield Acoustical Holography* (Academic, San Diego, 1999)
9. E.G. Williams, H.D. Dardy, R.G. Fink, Nearfield acoustical holography using an underwater automated scanner. J. Acoust. Soc. Am. **78**, 789–798 (1985)
10. J.D. Maynard, E.G. Williams, Y. Lee, Nearfield acoustic holography: I. Theory of generalized holography and the development of NAH. J. Acoust. Soc. Am. **78**, 1395–1413 (1985)
11. S. Hayek, T. Luce, Aperture effects in planar nearfield acoustical imaging. Trans. ASME J. Vib. Acoust. Stress. Reliab. Des. **110**, 91–96 (1988)
12. E.G. Williams, B.H. Houston, P.C. Herdic, Fast Fourier transform and singular value decomposition formulations for patch nearfield acoustical holography. J. Acoust. Soc. Am. **114**, 1322–1333 (2003)
13. A. Sarkissian, Method of superposition applied to patch near-field. J. Acoust. Soc. Am. **118**, 671–677 (2005)
14. M. Lee, J.S. Bolton, Patch near-field acoustical holography in cylindrical geometry. J. Acoust. Soc. Am. **118**, 3721–3732 (2005)
15. K. Saijyou, S. Yoshikawa, Reduction methods of the reconstruction error for large-scale implementation of near-field acoustical holography. J. Acoust. Soc. Am. **110**, 2007–2023 (2001)
16. F.J. Harris, On the use of windows for harmonic analysis with the discrete Fourier transform. Proc. IEEE **66**, 51–83 (1978)
17. J.W. Tukey, An introduction to the calculations of numerical spectrum analysis, in *Spectral Analysis of Time Series*, ed. by B. Harris (Wiley, New York, 1967), pp. 25–46

S.F. Wu, *The Helmholtz Equation Least Squares Method*, Modern Acoustics and Signal Processing, DOI 10.1007/978-1-4939-1640-5

18. B.K. Gardner, R.J. Bernhard, A noise source identification technique using an inverse Helmholtz integral equation method. ASME J. Vib. Acoust. Stress. Reliab. Des. **110**, 84–90 (1988)
19. T.W. Wu, A direct boundary element method for acoustic radiation and scattering from mixed regular and thin bodies. J. Acoust. Soc. Am. **97**, 84–91 (1995)
20. W.A. Veronesi, J.D. Maynard, Digital holographic reconstruction of sources with arbitrarily shaped surfaces. J. Acoust. Soc. Am. **85**, 588–598 (1989)
21. G.H. Golub, C.F. Loan, *Matrix Computations* (Johns Hopkins University Press, North Oxford, 1983)
22. G.V. Borgiotti, A. Sarkissian, E.G. Williams, L. Schuetz, Conformal generalized near-field acoustic holography for axisymmetric geometries. J. Acoust. Soc. Am. **88**, 199–209 (1990)
23. J.M. Varah, On the numerical solution of ill-conditioned linear systems with applications to ill-posed problems. SIAM J. Numer. Anal. **10**, 257–267 (1973)
24. G.-T. Kim, B.-T. Lee, 3-D sound source reconstruction and field reproduction using the Helmholtz integral equation. J. Sound Vib. **136**, 245–261 (1990)
25. M.R. Bai, Application of BEM (boundary element method)-based acoustic holography to radiation analysis of sound sources with arbitrarily shaped geometries. J. Acoust. Soc. Am. **92**, 533–549 (1992)
26. B.-K. Kim, J.-G. Ih, On the reconstruction of vibro-acoustic field over the surface enclosing an interior space using the boundary element method. J. Acoust. Soc. Am. **100**, 3003–3016 (1996)
27. Y.-K. Kim, Y.-H. Kim, Holographic reconstruction of active sources and surface admittance in an enclosure. J. Acoust. Soc. Am. **105**, 2377–2383 (1999)
28. Z. Zhang, N. Vlahopoulos, S.T. Raveendra, T. Allen, K.Y. Zhang, A computational acoustic field reconstruction process based on an indirect boundary element formulation. J. Acoust. Soc. Am. **108**, 2167–2178 (2000)
29. S.-C. Kang, J.-G. Ih, Use of non-singular boundary integral formulation for reducing errors due to near-field measurements in the boundary element method based near-field acoustic holography. J. Acoust. Soc. Am. **109**, 1320–1328 (2001)
30. A. Schuhmacher, J. Hald, K.B. Rasmussen, P.C. Hansen, Sound source reconstruction using boundary element calculations. J. Acoust. Soc. Am. **113**, 114–127 (2004)
31. K. Saijyou, H. Uchida, Data extrapolation method for boundary element method-based near-field acoustical holography. J. Acoust. Soc. Am. **115**, 785–796 (2004)
32. C. Langrenne, M. Melon, A. Garcia, Boundary element method for the acoustic characterization of a machine in bounded noisy environment. J. Acoust. Soc. Am. **121**, 2750–2757 (2007)
33. D.M. Photiadis, The relationship of singular value decomposition to wave-vector filtering in sound radiation problems. J. Acoust. Soc. Am. **88**, 1152–1159 (1990)
34. H.A. Schenck, Improved integral formulation for acoustic radiation problems. J. Acoust. Soc. Am. **44**, 41–58 (1968)
35. A.J. Burton, G.F. Miller, Application of the integral equation method to the numerical solution of some exterior boundary value problems. Proc. R. Soc. Lond. A **323**, 202–210 (1971)
36. Z. Wang, S.F. Wu, Helmholtz equation-least-squares method for reconstructing the acoustic pressure field. J. Acoust. Soc. Am. **102**, 2020–2032 (1997)
37. S.F. Wu, On reconstruction of acoustic pressure fields using the Helmholtz equation least squares method. J. Acoust. Soc. Am. **107**, 2511–2522 (2000)
38. S.F. Wu, Methods for reconstructing acoustic quantities based on acoustic pressure measurements. J. Acoust. Soc. Am. **124**, 2680–2697 (2008)
39. L.P. Eisenhart, Separable systems in Euclidean 3-space. Phys. Rev. **45**, 427–428 (1934)
40. M. Abramowitz, I.A. Stegun, *Handbook of Mathematical Functions* (Dover, New York, 1972)

41. I.S. Gradshteyn, I.M. Ryzhik, *Table of Integrals, Series and Products*, 4th edn. (Academic, New York, 1965)
42. A.D. Pierce, *Acoustics: An Introduction to Its Physical Principles and Applications* (McGraw-Hill, New York, 1981), pp. 165–194, Chap. 4
43. A.N. Tikhonov, V.Y. Arsenin, *Solutions of Ill-Posed Problems* (Wiley, New York, 1977), pp. 71–73, Chap. 2
44. D.L. Phillips, A technique for the numerical solution of certain integral equations of the first kind. J. Assoc. Comput. Mach. **9**, 84–97 (1962)
45. S. Twomey, On the numerical solution of Fredholm integral equations of the first kind by the inversion of the linear system produced by quadrature. J. Assoc. Comput. Mach. **10**, 97–101 (1962)
46. P.C. Hansen, *Rank-Deficient and Discrete Ill-Posed Problems* (SIAM, Philadelphia, 1998)
47. P.J. Davis, P. Rabinowitz, Advances in orthonormalizing computation, in *Advances in Computers*, ed. by F.L. Alt, vol. 2 (Academic, New York, 1961), pp. 55–133
48. L. K. Netarajan, Vibro-acoustic analysis of an arbitrarily shaped vibrating structure, Ph.D. dissertation, Wayne State University, 2013
49. E.G. Williams, Regularization methods for near-field acoustic holography. J. Acoust. Soc. Am. **110**, 1976–1988 (2001)
50. G.H. Golub, M. Heath, G. Wahba, Generalized cross-validation as a method for choosing a good ridge parameter. Technometrics **21**, 215–223 (1979)
51. L.K. Natarajan, S.F. Wu, Reconstruction of normal surface velocities on a baffled plate using Helmholtz equation least squares method. J. Acoust. Soc. Am. **131**, 4570–4583 (2012)
52. N.E. Rayess, S.F. Wu, Experimental validations of the HELS method for reconstructing acoustic radiation from a complex vibrating structure. J. Acoust. Soc. Am. **107**, 2955–2964 (2000)
53. S.F. Wu, N. Rayess, X. Zhao, Visualization of acoustic radiation from a vibrating bowling ball. J. Acoust. Soc. Am. **109**, 2771–2779 (2001)
54. S.F. Wu, J. Yu, Reconstructing interior acoustic pressure field via Helmholtz equation least-squares method. J. Acoust. Soc. Am. **104**, 2054–2060 (1998)
55. M. Moondra, S.F. Wu, Visualization of vehicle interior sound field using HELS based NAH. Noise Control Eng. J. **53**(4), 146–154 (2005)
56. S. F. Wu, in *An overview of reconstruction of radiated acoustic pressures from complex vibrating structures by using the HELS method*, The joint meeting of the meeting of the 136th Acoustical Society of America, 2nd Convention of the European Acoustics Association: Forum Acusticum 99, and 25th German Acoustics DAGA Conference, Berlin, Germany, March 1999
57. L. Rayleigh (J. W. Strutt), On the dynamical theory of gratings. Proc. R. Soc. Lond. A, **79**, 399–416 (1907)
58. L. Rayleigh (J. W. Strutt), *The Theory of Sound*, vol. 2 (Dover, New York, 1945), pp. 89–96
59. R.F. Millar, Rayleigh hypothesis in scattering problems. (With author's reply). Electron. Lett. **5**, 416–417 (1969)
60. R.F. Millar, On the Rayleigh assumption in scattering by a periodic surface. Proc. Camb. Philos. Soc. **65**, 773–791 (1969)
61. R.F. Millar, The Rayleigh hypothesis and a related least-squares solution to scattering problems for periodic surfaces and other scatters. Radio Sci. **8**, 785–796 (1973)
62. R.F. Millar, On the Rayleigh assumption in scattering by a periodic surface. II. Math. Proc. Camb. Philos. Soc. **69**, 217–225 (1971)
63. N.R. Hill, V. Celli, Limits of convergence of the Rayleigh method for surface scattering. Phys. Rev. B **17**, 2478–2481 (1987)
64. P.M. van den Berg, J.T. Fokkema, The Rayleigh hypothesis in the theory of diffraction by a perturbation in a plane surface. Radio Sci. **15**, 723–732 (1980)
65. P.M. van den Berg, J.T. Fokkema, The Rayleigh hypothesis in the theory of diffraction by a cylindrical obstacle. IEEE Trans. Antennas Propag. **AP-27**, 577–583 (1979)

66. B.V. Apel'tsin, On the method of nonorthogonal series in exterior problems in the theory of steady-state oscillations. Sov. Phys. Dokl. **26**, 831–833 (1982)
67. R.F. Millar, Singularities and the Rayleigh hypothesis for solutions to the Helmholtz equation. IMA J. Appl. Math. **37**, 155–171 (1986)
68. D. Maystre, M. Cadilhac, Singularities of the continuation of fields and validity of Rayleigh's hypothesis. J. Math. Phys. **26**, 2201–2204 (1985)
69. J.B. Keller, Singularities and Rayleigh's hypothesis for diffraction gratings. J. Opt. Soc. Am. **17**, 456–457 (2000)
70. P. Davis, The Schwarz function and its applications, in *The Carus Mathematical Monographs*, vol. 17 (Mathematical Association of America, Buffalo, 1974)
71. T. Semenova, On the behavior of HELS solutions for acoustic radiation and reconstruction, Ph.D. dissertation, Department of Mechanical Engineering, Wayne State University, May 2004
72. T. Semenova, S.F. Wu, The Helmholtz equation least squares method and the Rayleigh's hypothesis in nearfield acoustical holography. J. Acoust. Soc. Am. **115**(4), 1632–1640 (2004)
73. N. E. Rayess, Development of acoustic holography using the Helmholtz Equation-Least Squares (HELS) Method, Ph.D. dissertation, Department of Mechanical Engineering, Wayne State University, May 2001
74. V. Isakov, S.F. Wu, On theory and applications of the HELS method in inverse acoustics. Inverse Probl. **18**, 1147–1159 (2002)
75. V. Isakov, *Inverse Problems for Partial Differential Equations* (Springer, New York, 1990)
76. R. Pike, P. Sabatier (eds.), *Scattering* (Academic, New York, 2001), pp. 794–769
77. D. Colton, R. Kress, Inverse acoustic and electromagnetic scattering theory, in *Applied Mathematical Sciences*, vol. 93 (Springer, Berlin, 1992)
78. V. Isakov, *Inverse Source Problems* (AMS, Providence, 1990)
79. A.N. Tikhonov, On stability of inverse problems. Dokl. Akad. Nauk SSSR **39**, 176–179 (1943)
80. J. Prager, Approximate reconstruction of sound fields close to the source surface using spherical nearfield acoustic holography. J. Acoust. Soc. Am. **122**, 2067–2073 (2007)
81. The MathWorks, Inc., *MATLAB Documentation: MATLAB Function Reference*, (The MathWorks, Inc., MATLAB R2007b, 2007)
82. National Instruments Corporation, *LabVIEW Help: VI and Function Reference*, (National Instruments Corporation, LabVIEW 8.6, 2008)
83. H. Lu, Reconstruction of vibroacoustic responses using Helmholtz equation least squares method, Ph.D. dissertation, Wayne State University, Detroit, Michigan, 2007
84. S. F. Wu, *Noise Diagnostics Using Nearfield Acoustical Holography Seminar* (Society for Automobile Engineering International, Short Course ID# C0607, 2007)
85. H. Lu, S.F. Wu, Reconstruction of vibroacoustic responses of a highly nonspherical structure using Helmholtz equation least-squares method. J. Acoust. Soc. Am. **125**, 1538–1548 (2009)
86. C. E. Shannon, Communication in the presence of noise. Proc. Inst. Radio. Eng. **37**, 10–21 (1949). Reprint as classic paper in Proceedings of IEEE, vol. 86, February 1998. pii: S 0018-9219(98)01299-7
87. S.M. Pandit, *Modal and Spectrum Analysis: Data Dependent Systems in State Space* (Wiley, New York, 1991)
88. P. Mohanty, D.J. Rixen, Operational modal analysis in the presence of harmonic excitation. J. Sound Vib. **270**, 93–109 (2004)
89. H. Lord, W.S. Gatley, H.A. Evensen, *Noise Control for Engineers* (Krieger Publishing Company, Malabar, 1987)
90. E.G. Williams, H.D. Dardy, K.B. Washburn, Generalized nearfield acoustical holography for cylindrical geometry: theory and experiment. J. Acoust. Soc. Am. **81**, 389–407 (1987)
91. S.F. Wu, X. Zhao, Combined Helmholtz equation least squares (CHELS) method for reconstructing acoustic radiation. J. Acoust. Soc. Am. **112**, 179–188 (2002)

92. P. Morse, H. Feshbach, *Methods of Theoretical Physics*, vol. 1 (McGraw-Hill, New York, 1953)
93. P. Dirac, *Principles of Quantum Mechanics*, 4th edn. (Clarendon, Oxford, 1958)
94. M.E. Goldstein, *Aeroacoustics* (McGraw-Hill, New York, 1976), Chapter 1, 28–29, 45–47, and Chapter 3, 114–120
95. I.S. Gradshteyn, I.M. Ryzhik, The principal values of improper integrals, Section 3.05, in *Tables of Integrals, Series, and Products*, 6th edn. (Academic, San Diego, 2000), p. 248
96. F. Smithies, *Integral Equations* (Cambridge University Press, Cambridge, 1958), pp. 51–52
97. J.D. Pryce, *Numerical Solution of Sturm–Liouville Problems* (Clarendon, Oxford, 1993)
98. D.K. Arrowsmith, C.M. Place, *Dynamical Systems* (Chapman & Hall, London, 1992), Section 3.3
99. P.M. Morse, K.U. Ingard, *Theoretical Acoustics* (Princeton University Press, Princeton, 1986)
100. X. Zhao, S.F. Wu, Reconstruction of the vibro-acoustic fields using hybrid nearfield acoustical holography. J. Sound Vib. **282**, 1183–1199 (2005)
101. T.B. Hansen, Spherical expansions of time-domain acoustic fields: Application to near-field scanning. J. Acoust. Soc. Am. **98**, 1204–1215 (1995)
102. S.F. Wu, Hybrid nearfield acoustical holography. J. Acoust. Soc. Am. **115**(1), 207–217 (2004)
103. M. Moondra, S.F. Wu, Visualization of vehicle interior noise using HELS based NAH, (CD-ROM), in *Proceedings of the INTER-NOISE 2002*, August 2002
104. S.H. Yoon, P.A. Nelson, Estimation of acoustic source strength by inverse methods: part II, experimental investigation of methods for choosing regularization parameters. J. Sound Vib. **233**, 669–705 (2000)
105. L. Landweber, An iteration formula for Fredholm integral equations of the first kind. Am. J. Math. **73**, 615–624 (1951)
106. V.A. Morozov, The error principle in the solution of operational equations by the regularization method. USSR Comput. Math. Math. Phys. **8**, 63–87 (1968)
107. X. Zhao, S.F. Wu, Reconstruction of vibro-acoustic fields by using hybrid nearfield acoustical holography in half space. J. Acoust. Soc. Am. **117**, 555–565 (2005)
108. I.-Y. Jeon, J.-G. Ih, On the holographic reconstruction of vibroacoustic fields using equivalent sources and inverse boundary element method. J. Acoust. Soc. Am. **118**, 3475–3484 (2005)
109. G.H. Koopmann, L. Song, J.B. Fahnline, A method for computing acoustic fields based on the principle of wave superposition. J. Acoust. Soc. Am. **86**, 2433–2438 (1989)
110. L. Song, G.H. Koopmann, J.B. Fahnline, Numerical errors associated with the method of superposition for computing acoustic fields. J. Acoust. Soc. Am. **89**, 2625–2633 (1991)
111. J.B. Fahnline, G.H. Koopmann, A numerical solution for the general radiation problem based on the combined methods of superposition and singular-value decomposition. J. Acoust. Soc. Am. **90**, 2808–2819 (1991)
112. M.E. Johnson, S.J. Elliott, K.H. Baek, J. Garcia-Bonito, An equivalent source technique for calculating the sound field inside an enclosure containing scattering objects. J. Acoust. Soc. Am. **104**, 1221–1231 (1998)
113. Y.I. Bobrovnitskii, T.M. Tomilina, General properties and fundamental errors of the method of equivalent sources. Acoust. Phys. **41**, 649–660 (1995)
114. C.-X. Bi, X.-Z. Chen, J. Chen, Sound field separation technique based on equivalent source method and its application in nearfield acoustic holography. J. Acoust. Soc. Am. **123**, 1472–1478 (2008)
115. N.P. Valdivia, E.G. Williams, Study of the comparison of the methods of equivalent sources and boundary element methods for nearfield acoustic holography. J. Acoust. Soc. Am. **120**, 3694–3705 (2006)
116. M.R. Bai, J.H. Lin, Source identification system based on the time-domain nearfield equivalence source imaging: fundamental theory and implementation. J. Sound Vib. **307**, 202–225 (2007)

117. A. Sarkissian, Extension of measurement surface in near-field acoustic holography. J. Acoust. Soc. Am. **115**, 1593–1596 (2004)
118. M.R. Bai, C.-C. Chen, J.H. Lin, On optima retreat distance for the equivalent source method-based nearfield acoustical holography. J. Acoust. Soc. Am. **129**, 1407–1416 (2011)
119. K.H. Baek, S.J. Elliot, Natural algorithms for choosing source locations in active control systems. J. Sound Vib. **186**, 245–267 (1995)
120. D.C. Kammer, Sensor placement for on-orbit modal identification and correlation of large space structures. J. Guid. Contr. Dynam. **14**, 251–259 (1991)
121. T. Semenova, S.F. Wu, On the choice of expansion functions in the Helmholtz equation least-squares method. J. Acoust. Soc. Am. **117**, 701–710 (2005)
122. I.P. Lysanov, Theory of the scattering of waves at periodically uneven surfaces. Sov. Phys. Acoust. **4**, 1–10 (1958)
123. P.M. van den Berg, Reflection by a grating: Rayleigh methods. J. Opt. Soc. Am. **71**, 1224–1229 (1981)
124. A. Doicu, Y. Eremin, T. Wriedt, *Acoustic and Electromagnetic Scattering Analysis Using Discrete Sources* (Academic, New York, 2000)
125. Y.A. Eremin, Complete systems of functions for the study of boundary value problems in mathematical physics. Sov. Phys. Dokl. **32**, 635–637 (1987)
126. V.D. Kupradze, On the approximate solutions of problems in mathematical physics. Success Math. Sci. **22**, 58–108 (1967)
127. Y.A. Eremin, N.V. Orlov, A.G. Sveshnikov, The analysis of complex diffraction problems by the discrete source method. Comp. Maths Math. Phys. **35**, 731–743 (1995)
128. A. Wirgin, On Rayleigh's theory of sinusoidal diffraction gratings. Opt. Acta **27**, 1671–1692 (1980)
129. T. Namioka, T. Harada, K. Yasuura, Diffraction gratings in Japan. Opt. Acta **26**, 1021–1034 (1979)
130. D. Agassi, T.F. George, Convergent scheme for light scattering from an arbitrary deep metallic grating. Phys. Rev. B **33**, 2393–2400 (1986)
131. H. Engl, M. Hanke, A. Neubauer, Regularization of inverse problems, in *Mathematics and Its Applications*, vol. 375 (Kluwer, Dordrecht, 1996)
132. M. Hanke, P.C. Hansen, Regularization methods for large-scale problems. Surv. Math. Ind. **3**, 253–315 (1993)
133. A.B. Bakushinskii, Remarks on choosing a regularization parameter using the quasi-optimality and ration criterion. Comput. Math. Math. Phys. **24**, 181–182 (1984)
134. V.A. Morozov, On the solution of functional equations by the method of regularization. Sov. Math. Dokl. **7**, 414–417 (1966)
135. V.A. Morozov, *Methods for Solving Incorrectly Posed Problems* (Springer, New York, 1984)
136. C.L. Lawson, R.J. Hanson, *Solving Least-Squares Problems* (Prentice-Hall, Englewood Cliffs, 1974)
137. Z. Wang, Helmholtz equation-least-squares (HELS) method for inverse acoustic radiation problems, Ph.D. dissertation, Wayne State University, Detroit, Michigan, 1995
138. J. Hald, *Use of Non-Stationary STSF for the Analysis of Transient Engine Noise Radiation* (Brüel & Kjær, Nærum, 1999)
139. R. Bracewell, Heaviside's unit step function, H(x), in *The Fourier Transform and Its Applications*, 3rd edn. (McGraw-Hill, New York, 2000)
140. M.C. Junger, D. Feit, *Sound, Structures, and their Interactions* (MIT, Cambridge, 1972)
141. W.R. LePage, *Complex Variables and the Laplace Transform for Engineers* (Dover, New York, 1961)
142. S.F. Wu, Transient sound radiation from impulsively accelerated bodies. J. Acoust. Soc. Am. **94**, 542–553 (1993)
143. R.R. Craig Jr., *Structural Dynamics: An Introduction to Computer Methods* (Wiley, New York, 1981), Chap. 6, 123–127

144. S.F. Wu, H.-C. Lu, M.S. Bajwa, Reconstruction of transient acoustic radiation from a sphere. J. Acoust. Soc. Am. **117**, 2065–2077 (2005)
145. M. S. Bajwa, Investigation on transient acoustic radiation and reconstruction, Ph.D. dissertation, Wane State University, December 2008
146. S.F. Wu, L.K. Natarajan, Panel acoustic contribution analysis. J. Acoust. Soc. Am. **131**, 4570–4583 (2012)
147. D. De Vis, W. Hendricx, P. van der Linden, Development and integration of an advanced unified approach to structure borne noise analysis, in *2nd International Conference on Vehicle Comfort, ATA* (1992)
148. J. Hald, C. Blaabjerg, M. Kimura, Y. Ishii, M. Tsuchiya, and H. Ando, in Panel contribution analysis using a volume velocity source and a double layer array with the SONAH algorithm. Inter-Noise 2006, Honolulu, Hawaii, USA, December 3–6, 2006

Index

S.F. Wu, *The Helmholtz Equation Least Squares Method*, Modern Acoustics and Signal Processing, DOI 10.1007/978-1-4939-1640-5

U

V

W

MIX
Papier aus verantwortungsvollen Quellen
Paper from responsible sources
FSC® C105338

If you have any concerns about our products,
you can contact us on
ProductSafety@springernature.com

In case Publisher is established outside the EU,
the EU authorized representative is:
Springer Nature Customer Service Center GmbH
Europaplatz 3, 69115 Heidelberg, Germany

Printed by Libri Plureos GmbH
in Hamburg, Germany